J. Hamblin Smith

Elementary Algebra

J. Hamblin Smith

Elementary Algebra

ISBN/EAN: 9783743481688

Manufactured in Europe, USA, Canada, Australia, Japa

Cover: Foto ©berggeist007 / pixelio.de

Manufactured and distributed by brebook publishing software (www.brebook.com)

J. Hamblin Smith

Elementary Algebra

W. J. Gage & Co's Mathematical Series.

ELEMENTARY ALGEBRA,

—BY—

J. HAMBLIN SMITH, M.A.,

OF GONVILLE AND CAIUS COLLEGE, AND LATE LECTURER AT ST. PETER'S COLLEGE, CAMBRIDGE.

WITH APPENDIX BY

ALFRED BAKER, B.A.,

MATH. TUTOR UNIV. COL. TORONTO.

12th CANADIAN COPYRIGHT EDITION.

NEW REVISED EDITION.

Authorized by the Education Department, Ontario.
Authorized by the Council of Public Instruction, Quebec.
Recommended by the Senate of the Univ. of Halifax.

PRICE 60 CENTS.

TORONTO:
W. J. GAGE & CO.

Entered according to Act of the Parliament of Canada, in the year one thousand eight hundred and seventy-seven, in the office of the Minister of Agriculture, by ADAM MILLER & Co.

PREFACE

The design of this Treatise is to explain all that is commonly included in a First Part of Algebra. In the arrangement of the Chapters I have followed the advice of experienced Teachers. I have carefully abstained from making extracts from books in common use. The only work to which I am indebted for any material assistance is the Algebra of the late Dean Peacock, which I took as the model for the commencement of my Treatise. The Examples, progressive and easy, have been selected from University and College Examination Papers and from old English, French, and German works. Much care has been taken to secure accuracy in the Answers, but in a collection of more than 2300 Examples it is to be feared that some errors have yet to be detected. I shall be grateful for having my attention called to them.

I have published a book of Miscellaneous Exercises adapted to this work and arranged in a progressive order so as to supply constant practice for the student.

I have to express my thanks for the encouragement and advice received by me from many correspondents; and a special acknowledgment is due from me to Mr. E. J. Gross of Gonville and Caius College, to whom I am indebted for assistance in many parts of this work.

The Treatise on Algebra by Mr. E. J. Gross is a continuation of this work, and is in some important points supplementary to it.

J. HAMBLIN SMITH.

CAMBRIDGE, 1871.

CONTENTS.

CHAP.		PAGE
I.	Addition and Subtraction	1
II.	Multiplication	17
III.	Involution	29
IV.	Division	33
V.	On the Resolution of Expressions into Factors	43
VI.	On Simple Equations	57
VII.	Problems leading to Simple Equations	61
VIII.	On the Method of finding the Highest Common Factor	67
IX.	Fractions	76
X.	The Lowest Common Multiple	88
XI.	On Addition and Subtraction of Fractions	94
XII.	On Fractional Equations	105
XIII.	Problems in Fractional Equations	114
XIV.	On Miscellaneous Fractions	126
XV.	Simultaneous Equations of the First Degree	142
XVI.	Problems resulting in Simultaneous Equations	154
XVII.	On Square Root	163
XVIII.	On Cube Root	169
XIX.	Quadratic Equations	174
XX.	On Simultaneous Equations involving Quadratics	186
XXI.	On Problems resulting in Quadratic Equations	192
XXII.	Indeterminate Equations	196
XXIII.	The Theory of Indices	201
XXIV.	On Surds	213
XXV.	On Equations involving Surds	229

CONTENTS.

CHAP.		PAGE
XXVI.	On the Roots of Equations	234
XXVII.	On Ratio	243
XXVIII.	On Proportion	248
XXIX.	On Variation	258
XXX.	On Arithmetical Progression	264
XXXI.	On Geometrical Progression	273
XXXII.	On Harmonical Progression	282
XXXIII.	Permutations	287
XXXIV.	Combinations	291
XXXV.	The Binomial Theorem. Positive Integral Index	296
XXXVI.	The Binomial Theorem. Fractional and Negative Indices	307
XXXVII.	Scales of Notation	316
XXXVIII.	On Logarithms	328
	Appendix	344
	Answers	345

ELEMENTARY ALGEBRA.

I. ADDITION AND SUBTRACTION.

1. **Algebra** is the science which teaches the use of SYMBOLS to denote numbers and the operations to which numbers may be subjected.

2. The symbols employed in Algebra to denote numbers are, in addition to those of Arithmetic, the letters of some alphabet.

Thus a, b, c ...... x, y, z : α, β, γ ...... : a', b', c' ...... read a *dash*, b *dash*, c *dash* ...... : a_1, b_1, c_1 ...... read a *one*, b *one*, c *one* ...... are used as symbols to denote *numbers*.

3. The number *one*, or *unity*, is taken as the foundation of all numbers, and all other numbers are derived from it by the process of *addition*.

Thus *two* is defined to be the number that results from adding *one* to *one*;

three is defined to be the number that results from adding *one* to *two*;

four is defined to be the number that results from adding *one* to *three*;

and so on.

4. The symbol $+$, read *plus*, is used to denote the operation of **Addition**.

Thus $1+1$ symbolizes that which is denoted by 2,

$2+1$ 3,

and $a+b$ stands for the result obtained by adding b to a.

5. The symbol $=$ stands for the words "is equal to," or "the result is."

[S.A.]

Thus the definitions given in Art. 3 may be presented in an algebraical form thus:
$$1+1=2,$$
$$2+1=3,$$
$$3+1=4.$$

6. Since

$2 = 1+1$, where unity is written *twice*,
$3 = 2+1 = 1+1+1$, where unity is written *three* times,
$4 = 3+1 = 1+1+1+1$ .... *four* times,

it follows that

$a = 1+1+1 \ldots\ldots +1+1$ with unity written a times,
$b = 1+1+1 \ldots\ldots +1+1$ with unity written b times.

7. The process of addition in Arithmetic can be presented in a *shorter* form by the use of the sign $+$. Thus if we have to add 14, 17, and 23 together we can represent the process thus:
$$14+17+23=54.$$

8. When several numbers are added together, it is indifferent in what order the numbers are taken. Thus if 14, 17, and 23 be added together, their sum will be the same in whatever order they be set down in the common arithmetical process:

14	14	17	17	23	23
17	23	14	23	14	17
23	17	23	14	17	14
—	—	—	—	—	—
54	54	54	54	54	54

So also in Algebra, when any number of *symbols* are added together, the result will be the same in whatever order the symbols succeed each other. Thus if we have to add together the numbers symbolized by a and b, the result is represented by $a+b$, and this result is the same number as that which is represented by $b+a$.

Similarly the result obtained by adding together a, b, c might be expressed algebraically by

$a+b+c$, or $a+c+b$, or $b+a+c$, or $b+c+a$, or $c+a+b$, or $c+b+a$.

9. When a number denoted by a is added to itself the result is represented algebraically by $a+a$. This result is for

the sake of brevity represented by $2a$, the figure prefixed to the symbol expressing the number of times the number denoted by a is repeated.

Similarly $a+a+a$ is represented by $3a$.

Hence it follows that

$2a+a$ will be represented by $3a$,
$3a+a$ .... by $4a$.

10. The symbol $-$, read *minus*, is used to denote the operation of **Subtraction.**

Thus the operation of subtracting 15 from 26 and its connection with the result may be briefly expressed thus:

$$26-15=11.$$

11. The result of subtracting the number b from the number a is represented by

$$a-b.$$

Again $a-b-c$ stands for the number obtained by taking c from $a-b$.

Also $a-b-c-d$ stands for the number obtained by taking d from $a-b-c$.

Since we cannot take away a greater *number* from a smaller, the expression $a-b$, where a and b represent *numbers*, can denote a possible result only when a is not less than b.

So also the expression $a-b-c$ can denote a possible result only when the number obtained by taking b from a is not less than c.

12. A combination of symbols is termed an algebraical *expression*.

The parts of an expression which are connected by the symbols of operation $+$ and $-$ are called TERMS.

Compound expressions are those which have more than one term.

Thus $a-b+c-d$ is a compound expression made up of four terms.

When a compound expression contains

two terms it is called a *Binomial*,
three .... *Trinomial*,
four or more .... *Multinomial*.

Terms which are preceded by the symbol + are called *positive* terms. Terms which are preceded by the symbol − are called *negative* terms. When no symbol precedes a term the symbol + is understood.

Thus in the expression $a-b+c-d+e-f$

a, c, e are called positive terms,
b, d, f .... negative ......

The symbols of operation + and − are usually called positive and negative SIGNS.

13. If the number 6 be added to the number 13, and if 6 be taken from the result, the final result will plainly be 13.

So also if a number b be added to a number a, and if b be taken from the result, the final result will be a: that is,
$$a+b-b=a.$$

Since the operations of addition and subtraction when performed by the same number neutralize each other, we conclude that we may obliterate the same symbol when it presents itself as a positive term and also as a negative term in the same expression.

Thus $\qquad a-a=0,$
and $\qquad a-a+b=b.$

14. If we have to add the numbers 54, 17, and 23 we may first add 17 and 23, and add their sum 40 to the number 54, thus obtaining the final result 94. This process may be represented algebraically by enclosing 17 and 23 in a BRACKET ( ), thus:
$$54+(17+23)=54+40=94.$$

15. If we have to subtract from 54 the sum of 17 and 23 the process may be represented algebraically thus:
$$54-(17+23)=54-40=14.$$

16. If we have to add to 54 the difference between 23 and 17, the process may be represented algebraically thus:
$$54+(23-17)=54+6=60.$$

17. If we have to subtract from 54 the difference between 23 and 17, the process may be represented algebraically thus:
$$54-(23-17)=54-6=48,$$

18. The use of brackets is so frequent in Algebra, that the rules for their removal and introduction must be carefully considered.

We shall first treat of the *removal* of brackets in cases where symbols supply the places of numbers corresponding to the arithmetical examples considered in Arts. 14, 15, 16, 17.

Case I. To add to a the sum of b and c.
This is expressed thus: $a+(b+c)$.
First add b to a, the result will be
$$a+b.$$

This result is *too small*, for we have to add to a a number *greater* than b, and greater by c. Hence our final result will be obtained by adding c to $a+b$, and it will be
$$a+b+c.$$

Case II. To take from a the sum of b and c.
This is expressed thus: $a-(b+c)$.
First take b from a, the result will be
$$a-b.$$

This result is *too large*, for we have to take from a a number *greater* than b, and greater by c. Hence our final result will be obtained by taking c from $a-b$, and it will be
$$a-b-c.$$

Case III. To add to a the difference between b and c.
This is expressed thus: $a+(b-c)$.
First add b to a, the result will be
$$a+b.$$

This result is *too large*, for we have to add to a a number *less* than b, and less by c. Hence our final result will be obtained by taking c from $a+b$, and it will be
$$a+b-c.$$

Case IV. To take from a the difference between b and c.
This is expressed thus: $a-(b-c)$.
First take b from a, the result will be
$$a-b.$$

This result is *too small*, for we have to take from a a number *less* than b, and less by c. Hence our final result will be obtained by adding c to $a-b$, and it will be
$$a-b+c.$$

Note. We assume that a, b, c represent such numbers that in CASE II. a is not less than the sum of b and c, in CASE III. b is not less than c, and in CASE IV. b is not less than c, and a is not less than b.

19. Collecting the results obtained in Art. 18, we have

$$a+(b+c)=a+b+c,$$
$$a-(b+c)=a-b-c,$$
$$a+(b-c)=a+b-c,$$
$$a-(b-c)=a-b+c.$$

From which we obtain the following rules for the removal of a bracket.

Rule I. When a bracket is preceded by the sign $+$, remove the bracket and leave the signs of the terms in it *unchanged*.

Rule II. When a bracket is preceded by the sign $-$, remove the bracket and *change the sign of each term in it*.

These rules apply to cases in which any number of terms are included in the bracket.

Thus

$$a+b+(c-d+e-f)=a+b+c-d+e-f,$$

and

$$a+b-(c-d+e-f)=a+b-c+d-e+f.$$

20. The rules given in the preceding Article for the removal of brackets furnish corresponding rules for the *introduction* of brackets.

Thus if we enclose two or more terms of an expression in a bracket.

  I. The sign of each term remains the same if $+$ precedes the bracket :

  II. The sign of each term is changed if $-$ precedes the bracket.

Ex. $a-b+c-d+e-f=a-b+(c-d)+(e-f),$
    $a-b+c-d+e-f=a-(b-c)-(d-e+f).$

ADDITION AND SUBTRACTION. 7

21. We may now proceed to give rules for the Addition and Subtraction of algebraical expressions.

Suppose we have to *add* to the expression $a+b-c$ the expression $d-e+f$.

$$\text{The Sum} = a+b-c+(d-e+f)$$
$$= a+b-c+d-e+f \text{ (by Art. 19, Rule I.).}$$

Also, if we have to *subtract* from the expression $a+b-c$ the expression $d-e+f$.

$$\text{The Difference} = a+b-c-(d-e+f)$$
$$= a+b-c-d+e-f \text{ (by Art. 19, Rule II.).}$$

We might arrange the expressions in each case under each other as in Arithmetic: thus

To $a+b-c$	From $a+b-c$
Add $d-e+f$	Take $d-e+f$
Sum $a+b-c+d-e+f$	Difference $a+b-c-d+e-f$

and then the rules may be thus stated.

I. In Addition attach the lower line to the upper with the signs of both lines unchanged.

II. In Subtraction attach the lower line to the upper with *the signs of the lower line changed*, the signs of the upper line being unchanged.

The following are examples.

(1)
$$\text{To } a+b+9$$
$$\text{Add } a-b-6$$
$$\text{Sum } a+b+9+a-b-6$$

and this sum $= a+a+b-b+9-6$
$= 2a+3.$

For it has been shown, Art. 9, that $a+a=2a$, and, Art. 13, that $b-b=0$.

(2)
$$\text{From } a+b+9$$
$$\text{Take } a-b-6$$
$$\text{Remainder } a+b+9-a+b+6$$

and this remainder $= 2b+15.$

22. We have worked out the examples in Art. 21 at full length, but in practice they may be abbreviated, by combining the symbols or digits by a mental process, thus

$$\begin{array}{ll} \text{To } c+d+10 & \text{From } c+d+10 \\ \text{Add } c-d-7 & \text{Take } c-d-7 \\ \hline \text{Sum } 2c\ \ +3 & \text{Remainder } 2d+17 \end{array}$$

23. We have said that

$$\text{instead of } a+a \text{ we write } 2a,$$
$$\ldots\ldots\ a+a+a\ \ldots\ldots\ 3a,$$

and so on.

The digit thus prefixed to a symbol is called the *coefficient* of the term in which it appears.

24. Since
$$3a = a+a+a,$$
$$\text{and } 5a = a+a+a+a+a,$$
$$3a + 5a = a+a+a+a+a+a+a+a$$
$$= 8a.$$

Terms which have the same symbol, whatever their coefficients may be, are called *like* terms: those which have different symbols are called *unlike* terms.

Like terms, when positive, may be combined into one by adding their coefficients together and subjoining the common symbol; thus

$$2x + 5x = 7x,$$
$$3y + 5y + 8y = 16y.$$

25. If a term appears without a coefficient, *unity* is to be taken as its coefficient.

Thus $$x + 5x = 6x.$$

26. Negative terms, when like, may be combined into one term with a negative sign prefixed to it by adding the coefficients and subjoining to the result the common symbol.

Thus
$$2x - 3y - 5y = 2x - 8y,$$
$$\text{for } 2x - 3y - 5y = 2x - (3y + 5y)$$
$$= 2x - 8y.$$

So again: $$3x - y - 4y - 6y = 3x - 11y.$$

27. If an expression contain two or more like terms, some being positive and others negative, we must first collect all the positive terms into one positive term, then all the negative terms into one negative term, and finally combine the two remaining terms into one by the following process. Subtract the smaller *coefficient* from the greater, and set down the remainder with the sign of the greater prefixed and the common symbol attached to it.

Ex. $8x - 3x = 5x$,
$7x - 4x + 5x - 3x = 12x - 7x = 5x$,
$a - 2b + 5b - 4b = a + 5b - 6b = a - b$.

28. The rules for the combination of any number of like terms into one single term enable us to extend the application of the rules for Addition and Subtraction in Algebra, and we proceed to give some Examples.

ADDITION.

(1) $a - 2b + 3c$
$3a - 4b - 5c$
$\overline{4a - 6b - 2c}$

(2) $5a + 7b - 3c - 4d$
$6a - 7b + 9c + 4d$
$\overline{11a + 6c}$

The terms containing b and d in Ex. (2) destroying one another.

(3) $7x - 5y + 4z$
$x + 2y - 11z$
$3x - y + 5z$
$5x - 3y - z$
$\overline{16x - 7y - 3z}$

(4) $6m - 13n + 5p$
$8m + n - 9p$
$m - n - p$
$m + 2n + 5p$
$\overline{16m - 11n}$

SUBTRACTION.

(1) $5a - 3b + 6c$
$2a + 5b - 4c$
$\overline{3a - 8b + 10c}$

(2) $3a + 7b - 8c$
$3a - 7b + 4c$
$\overline{14b - 12c}$

(2) $5a - 6b + 2c$
$2a - 6b + 2c$
$\overline{3a}$

(4) $x - y + z$
$x - y - z$
$\overline{2z}$

(5) $3x + 7y + 12z$
$5y - 2z$
$\overline{3x + 2y + 14z}$

(6) $7x - 19y - 14z$
$6x - 24y + 9z$
$\overline{x + 5y - 23z}$

29. We have placed the expressions in the examples given in the preceding Article under each other, as in Arithmetic, for the sake of clearness, but the same operations might be exhibited by means of signs and brackets, thus Examples (2) of each rule might have been worked thus, in Addition,

$$5a + 7b - 3c - 4d + (6a - 7b + 9c + 4d)$$
$$= 5a + 7b - 3c - 4d + 6a - 7b + 9c + 4d$$
$$= 11a + 6c \;;$$

and, in Subtraction,

$$3a + 7b - 8c - (3a - 7b + 4c)$$
$$= 3a + 7b - 8c - 3a + 7b - 4c$$
$$= 14b - 12c.$$

EXAMPLES.—i.

Simplify the following expressions, by combining like symbols in each.

1. $3a + 4b + 5c + 2a + 3b + 7c.$ 2. $4a + 5b + 6c - 3a - 2b - 4c.$
3. $6a - 3b - 4c - 4a + 5b + 6c.$
4. $8a - 5b + 3c - 7a - 2b + 6c - 3a + 9b - 7c + 10a.$
5. $5x - 3a + b + 7 + 2b - 3x - 4a - 9.$
6. $a - b - c + b + c - d + d - a.$
7. $5a + 10b - 3c + 2b - 3a + 2c - 2a + 4c.$

EXAMPLES.—ii. ADDITION.

Add together

1. $a + x$ and $a - x.$ 2. $a + 2x$ and $a + 3x.$
3. $a - 2x$ and $2a - x.$ 4. $3x + 7y$ and $5x - 2y.$
5. $a + 3b + 5c$ and $3a - 2b - 3c.$
6. $a - 2b + 3c$ and $a + 2b - 3c.$ 7. $1 + x - y$ and $3 - x + y.$
8. $2x - 3y + 4z, 5x - 7y - 2z,$ and $6x + 9y - 8z.$
9. $2a + b - 3x, 3a - 2b + x, a + b - 5x,$ and $4a - 7b + 6x.$

EXAMPLES.—iii. SUBTRACTION.

1. From $a + b$ take $a - b.$
2. ...... $3x + y$ ...... $2x - y.$
3. ...... $2a + 3c + 4d$ ...... $2c + 3d.$
4. ...... $x + y$ ...... $y - z.$

5. From $m-n+r$ take $m-n-r$.
6. ...... $a+b+c$ ...... $a-b-c$.
7. ...... $3a+4b+5c$ ...... $2a+7b+6c$.
8. ...... $3x+5y-4z$ ...... $3x+2y-5z$.

30. We have given examples of the use of a bracket. The methods of denoting a bracket are various; thus, besides the marks ( ), the marks [ ], or { }, are often employed. Sometimes a mark called "The Vinculum" is drawn over the symbols which are to be connected, thus $a-\overline{b+c}$ is used to represent the same expression as that represented by $a-(b+c)$.

Often the brackets are made to enclose one another, thus
$$a-[b+\{c-(d-\overline{e-f})\}].$$

In removing the brackets from an expression of this kind it is best to commence with the *innermost*, and to remove the brackets one by one, the outermost last of all.

Thus
$$a-[b+\{c-(d-\overline{e-f})\}]$$
$$=a-[b+\{c-(d-e+f)\}]$$
$$=a-[b+\{c-d+e-f\}]$$
$$=a-[b+c-d+e-f]$$
$$=a-b-c+d-e+f.$$

Again
$$5x-(3x-7)-\{4-2x-(6x-3)\}$$
$$=5x-3x+7-\{4-2x-6x+3\}$$
$$=5x-3x+7-4+2x+6x-3$$
$$=10x.$$

EXAMPLES.—iv. BRACKETS.

Simplify the following expressions, combining all like quantities in each.

1. $a+b+(3a-2b)$.
2. $a+b-(a-3b)$.
3. $3a+5b-6c-(2a+4b-2c)$.
4. $a+b-c-(a-b-c)$.
5. $14x-(5x-9)-\{4-3x-(2x-3)\}$.
6. $4x-\{3x-(2x-x-a)\}$.
7. $15x-\{7x+(3x+a-x)\}$.

8. $a-[b+\{a-(b+a)\}]$.
9. $6a+[4a-\{8b-(2a+4b)-22b\}-7b]-[7b+\{8a-(3b+4a)+8b\}+6a]$.
10. $b-[b-(a+b)-\{b-(b-a-b)\}]$.
11. $2c-(6a-b)-\{c-(5a+2b)-(a-3b)\}$.
12. $2x-\{a-(2a-[3a-(4a-[5a-(6a-x)])])\}$.
13. $25a-19b-[3b-\{4a-(5b-6c)\}]$.

31. We have hitherto supposed the symbols in every expression used for illustration to represent *such* numbers that the expressions symbolize results which would be arithmetically possible.

Thus $a-b$ symbolizes a possible result, so long as a is not less than b.

If, for instance, a stands for 10 and b for 6,
$$a-b \text{ will stand for } 4.$$
But if a stands for 6 and b for 10,
$a-b$ denotes no possible result, because we cannot take the number 10 from the number 6.

But though there can be no such a thing as *a negative number*, we can conceive the real existence of *a negative quantity*.

To explain this we must consider

    I. What we mean by Quantity.
    II. How Quantities are measured.

32. A Quantity is anything which may be regarded as being made up of parts like the whole.

Thus a distance is *a quantity*, because we may regard it as made up of parts each of themselves a distance.

Again a sum of money is *a quantity*, because we may regard it as made up of parts like the whole.

33. To *measure* any quantity we fix upon some known quantity of the same kind for our standard, or *unit*, and then any quantity of that kind is measured by saying how many times it contains this unit, and this number of times is called *the measure* of the **quantity.**

For example, to measure any distance along a road we fix upon a known distance, such as a mile, and express all distances by saying how many times they contain this unit. Thus 16 is the measure of a distance containing 16 miles.

Again, to measure a man's income we take one pound as our unit, and thus if we said (as we often do say) that a man's income is 500 a year, we should mean 500 times the unit, that is, £500. Unless we knew what the unit was, to say that a man's income was 500 would convey no definite meaning: all we should know would be that, whatever our unit was, a pound, a dollar, or a franc, the man's income would be 500 times that unit, that is, £500, 500 dollars, or 500 francs.

N.B. Since the unit contains itself *once*, its measure is *unity*, and hence its name.

34. Now we can conceive a quantity to be such that when put to another quantity of the same kind it will entirely or in part neutralize its effect.

Thus, if I walk 4 miles towards a certain object and then return along the same road 2 miles, I may say that the latter distance is such a quantity that it neutralizes part of my first journey, so far as regards my position with respect to the point from which I started.

Again, if I gain £500 in trade and then lose £400, I may say that the latter sum is such a quantity that it neutralizes part of my first gain.

If I gain £500 and then lose £700, I may say that the latter sum is such a quantity that it neutralizes all my first gain, and not only that, but also a quantity of which the absolute value is £200 remains *in readiness to neutralize some future gain*. Regarding this £200 by itself we call it *a quantity* which will have a *subtractive* effect on subsequent profits.

Now, since Algebra is intended to deal with such questions in a general way, and to teach us how to put quantities, alike or opposite in their effect, together, a convention is adopted, founded on the *additive* or *subtractive* effect of of the quantities in question, and stated thus:

"To the quantities to be *added* prefix the sign $+$, and to the quantities to be *subtracted* prefix the sign $-$, and then write down all the quantities involved in such a question connected with these signs.

Thus, suppose a man to trade for 4 years, and to gain a pounds the first year, to lose b pounds the second year, to gain c pounds the third year, and to lose d pounds the fourth year.

The *additive* quantities are here a and c, which we are to write $+a$ and $+c$,

The *subtractive* quantities are here b and d, which we are to write $-b$ and $-d$,

$\therefore$ Result of trading $= +a-b+c-d$.

35. Let us next take the case in which the gain for the first year is a pounds, and the loss for each of three subsequent years is a pounds.

Result of trading $= +a-a-a-a$
$= -2a$.

Thus we arrive at an isolated quantity of a *subtractive* nature.

Arithmetically we *interpret* this result as a *loss* of £2a.

Algebraically we call the result a *negative quantity*.

When once we have admitted the possibility of the independent existence of such quantities as this we may extend the application of the rules for Addition and Subtraction, for

I. A negative quantity may stand by *itself*, and we may then add it to or take it from some other quantity or expression.

II. A negative quantity may stand *first* in an expression which we may have to add to or subtract from any other expression.

The Rules for Addition and Subtraction given in Art. 21 will be applicable to these expressions, as in the following Examples.

ADDITION.

(1) $5a-7a = -2a$.
(2) $4a-3b-6a+7b = -2a+4b$.
(3) To $4a$ To $5a-3b$
     Add $-3a$ Add $-2a-2b$
     Sum a Sum $3a-5b$

(4) $6a - 5b - 4c + 6$
     $-5a + 7b - 12c - 17$
     $- a - 8b + 19c + 4$
     ───────────────────
     $-6b + 3c - 7$

(5) $7x - 5y + 9z$
     $-18x + 9y - 5z$
     $- 3x - 8y + z$
     ───────────────
     $-14x - 4y + 5z$

SUBTRACTION.

(1)   From x
      Take $-y$
      Remainder $x+y$

or we might represent the operation thus,

$$x-(-y)=x+y.*$$

(2) $a+b-(-a+b)=a+b+a-b=2a.$

(3) $-a-b-(a-b)= -a-b-a+b= -2a.$

(4) $-3a + 4b - 7c + 10$
          $5a - 9b + 8c + 19$
         ──────────────────
         $-8a + 13b - 15c - 9$

(5) $x-y-[3x-\{-5x-(-4y+7x)\}]$
   $=x-y-[3x-\{-5x+4y-7x\}]$
   $=x-y-[3x+5x-4y+7x]$
   $=x-y-3x-5x+4y-7x$
   $= -14x+3y.$

(6) $7a + 5b + 9c - 12d$
            $-3b - 12c - 8d + 6e$
         ─────────────────────
          $7a + 8b + 21c - 4d - 6e$

In this example we have deviated from our previous practice of placing *like terms* under each other. This arrangement is useful to facilitate the calculation, but is not absolutely necessary; for the terms which are alike can be combined independently of it.

* NOTE.—The meaning of Subtraction is here extended so that the result in Art. 18, CASE IV. may be true when b is less than c.

EXAMPLES.—V.

(1.) ADDITION.

Add together

1. $6a+7b$, $-2a+4b$, and $3a-5b$.
2. $-5a+6b-7c$, $-2a+13b+9c$, and $7a-29b+4c$.
3. $2x-3y+4z$, $-5x+4y-7z$, and $-8x-9y-3z$.
4. $-a+b-c+d$, $a-2b-3c+d$, $-5b+4c$, and $-5c+d$.
5. $a+b-c+7$, $-2a-3b-4c+9$, and $3a+2b+5c-10$.
6. $5x-3z-4b$, $6y-2a$, $3a-2y$, and $5b-7x$.
7. $a+b-c$, $c-a+b$, $2b-c+3a$, and $4a-3c$.
8. $7a-3b-5c+9d$, $2b-3c-5d$, and $-4d+15c$.
9. $-12x-5y+4z$, $3x+2y-3z$, and $9x-3y+z$.

(2.) SUBTRACTION.

1. From $a+b$ take $-a-b$.
2. From $a-b$ take $-b+c$.
3. From $a-b+c$ take $-a+b-c$.
4. From $6x-8y+3$ take $-2x+9y-2$.
5. From $5a-12b+17c$ take $-2a+4b-3c$.
6. From $2a+b-3x$ take $4b-3a+5x$.
7. From $a+b-c$ take $3c-2b+4a$.
8. From $a+b+c-7$ take $8\ c-b+a$.
9. From $12x-3y-z$ take $4y-5z+x$.
10. From $8a-5b+7c$ take $2c-4b+2a$.
11. From $9p-4q+3r$ take $5q-3p+r$.

II. MULTIPLICATION.

36. THE operation of finding the sum of a numbers each equal to b is called **Multiplication**.

The number a is called the Multiplier.
.... b .... Multiplicand.

This Sum is called the PRODUCT of the multiplication of b by a.

This Product is represented in Algebra by three distinct symbols:

I. By writing the symbols side by side, with no sign between them, thus, ab;

II. By placing a small dot between the symbols, thus, $a.b$;

III. By placing the sign $\times$ between the symbols, thus, $a \times b$; and all these are read thus, "a into b," or "a times b."

In Arithmetic we chiefly use the *third* way of expressing a Product, for we cannot symbolize the product of 5 into 7 by 57, which means the sum of fifty and seven, nor can we well represent it by 5.7, because it might be confounded with the notation used for decimal fractions, as 5·7.

37. In Arithmetic

2×7 stands for the same as $7 + 7$.
3×4 .... $4 + 4 + 4$.

In Algebra

ab stands for the same as $b + b + b + \ldots$ with b written a times.

$(a + b) c$ stands for the same as $c + c + c \ldots$ with c written $a + b$ times.

[S.A.]

38. *To shew that* $3 \text{ times } 4 = 4 \text{ times } 3$.

$$\begin{aligned} 3 \text{ times } 4 &= 4+4+4 \\ &= \left. \begin{array}{l} 1+1+1+1 \\ +1+1+1+1 \\ +1+1+1+1 \end{array} \right\} \ldots\ldots\ldots \text{I.} \end{aligned}$$

$$\begin{aligned} 4 \text{ times } 3 &= 3+3+3+3 \\ &= \left. \begin{array}{l} 1+1+1 \\ +1+1+1 \\ +1+1+1 \\ +1+1+1 \end{array} \right\} \ldots\ldots\ldots \text{II.} \end{aligned}$$

Now the results obtained from I. and II. must be the same, for the horizontal columns of one are identical with the vertical columns of the other.

39. *To prove that* $ab = ba$.

ab means that the sum of a numbers each equal to b is to be taken.

$$\begin{aligned} \therefore ab &= b+b+\ldots\ldots \text{with } b \text{ written } a \text{ times} \\ &= \begin{array}{l} b \\ +b \\ + \\ \ldots\ldots \\ \text{to } a \text{ lines} \end{array} \\ &= \left. \begin{array}{l} 1+1+1+\ldots\ldots \text{to } b \text{ terms} \\ +1+1+1+\ldots\ldots \text{to } b \text{ terms} \\ +\ldots\ldots\ldots\ldots\ldots\ldots\ldots\ldots \\ \text{to } a \text{ lines.} \end{array} \right\} \ldots\ldots \text{I.} \end{aligned}$$

Again,

$$\begin{aligned} ba &= a+a+\ldots\ldots \text{with } a \text{ written } b \text{ times} \\ &= \begin{array}{l} a \\ +a \\ + \\ \ldots\ldots \\ \text{to } b \text{ lines} \end{array} \\ &= \left. \begin{array}{l} 1+1+1+\ldots\ldots \text{to } a \text{ terms} \\ +1+1+1+\ldots\ldots \text{to } a \text{ terms} \\ +\ldots\ldots\ldots\ldots\ldots\ldots\ldots\ldots \\ b \text{ lines} \end{array} \right\} \ldots\ldots \text{II.} \end{aligned}$$

Now the results obtained from I. and II. must be the same, for the horizontal columns of one are clearly the same as the vertical columns of the other.

40. Since the expressions ab and ba are the same in meaning, we may regard either a or b as the multiplier in forming the product of a and b, and so we may read ab in two ways:

(1) a into b,
(2) a multiplied by b.

41. The expressions abc, acb, bac, bca, cab, cba are all the same in meaning, denoting that the three numbers symbolized by a, b, and c are to be multiplied together. It is, however, generally desirable that the alphabetical order of the letters representing a product should be observed.

42. Each of the numbers a, b, c is called a FACTOR of the product abc.

43. When a number expressed in figures is one of the factors of a product it always stands first in the product.

Thus the product of the factors x, y, z and 9 is represented by $9xyz$.

44. Any one or more of the factors that make up a product is called the COEFFICIENT of the other factors.

Thus in the expression $2ax$, $2a$ is called the coefficient of x.

45. When a factor a is repeated *twice* the product would be represented, in accordance with Art. 36, by aa; when *three* times, by aaa. In such cases these products are, for the sake of brevity, expressed by writing the symbol with a number *placed above it on the right*, expressing the number of times the symbol is repeated; thus

instead of aa we write a^2
.... aaa .... a^3
.... $aaaa$ .... a^4

These expressions a^2, a^3, a^4...... are called the second, third, fourth...... POWERS of a.

The number placed over a symbol to express the power of the symbol is called the INDEX or EXPONENT.

a^2 is generally called the *square* of a.
a^3 ... . the *cube* of a.

46. The product of a^2 and $a^3 = a^2 \times a^3$
$$= aa \times aaa = aaaaa = a^5.$$

Thus the index of the resulting power is the *sum* of the indices of the two factors.

Similarly $\quad a^4 \times a^6 = aaaa \times aaaaaa$
$$= aaaaaaaaaa = a^{10} = a^{4+6}.$$

If one of the factors be a symbol without an index, we may assume it to have an index[1], that is
$$a = a^1.$$

Examples in multiplying powers of the same symbol are

(1) $a \times a^2 = a^{1+2} = a^3.$

(2) $7a^3 \times 5a^7 = 7 \times 5 \times a^3 \times a^7 = 35a^{3+7} = 35a^{10}.$

(3) $a^3 \times a^6 \times a^9 = a^{3+6+9} = a^{18}.$

(4) $x^2y \times xy^2 = x^2 . y . x . y^2 = x^2 . x . y . y^2 = x^{2+1} . y^{1+2} = x^3y^3.$

(5) $a^2b \times ab^3 \times a^5b^7 = a^{2+1+5} . b^{1+3+7} = a^8 . b^{11}.$

EXAMPLES.—vi.

Multiply

1. x into $3y$.
2. $3x$ into $4y$.
3. $3xy$ into $4xy$.
4. $3abc$ into ac.
5. a^3 into a^4.
6. a^7 into a.
7. $3a^2b$ into $4a^3b^2$.
8. $7a^4c$ into $5a^2bc^3$.
9. $15ab^4c^3$ by $12a^3bc$.
10. $7a^5c^7$ by $4a^2bc^3$.
11. a^8 by $3a^3$.
12. $4a^3bx$ by $5ab^2y$.
13. $19x^3yz$ by $4xy^3z^2$.
14. $17ab^3z$ by $3bc^2y$.
15. $6x^5y^8z^3$ by $8x^3y^2z^3$.
16. $3abc$ by $4axy$.
17. a^7b^2c by $8a^7b^3c$.
18. $9m^2np$ by $m^3n^2p^2$.
19. ay^2z by bx^2z^3.
20. $11a^3bx$ by $3a^{17}b^{15}m^2$.

47. The rules for the addition and subtraction of powers are similar to those laid down in Chap. I. for simple quantities.

Thus the sum of the second and third powers of x is represented by
$$x^2 + x^3,$$
and the remainder after taking the fourth power of y from the fifth power of y is represented by
$$y^5 - y^4,$$
and these expressions cannot be abridged.

But when we have to add or subtract the same powers of the same quantities the terms may be combined into one: thus

$$x^3 + x^3 = 2x^3,$$
$$3y^3 + 5y^3 + 7y^3 = 15y^3,$$
$$8x^4 - 5x^4 = 3x^4,$$
$$9y^5 - 3y^5 - 2y^5 = 4y^5.$$

Again, whenever two or more terms are entirely the same with respect to the symbols they contain, their sum may be abridged.

Thus
$$ad + ad = 2ad,$$
$$3a^2b - 2a^2b = a^2b,$$
$$5a^3b^3 + 6a^3b^3 - 9a^3b^3 = 2a^3b^3,$$
$$7a^2x - 10a^2x - 12a^2x = -15a^2x.$$

48. From the multiplication of *simple* expressions we pass on to the case in which *one* of the quantities whose product is to be found is a *compound* expression.

To shew that $(a + b)\,c = ac + bc$.

$(a + b)\,c = c + c + c + \ldots$ with c written $a + b$ times,
$\qquad = (c + c + c + \ldots$ with c written a times$)$
$\qquad\quad + (c + c + c \ldots$ with c written b times$)$,
$\qquad = ac + bc.$

49. To shew that $(a - b)\,c = ac - bc$.

$(a - b)c = c + c + c + \ldots$ with c written $a - b$ times,
$\qquad = (c + c + c + \ldots$ with c written a times$)$
$\qquad\quad - (c + c + c \ldots$ with c written b times$)$,
$\qquad = ac - bc.$

Note. We assume that a is greater than b.

50. Similarly it may be shewn that
$$(a + b + c)\,d = ad + bd + cd,$$
$$(a - b - c)\,d = ad - bd - cd,$$

and hence we obtain the following general rule for finding the product of a *single* symbol and an expression consisting of two or more terms.

"Multiply each of the terms by the single symbol, and connect the terms of the result by the signs of the several terms of the *compound* expression."

EXAMPLES.—vii.

Multiply

1. $a+b-c$ by a.
2. $a+3b-4c$ by $2a$.
3. a^3+3a^2+4a by a.
4. $3a^3-5a^2-6a+7$ by $3a^2$.
5. $a^2-2ab+b^2$ by ab.
6. $a^3-3a^2b^2+b^3$ by $3a^2b$.
7. $8m^2+9mn+10n^2$ by mn.
8. $9a^5+4a^4b-3a^3b^2+4a^2b^3$ by $2ab$.
9. $x^3y^3-x^2y^2+xy-7$ by xy.
10. $m^3-3m^2n+3mn^2-n^3$ by n.
11. $12a^3b-6a^2b^2+5ab^3$ by $12a^2b^3$.
12. $13x^3-17x^2y+5xy^2-y^3$ by $8xy$.

51. We next proceed to the case in which both multiplier and multiplicand are *compound expressions*.

First to multiply $a+b$ into $c+d$.

Represent $c+d$ by x.

Then $(a+b)(c+d) = (a+b)x$
$\qquad = ax + bx$, by Art. 48,
$\qquad = a(c+d) + b(c+d)$
$\qquad = ac + ad + bc + bd$, by Art. 48.

The same result is obtained by the following process:

$$\begin{array}{r} c+d \\ a+b \\ \hline ac+ad \\ +bc+bd \\ \hline ac+ad+bc+bd \end{array}$$

which may be thus described:

Write $a+b$ considered as the *multiplier* under $c+d$ considered as the *multiplicand*, as in common Arithmetic. Then multiply each term of the multiplicand by a, and set down the result. Next multiply each term of the multiplicand by b, and set down the result under the result obtained before. The sum of the two results will be the product required.

Note. The second result is shifted one place to the right The object of this will be seen in Art. 56.

52. Next, to multiply $a+b$ into $c-d$.

Represent $c-d$ by x.

Then
$$\begin{aligned}(a+b)(c-d)&=(a+b)x\\&=ax+bx\\&=a(c-d)+b(c-d)\\&=ac-ad+bc-bd,\end{aligned}$$
by Art. 49.

From a comparison of this result with the factors from which it is produced it appears that if we regard the terms of the multiplicand $c-d$ as *independent* quantities, and call them $+c$ and $-d$, the effect of multiplying the positive terms $+a$ and $+b$ into the positive term $+c$ is to produce two positive terms $+ac$ and $+bc$, whereas the effect of multiplying the positive terms $+a$ and $+b$ into the *negative term* $-d$ is to produce *two negative terms* $-ad$ and $-bd$.

The same result is obtained by the following process:

$$\begin{array}{r}c-d\\a+b\\\hline ac-ad\\+bc-bd\\\hline ac-ad+bc-bd\end{array}$$

This process may be described in a similar manner to that in Art. 51, it being assumed that a positive term multiplied into a negative term gives a negative result.

Similarly we may shew that $a-b$ into $c+d$ gives
$$ac+ad-bc-bd.$$

53. Next to multiply $a-b$ into $c-d$.

Represent $c-d$ by x.

Then
$$\begin{aligned}(a-b)(c-d)&=(a-b)x\\&=ax-bx\\&=a(c-d)-b(c-d)\\&=(ac-ad)-(bc-bd),\text{ by Art. 49,}\\&=ac-ad-bc+bd.\end{aligned}$$

When we compare this result with the factors from which it is produced, we see that

> The product of the positive term a into the positive term c is the positive term ac.

The product of the positive term a into the negative term $-d$ is the negative term $-ad$.

The product of the negative term $-b$ into the positive term c is the negative term $-bc$.

The product of the negative term $-b$ into the negative term $-d$ is the positive term bd.

The multiplication of $c-d$ by $a-b$ may be written thus:

$$\begin{array}{r} c - d \\ a - b \\ \hline ac - ad \\ -bc + bd \\ \hline ac - ad - bc + bd \end{array}$$

54. The results obtained in the preceding Article enable us to state what is called the RULE OF SIGNS in Multiplication, which is

"*The product of two positive terms or of two negative terms is positive: the product of two terms, one of which is positive and the other negative, is negative.*"

55. The following more concise proof may now be given of the RULE OF SIGNS.

To shew that $(a-b)(c-d) = ac - ad - bc + bd$.

First, $(a-b)M = M + M + M + \ldots$ with M written $a-b$ times,
$= (M + M + M + \ldots$ with M written a times$)$
$- (M + M + M + \ldots$ with M written b times$)$,
$= aM - bM$.

Next, let $M = c - d$.

Then $aM = a(c-d)$
$= (c-d)a$ Art. 39.
$= ca - da$.   Art. 49.

Similarly, $bM = cb - db$.

$\therefore (a-b)(c-d) = (ca - da) - (cb - db)$.

Now to subtract $(cb - db)$ from $(ca - da)$, if we take away cb we take away db too much, and we must therefore add db to the result,

$\therefore$ we get $ca - da - cb + db$,
which is the same as $ac - ad - bc + bd$.   Art. 39.

So it appears that in multiplying $(a-b)(c-d)$ we must multiply each term in one factor by each term in the other and prefix the sign according to this law :—

When the factors multiplied have like signs prefix $+$, when unlike $-$ to the product.

This is the RULE OF SIGNS

56. We shall now give some examples in illustration of the principles laid down in the last five Articles.

Examples in Multiplication worked out.

(1) Multiply $x+5$ by $x+7$.

$$\begin{array}{r} x+5 \\ x+7 \\ \hline x^2+5x \\ +7x+35 \\ \hline x^2+12x+35 \end{array}$$

(2) Multiply $x-5$ by $x+7$.

$$\begin{array}{r} x-5 \\ x+7 \\ \hline x^2-5x \\ +7x-35 \\ \hline x^2+2x-35 \end{array}$$

The reason for shifting the second result one place to the right is that it enables us generally to place *like terms* under each other.

(3) Multiply $x+5$ by $x-7$.

$$\begin{array}{r} x+5 \\ x-7 \\ \hline x^2+5x \\ -7x-35 \\ \hline x^2-2x-35 \end{array}$$

(4) Multiply $x-5$ by $x-7$.

$$\begin{array}{r} x-5 \\ x-7 \\ \hline x^2-5x \\ -7x+35 \\ \hline x^2-12x+35 \end{array}$$

(5) Multiply x^2+y^2 by x^2-y^2.

$$\begin{array}{r} x^2+y^2 \\ x^2-y^2 \\ \hline x^4+x^2y^2 \\ -x^2y^2-y^4 \\ \hline x^4-y^4 \end{array}$$

(6) Multiply $3ax-5by$ by $7ax-2by$

$$\begin{array}{r} 3ax-5by \\ 7ax-2by \\ \hline 21a^2x^2-35abxy \\ -6abxy+10b^2y^2 \\ \hline 21a^2x^2-41abxy+10b^2y^2 \end{array}$$

57. The process in the multiplication of factors, one or both of which contains more than two terms, is similar to the processes which we have been describing, as may be seen from the following examples:

Multiply

(1) $x^2 + xy + y^2$ by $x - y$.

$$\begin{array}{l} x^2 + xy + y^2 \\ \underline{x - y} \\ x^3 + x^2y + xy^2 \\ - x^2y - xy^2 - y^3 \\ \overline{x^3 - y^3} \end{array}$$

(2) $a^2 + 6a + 9$ by $a^2 - 6a + 9$.

$$\begin{array}{l} a^2 + 6a + 9 \\ \underline{a^2 - 6a + 9} \\ a^4 + 6a^3 + 9a^2 \\ - 6a^3 - 36a^2 - 54a \\ + 9a^2 + 54a + 81 \\ \overline{a^4 - 18a^2 + 81} \end{array}$$

(3) Multiply $3x^2 + 4xy - y^2$ by $3x^2 - 4xy + y^2$.

$$\begin{array}{l} 3x^2 + 4xy - y^2 \\ \underline{3x^2 - 4xy + y^2} \\ 9x^4 + 12x^3y - 3x^2y^2 \\ - 12x^3y - 16x^2y^2 + 4xy^3 \\ + 3x^2y^2 + 4xy^3 - y^4 \\ \overline{9x^4 - 16x^2y^2 + 8xy^3 - y^4} \end{array}$$

(4) To find the continued product of $x+3$, $x+4$, and $x+6$.

To effect this we must multiply $x+3$ by $x+4$, and then multiply the result by $x+6$.

$$\begin{array}{l} x + 3 \\ \underline{x + 4} \\ x^2 + 3x \\ + 4x + 12 \\ \overline{x^2 + 7x + 12} \\ \underline{x + 6} \\ x^3 + 7x^2 + 12x \\ + 6x^2 + 42x + 72 \\ \overline{x^3 + 13x^2 + 54x + 72} \end{array}$$

Note. The numbers 13 and 54 are called the *coefficients* of x^2 and x in the expression $x^3 + 13x^2 + 54x + 72$, in accordance with Art. 44.

(5) Find the continued product of $x+a$, $x+b$, and $x+c$.

$$x+a$$
$$x+b$$
$$\overline{x^2+ax}$$
$$+bx+ab$$
$$\overline{x^2+ax+bx+ab}$$
$$x+c$$
$$\overline{x^3+ax^2+bx^2+abx}$$
$$+cx^2+acx+bcx+abc$$
$$\overline{x^3+(a+b+c)x^2+(ab+ac+bc)x+abc}$$

Note. The coefficients of x^2 and x in the expression just obtained are $a+b+c$ and $ab+ac+bc$ respectively.

When a coefficient is expressed in *letters*, as in this example, it is called a *literal* coefficient.

EXAMPLES.—viii.

Multiply

1. $x+3$ by $x+9$.   2. $x+15$ by $x-7$.   3. $x-12$ by $x+10$.
4. $x-8$ by $x-7$.   5. $a-3$ by $a-5$.   6. $y-6$ by $y+13$.
7. x^2-4 by x^2+5.   8. x^2-6x+9 by x^2-6x+5.
9. x^2+5x-3 by x^2-5x-3.   10. a^3-3a+2 by a^3-3a^2+2.
11. x^2-x+1 by x^2+x-1.   12. x^2+xy+y^2 by x^2-xy+y^2.
13. x^2+xy+y^2 by $x-y$.   14. a^2-x^2 by $a^4+a^2x^2+x^4$.
15. x^3-3x^2+3x-1 by x^2+3x+1.
16. $x^3+3x^2y+9xy^2+27y^3$ by $x-3y$.
17. $a^3+2a^2b+4ab^2+8b^3$ by $a-2b$.
18. $8a^3+4a^2b+2ab^2+b^3$ by $2a-b$.
19. $a^3-2a^2b+3ab^2+4b^3$ by $a^2-2ab-3b^2$.
20. $a^3+3a^2b-2ab^2+3b^3$ by $a^2+2ab-3b^2$.
21. $a^2-2ax+4x^2$ by $a^2+2ax+4x^2$.
22. $9a^2+3ax+x^2$ by $9a^2-3ax+x^2$.
23. $x^4-2ax^2+4a^2$ by $x^4+2ax^2+4a^2$.
24. $a^2+b^2+c^2-ab-ac-bc$ by $a+b+c$.
25. $x^2+4xy+5y^2$ by $x^3-3x^2y-2xy^2+3y^3$.
26. $ab+cd+ac+bd$ by $ab+cd-ac-bd$.

Find the continued product of the following expressions:

27. $x-a$, $x+a$, x^2+a^2, x^4+a^4.   28. $x-a$, $x+b$, $x-c$.

29. $1-x$, $1+x$, $1+x^2$, $1+x^4$.
30. $x-y$, $x+y$, x^2-xy+y^2, x^2+xy+y^2.
31. $a-x$, $a+x$, a^2+x^2, a^4+x^4, a^8+x^8.

Find the coefficient of x in the following expansions:

32. $(x-5)(x-6)(x+7)$ 33. $(x+8)(x+3)(x-2)$.
34. $(x-2)(x-3)(x+4)$ 35. $(x-a)(x-b)(x-c)$.
36. $(x^2+3x-2)(x^2-3x+2)(x^4-5)$.
37. $(x^2-x+1)(x^2+x-1)(x^4-x^2+1)$.
38. $(x^2-mx+1)(x^2-mx-1)(x^4-m^2x-1)$.

58. Our proof of the Rule of Signs in Art. 55 is founded on the supposition that a is greater than b and c is greater than d.

To include cases in which the *multiplier* is an isolated negative quantity we must *extend* our definition of Multiplication. For the definition given in Art. 36 does not cover this case, since we cannot say that c shall be taken $-d$ times.

We give then the following definition: "*The operation of Multiplication is such that the product of the factors $a-b$ and $c-d$ will be equivalent to $ac-ad-bc+bd$, whatever may be the values of a, b, c, d.*"

Now since
$$(a-b)(c-d) = ac-ad-bc+bd.$$
make $a=0$ and $d=0$.

Then $(0-b)(c-0) = 0 \times c - 0 \times 0 - bc + b \times 0$.
or $-b \times c = -bc$.

Similarly it may be shewn that
$$-b \times -d = +bd.$$

EXAMPLES.—IX.

Multiply

1. a^2 by $-b$.   2. a^2 by $-a^3$ 3. a^2b by $-ab^2$.
4. $-4a^2b$ by $-3ab^2$.   5. $5x^3y$ by $-6xy^2$.
6. a^2ab+b^2 by $-a$.   7. $3a^3+4a^2-5ab$ by $-2a^2$.
8. $-a^3-a^2-a$ by $-a-1$.
9. $3x^2y-5xy^2+4y^3$ by $2x-3y$.
10. $-5m^2-6mn+7n^2$ by $-m+n$.
11. $13r^2-17r-45$ by $-r-3$.
12. $7x^3-8x^2z-9z^2$ by $-x-z$.
13. $-x^5+x^4y-x^3y^2$ by $-y-x$.
14. $-y^3-xy^2-x^2y-x^3$ by $-x-y$.

III. INVOLUTION.

59. To this part of Algebra belongs the process called Involution. This is the operation of multiplying a quantity *by itself* any number of times.

The power to which the quantity is *raised* is expressed by the number of times the quantity has been employed as a factor in the operation.

Thus, as has been already stated in Art. 45,
a^2 is called the second power of a,
a^3 is called the third power of a.

60. When we have to raise negative quantities to certain powers we symbolize the operation by putting the quantity in a bracket with the number denoting the *index* (Art. 45) placed over the bracket on the right hand.

Thus $(-a)^3$ denotes the third power of $-a$,
$(-2x)^4$ denotes the fourth power of $-2x$.

61. The signs of all *even* powers of a negative quantity will be *positive*, and the signs of the *odd* powers will be *negative*.

Thus $(-a)^2 = (-a) \times (-a) = a^2$,
$(-a)^3 = (-a).(-a)(-a) = a^2.(-a) = -a^3$.

62. To raise a simple quantity to any power we multiply the index of the quantity by the number denoting the power to which it is to be raised, and prefix the proper sign.

Thus the square of a^3 is a^6,
the cube of a^3 is a^9,
the cube of $-x^2yz^3$ is $-x^6y^3z^9$.

63. We form the second, third and fourth powers of $a+b$ in the following manner:

$$a+b$$
$$a+b$$
$$\overline{a^2+ab}$$
$$+ab+b^2$$
$$(a+b)^2 = \overline{a^2+2ab+b^2}$$
$$a+b$$
$$\overline{a^3+2a^2b+ab^2}$$
$$+\ a^2b+2ab^2+b^3$$
$$(a+b)^3 = \overline{a^3+3a^2b+3ab^2+b^3}$$
$$a+b$$
$$\overline{a^4+3a^3b+3a^2b^2+ab^3}$$
$$+\ a^3b+3a^2b^2+3ab^3+b^4$$
$$(a+b)^4 = \overline{a^4+4a^3b+6a^2b^2+4ab^3+b^4}.$$

Here observe the following laws:

I. The indices of a *decrease* by unity in each term.
II. The indices of b *increase* by unity in each term.
III. The numerical coefficient of the *second* term is always the same as the index of the power to which the binomial is raised.

64. We form the second, third and fourth powers of $a-b$ in the following manner:

$$a-b$$
$$a-b$$
$$a^2-ab$$
$$-ab+b^2$$
$$(a-b)^2 = \overline{a^2-2ab+b^2}$$
$$a-b$$
$$a^3-2a^2b+ab^2$$
$$-\ a^2b+2ab^2-b^3$$
$$(a-b)^3 = a^3-3a^2b+3ab^2-b^3$$
$$a-b$$
$$\overline{a^4-3a^3b+3a^2b^2-ab^3}$$
$$-\ a^3b+3a^2b^2-3ab^3+b^4$$
$$(a-b)^4 = a^4-4a^3b+6a^2b^2-4ab^3+b^4.$$

Now observe that the powers of $a-b$ do not differ from the powers of $a+b$ except that the terms, in which the *odd* powers of b, as b^1, b^3, occur have the sign $-$ prefixed.

Hence if any power of $a+b$ be *given* we can write the corresponding power of $a-b$: thus

since
$$(a+b)^5 = a^5 + 5a^4b + 10a^3b^2 + 10a^2b^3 + 5ab^4 + b^5,$$
$$(a-b)^5 = a^5 - 5a^4b + 10a^3b^2 - 10a^2b^3 + 5ab^4 - b^5.$$

65. Since $(a+b)^2 = a^2 + b^2 + 2ab$ and $(a-b)^2 = a^2 + b^2 - 2ab$, it appears that the square of a binomial is formed by the following process:

"To the sum of the squares of each term add twice the product of the terms."

Thus
$$(x+y)^2 = x^2 + y^2 + 2xy,$$
$$(x+3)^2 = x^2 + 9 + 6x,$$
$$(x-5)^2 = x^2 + 25 - 10x,$$
$$(2x-7y)^2 = 4x^2 + 49y^2 - 28xy.$$

66. To form the square of a trinomial:

$$\begin{array}{l}a+b+c\\a+b+c\\\hline a^2+ab+ac\\\quad+ab+b^2+bc\\\quad\quad+ac+bc+c^2\\\hline a^2+2ab+b^2+2ac+2bc+c^2.\end{array}$$

Arranging this result thus $a^2 + b^2 + c^2 + 2ab + 2ac + 2bc$, we see that it is composed of two sets of quantities:

I. The squares of the quantities a, b, c.

II. The double products of a, b, c taken two and two.

Now, if we form the square of $a-b-c$, we get

$$\begin{array}{l}a-b-c\\a-b-c\\\hline a^2-ab-ac\\\quad-ab+b^2+bc\\\quad\quad-ac+bc+c^2\\\hline a^2-2ab+b^2-2ac+2bc+c^2.\end{array}$$

The law of formation is the same as before, for we have

I. The squares of the quantities.

II. The double products of $a, -b, -c$ taken two by two: the sign of each result being $+$ or $-$, according as the signs of the algebraical quantities composing it are like or unlike.

67. The same law holds good for expressions containing more than three terms, thus

$(a+b+c+d)^2 = a^2+b^2+c^2+d^2$
$\qquad +2ab+2ac+2ad+2bc+2bd+2cd,$
$(a-b+c-d)^2 = a^2+b^2+c^2+d^2$
$\qquad -2ab+2ac-2ad-2bc+2bd-2cd.$

And generally, the square of an expression containing 2, 3, 4 or more terms will be formed by the following process:

"To the sum of the squares of each term add twice the product of each term into each of the terms that follow it."

EXAMPLES.—X.

Form the square of each of the following expressions:

1. $x+a$. 2. $x-a$. 3. $x+2$. 4. $x-3$. 5. x^2+y^2.
6. x^2-y^2. 7. a^3+b^3. 8. a^3-b^3. 9. $x+y+z$. 10. $x-y+z$.
11. $m+n-p-r$. 12. x^2+2x-3. 13. x^2-6x+7.
14. $2x^2-7x+9$. 15. $x^2+y^2-z^2$. 16. $x^4-4x^2y^2+y^4$.
17. $a^3+b^3+c^3$. 18. $x^3-y^3-z^3$. 19. $x+2y-3z$.
20. $x^2-2y^2+5z^2$.

Expand the following expressions:

21. $(x+a)^3$. 22. $(x-a)^3$. 23. $(x+1)^3$. 24. $(x-1)^3$.
25. $(x+2)^3$. 26. $(a^2-b^2)^3$. 27. $(a+b+c)^3$. 28. $(a-b-c)^3$.
29. $(m+n)^2.(m-n)^2$. 30. $(m+n)^2.(m^2-n^2)$.

68. An algebraical product is said to be of 2, 3......dimensions, when the sum of the indices of the quantities composing the product is 2, 3..........

Thus $\quad ab$ is an expression of 2 dimensions,
$\qquad a^2b^2c$ is an expression of 5 dimensions.

69. An algebraical expression is called *homogeneous* when each of its terms is of the same dimensions.

Thus $x^2 + xy + y^2$ is homogeneous, for each term is of 2 dimensions.

Also $3x^3 + 4x^2y + 5y^3$ is homogeneous, for each term is of 3 dimensions, the numerical coefficients not affecting the dimensions of each term.

70. An expression is said to be *arranged* according to powers of some letter, when the indices of that letter occur in the order of their magnitudes, either increasing or decreasing.

Thus the expression $a^3 + a^2x + ax^2 + x^3$ is arranged according to *descending* powers of a, and *ascending* powers of x.

71. One expression is said to be of a *higher* order than another when the former contains a higher power of some distinguishing letter than the other.

Thus $a^3 + a^2x + ax^2 + x^3$ is said to be of a higher order than $a^2 + ax + x^2$, with reference to the index of a.

IV. DIVISION.

72. **Division** is the process by which, when a *product* is given and we know *one* of the factors, the *other* factor is determined.

The product is, with reference to this process, called the DIVIDEND.

The given factor is called the DIVISOR.

The factor which has to be found is called the QUOTIENT.

73. The operation of Division is denoted by the sign $\div$.

Thus $ab \div a$ signifies that ab is to be divided by a.

The same operation is denoted by writing the dividend over the divisor with a line drawn between them, thus $\dfrac{ab}{a}$.

In this chapter we shall treat only of cases in which the dividend contains the divisor an exact number of times.

Case I.

74. When the dividend and divisor are each included in a single term, we can usually tell by inspection the factors of which each is composed. The quotient will in this case be represented by the factors which remain in the dividend, when those factors which are common to the dividend and the divisor have been removed from the dividend.

Thus
$$\frac{ab}{b} = a,$$
$$\frac{3a^2}{a} = \frac{3aa}{a} = 3a,$$
$$\frac{a^5}{a^3} = \frac{aaaaa}{aaa} = aa = a^2.$$

Thus, when one power of a number is divided by a smaller power of the same number, the quotient is that power of the number whose index is *the difference between the indices of the dividend and the divisor.*

Thus
$$\frac{a^{12}}{a^5} = a^{12-5} = a^7,$$
$$\frac{15a^3b^2}{3ab} = 5a^2b.$$

75. The quotient is *unity* when the dividend and the divisor are equal.

Thus
$$\frac{a}{a} = 1; \quad \frac{x^2y^2}{x^2y^2} = 1;$$

and this will hold true when the dividend and the divisor are *compound* quantities.

Thus
$$\frac{a+b}{a+b} = 1; \quad \frac{x^2-y^2}{x^2-y^2} = 1.$$

EXAMPLES.—xi.

Divide

1. x^6 by r^3.
2. x^{10} by x^2.
3. x^4y^2 by xy.
4. $x^7y^{3.6}$ by xy^2z.
5. $24ab^2c$ by $4ab$.
6. $72a^2b^3c^3$ by $9a^2b^2c$.
7. $256a^3b^7c^9$ by $16abc^3$.
8. $1331m^{10}n^{11}p^{12}$ by $11m^2n^3p^4$.
9. $60a^3x^2y^5$ by $5xy$.
10. $96a^4b^2c^3$ by $12bc$.

Case II.

76. If the divisor be a single term, while the dividend contains two or more terms, the quotient will be found by dividing each term of the dividend separately by the divisor and connecting the results with their proper signs.

Thus
$$\frac{ax + bx}{x} = a + b,$$

$$\frac{a^3x^3 + a^2x^2 + ax}{ax} = a^2x^2 + ax + 1,$$

$$\frac{12x^3y^4 + 16x^2y^3 - 8xy^2}{4xy^2} = 3x^2y^2 + 4xy - 2.$$

EXAMPLES.—xii.

Divide

1. $x^3 + 2x^2 + x$ by x.
2. $y^5 - y^4 + y^3 - y^2$ by y^2.
3. $8a^3 + 16a^2b + 24ab^2$ by $8a$.
4. $mpx^4 + m^2p^2x^2 + m^3p^3$ by mp.
5. $16a^3xy - 28a^2x^2 + 4a^2x^3$ by $4a^2x$.
6. $72x^5y^6 - 36x^4y^3 - 18x^2y^2$ by $9x^2y$.
7. $81m^8n^7 - 54m^5n^6 + 27m^3n^2p$ by $3m^2n^2$.
8. $12x^5y^2 - 8x^4y^3 - 4x^3y^4$ by $4x^3$.
9. $169a^4b - 117a^3b^2 + 91a^2b$ by $13a^2$.
10. $361b^5c^3 + 228b^4c^4 - 133b^3c^5$ by $19b^2c$.

77. Admitting the possibility of the independent existence of a term affected with the sign −, we can extend the Examples in Arts. 74—76, by taking the first term of the dividend or the divisor, or both, *negative*. In such cases we apply the Rule of Signs in Multiplication to form a Rule of Signs in Division.

Thus since $-a \times b = -ab$, we conclude that $\dfrac{-ab}{b} = -a,$

...... $a \times -b = -ab,$ .... $\dfrac{-ab}{-b} = a,$

...... $-a \times -b = ab,$ .... $\dfrac{ab}{-b} = -a;$

and hence the rules

    I. When the dividend and the divisor have the *same sign* the quotient is *positive*.

    II. When the dividend and the divisor have *different signs* the quotient is *negative*.

78. The following Examples illustrate the conclusions just obtained:

(1) $\dfrac{abx^2}{-x} = -abx.$ (3) $\dfrac{-27a^3y^3}{-3x^2y} = 9xy^2.$

(2) $\dfrac{-12a^2b^3c^4}{4abx^2} = -3ab^2x^2.$ (4) $\dfrac{ax-bx}{-x} = -a+b.$

(5) $\dfrac{ab^4 - a^2b^3 + a^3b^2 - a^4b}{-ab} = -b^3 + ab^2 - a^2b + a^3.$

(6) $\dfrac{-12x^3y^4 + 16x^2y^3 - 8xy^2}{-4xy^2} = 3x^2y^2 - 4xy + 2.$

EXAMPLES.—xiii.

Divide

1. $72ab$ by $-9ab$.
2. $-60a^8$ by $-4a^3$.
3. $-84x^8y^9$ by $4x^5y^3$.
4. $-18m^3n^2$ by $3mn$.
5. $-128a^3b^2c$ by $-8bc$.
6. $-a^3x^3 - a^2x^2 - ax$ by $-ax$.
7. $-34a^3 + 51a^2 - 17ax^2$ by $17a$.
8. $-8a^3b^2 - 24a^5b^3 + 32a^7b^8$ by $-4a^3b^2$.
9. $-144x^3 + 108x^2y - 96xy^2$ by $12x$.
10. $b^2x^3z^2 - b^5x^7z^4 - b^3y^4z^2$ by $-b^2z^2$.

Case III.

79. The third case of the operation of Division is that in which the divisor and the dividend contain more terms than one. The operation is conducted in the following way:

> Arrange the divisor and dividend according to the powers of some one symbol, and place them in the same line as in the process of Long Division in Arithmetic.
>
> Divide the first term of the dividend by the first term of the divisor.
>
> Set down the result as the first term of the quotient.
>
> Multiply all the terms of the divisor by the first term of the quotient.
>
> Subtract the resulting product from the dividend. If there be a remainder, consider it as a *new dividend*, and proceed as before.

DIVISION. 37

The process will best be understood by a careful study of the following Examples:

(1) Divide $a^2 + 2ab + b^2$ by $a + b$.

$$a + b \,)\, a^2 + 2ab + b^2 \,(\, a + b$$
$$\underline{a^2 + ab}$$
$$ab + b^2$$
$$ab + b^2$$

(2) Divide $a^2 - 2ab + b^2$ by $a - b$.

$$a - b \,)\, a^2 - 2ab + b^2 \,(\, a - b$$
$$\underline{a^2 - ab}$$
$$-ab + b^2$$
$$-ab + b^2$$

(3) Divide $x^6 - y^6$ by $x^2 - y^2$.

$$x^2 - y^2 \,)\, x^6 - y^6 \,(\, x^4 + x^2y^2 + y^4$$
$$\underline{x^6 - x^4y^2}$$
$$x^4y^2 - y^6$$
$$\underline{x^4y^2 - x^2y^4}$$
$$x^2y^4 - y^6$$
$$x^2y^4 - y^6$$

(4) Divide $x^6 - 4a^2x^4 + 4a^4x^2 - a^6$ by $x^2 - a^2$.

$$x^2 - a^2 \,)\, x^6 - 4a^2x^4 + 4a^4x^2 - a^6 \,(\, x^4 - 3a^2x^2 + a^4$$
$$\underline{x^6 - a^2x^4}$$
$$-3a^2x^4 + 4a^4x^2 - a^6$$
$$\underline{-3a^2x^4 + 3a^4x^2}$$
$$a^4x^2 - a^6$$
$$a^4x^2 - a^6$$

(5) Divide $3xy + x^3 + y^3 - 1$ by $y + x - 1$.

Arranging the divisor and dividend by *descending powers* of x,

$$x + y - 1 \,)\, x^3 + 3xy + y^3 - 1 \,(\, x^2 - xy + x + y^2 + y + 1$$
$$\underline{x^3 + x^2y - x^2}$$
$$-x^2y + x^2 + 3xy + y^3 - 1$$
$$\underline{-x^2y - xy^2 + xy}$$
$$x^2 + xy^2 + 2xy + y^3 - 1$$
$$\underline{x^2 + xy - x}$$
$$xy^2 + xy + x + y^3 - 1$$
$$\underline{xy^2 + y^3 - y^2}$$
$$xy + x + y^2 - 1$$
$$\underline{xy + y^2 - y}$$
$$x + y - 1$$
$$x + y - 1$$

80. We must now direct the attention of the student to two points of great importance in Division.

> I. The dividend and divisor must be arranged according to the order of the powers of one of the symbols involved in them. This order may be *ascending* or *descending*. In the Examples given above we have taken the *descending* order, and in the Examples worked out in the next Article we shall take an *ascending* order of arrangement.
>
> II. In each remainder the terms must be arranged in the same order, ascending or descending, as that in which the dividend is arranged at first.

81. To divide (1) $1 - x^4$ by $x^3 + x^2 + x + 1$, arrange the dividend and divisor by ascending powers of x, thus:

$$1 + x + x^2 + x^3 \,)\, 1 - x^4 \,(\, 1 - x$$
$$\underline{1 + x + x^2 + x^3}$$
$$-x - x^2 - x^3 - x^4$$
$$\underline{-x - x^2 - x^3 - x^4}$$

(2) $48x^2 + 6 - 35x^5 + 58x^4 - 70x^3 - 23x$ by $6x^2 - 5x + 2 - 7x^3$, arrange the dividend and divisor by ascending powers of x, thus:

$$2 - 5x + 6x^2 - 7x^3 \,)\, 6 - 23x + 48x^2 - 70x^3 + 58x^4 - 35x^5 \,(\, 3 - 4x + 5x^2$$
$$\underline{6 - 15x + 18x^2 - 21x^3}$$
$$-8x + 30x^2 - 49x^3 + 58x^4$$
$$\underline{-8x + 20x^2 - 24x^3 + 28x^4}$$
$$10x^2 - 25x^3 + 30x^4 - 35x^5$$
$$\underline{10x^2 - 25x^3 + 30x^4 - 35x^5}$$

EXAMPLES.—XIV.

Divide

1. $x^2 + 15x + 50$ by $x + 10$.
2. $x^2 - 17x + 70$ by $x - 7$.
3. $x^2 + x - 12$ by $x - 3$.
4. $x^2 + 13x + 12$ by $x + 1$.
5. $x^3 + 13x^2 + 54x + 72$ by $x + 6$.
6. $x^3 + x^2 - x - 1$ by $x + 1$.
7. $x^3 + 2x^2 + 2x + 1$ by $x + 1$.
8. $x^5 - 5x^3 + 7x^2 + 6x + 1$ by $x^2 + 3x + 1$.
9. $x^4 - 4x^3 + 2x^2 + 4x + 1$ by $x^2 - 2x - 1$.
10. $x^4 - 4x^3 + 6x^2 - 4x + 1$ by $x^2 - 2x + 1$.

11. $x^4 - x^2 + 2x - 1$ by $x^2 + x - 1$. 12. $x^4 - 4x^2 + 8x + 16$ by $x + 2$.
13. $x^3 + 4x^2y + 3xy^2 + 12y^3$ by $x + 4y$.
14. $a^4 + 4a^3b + 6a^2b^2 + 4ab^3 + b^4$ by $a + b$.
15. $a^5 - 5a^4b + 10a^3b^2 - 10a^2b^3 + 5ab^4 - b^5$ by $a - b$.
16. $x^4 - 12x^3 + 50x^2 - 84x + 45$ by $x^2 - 6x + 9$.
17. $a^5 - 4a^4b + 4a^3b^2 + 4a^2b^3 - 17ab^4 - 12b^5$ by $a^2 - 2ab - 3b^2$.
18. $4a^2x^4 - 12a^3x^3 + 13a^4x^2 - 6a^5x + a^6$ by $2ax^2 - 3a^2x + a^3$.
19. $x^4 - x^2 + 2x - 1$ by $x^2 + x - 1$.
20. $x^4 + a^2x^2 - 2a^4$ by $x^2 + 2a^2$.   23. $x^6 - y^6$ by $x - y$.
21. $x^2 - 13xy - 30y^2$ by $x - 15y$.   24. $a^2 - b^2 + 2bc - c^2$ by $a - b + c$.
22. $x^5 + y^5$ by $x + y$.   25. $b - 3b^2 + 3b^3 - b^4$ by $b - 1$.
26. $a^2 - b^2 - c^2 + d^2 - 2(ad - bc)$ by $a + b - c - d$.
27. $x^3 + y^3 + z^3 - 3xyz$ by $x + y + z$.   28. $x^{15} + y^{10}$ by $x^3 + y^2$.
29. $p^2 + pq + 2pr - 2q^2 + 7qr - 3r^2$ by $p - q + 3r$.
30. $a^8 + a^6b^2 + a^4b^4 + a^2b^6 + b^8$ by $a^4 + a^3b + a^2b^2 + ab^3 + b^4$.
31. $x^8 + x^6y^2 + x^4y^4 + x^2y^6 + y^8$ by $x^4 - x^3y + x^2y^2 - xy^3 + y^4$.
32. $4x^5 - x^3 + 4x$ by $2x^2 + 3x + 2$.   33. $a^5 - 243$ by $a - 3$.
34. $k^{10} - k$ by $k^3 - 1$.   35. $x^3 - 5x^2 - 46x - 40$ by $x + 4$.
36. $48x^3 - 76ax^2 - 64a^2x + 105a^3$ by $2x - 3a$.
37. $18x^4 - 45x^3 + 82x^2 - 67x + 40$ by $3x^2 - 4x + 5$.
38. $16x^4 - 72a^2x^2 + 81a^4$ by $2x - 3a$.
39. $81x^4 - 256a^4$ by $3x + 4a$.   41. $x^3 + 2ax^2 - a^2x - 2a^3$ by $x^2 - a^2$.
40. $2a^3 + 3a^2b - 2ab^2 - 3b^3$ by $a^2 - b^2$.   42. $a^4 - a^2b^2 - 12b^4$ by $a^2 + 3b^2$.
43. $x^4 - 9x^2 - 6xy - y^2$ by $x^2 + 3x + y$.
44. $x^4 - 6x^3y + 9x^2y^2 - 4y^4$ by $x^2 - 3xy + 2y^2$.
45. $x^4 - 81y^4$ by $x - 3y$.   47. $81a^4 - 16b^4$ by $3a + 2b$.
46. $a^4 - 16b^4$ by $a - 2b$.   48. $16x^4 - 81y^4$ by $2x + 3y$.
49. $3a^2 + 8ab + 4b^2 + 10ac + 8bc + 3c^2$ by $a + 2b + 3c$.
50. $a^4 + 4a^2x^2 + 16x^4$ by $a^2 + 2ax + 4x^2$.
51. $x^4 + x^2y^2 + y^4$ by $x^2 - xy + y^2$.
52. $256c^4 + 16x^2y^2 + y^4$ by $16x^2 + 4xy + y^2$.
53. $x^5 + x^4y - x^3y^2 + x^3 - 2xy^2 + y^3$ by $x^3 + x - y$.

54. $ax^3 + 3a^2x^2 - 2a^3x - 2a^4$ by $x - a$.   55. $a^2 - x^2$ by $x + a$.
56. $2c^2 + xy - 3y^2 - 4yz - xz - z^2$ by $2x + 3y + z$.
57. $9x + 3x^4 + 14x^3 + 2$ by $1 + 5x + x^2$.
58. $12 - 38x + 82x^2 - 112x^3 + 106x^4 - 70x^5$ by $7x^2 - 5x + 3$.
59. $x^5 + y^5$ by $x^4 - x^3y + x^2y^2 - xy^3 + y^4$.
60. $(a^2x^2 + b^2y^2) - (a^2b^2 + x^2y^2)$ by $ax + by + ab + xy$.
61. $ab(x^2 + y^2) + xy(a^2 + b^2)$ by $ax + by$.
62. $x^4 + (2b^2 - a^2)x^2 + b^4$ by $x^2 + ax + b^2$.

82. The process may in some cases be shortened by the use of brackets, as in the following Example.

$x + b\,) \; x^3 + (a + b + c)\, x^2 + (ab + ac + bc)\, x + abc \;(\; x^2 + (a + c)\, x + ac$
$ x^3 + bx^2$

$ (a + c)\, x^2 + (ab + ac + bc)\, x$
$ (a + c)\, x^2 + (ab + bc)\, x$

$ acx + abc$
$ acx + abc$

$x - 1\,) \; x^5 - mx^4 + nx^3 - nx^2 + mx - 1 \;(\; x^4 - (m - 1)\, x^3$
$ x^5 - x^4 - (m - n - 1)\, x^2 - (m - 1)\, x + 1.$

$ - (m - 1)\, x^4 + nx^3$
$ - (m - 1)\, x^4 + (m - 1)\, x^3$

$ - (m - n - 1)\, x^3 - nx^2$
$ - (m - n - 1)\, x^3 + (m - n - 1)\, x^2$

$ - (m - 1)\, x^2 + mx$
$ - (m - 1)\, x^2 + (m - 1)\, x$

$ x - 1$
$ x - 1$

EXAMPLES.—XV.

Divide

1. $x^4 - (a^2 - b - c)\, x^2 - (b - c)\, ax + bc$ by $x^2 - ax + c$.
2. $y^3 - (l + m + n)\, y^2 + (lm + ln + mn)\, y - lmn$ by $y - n$.
3. $x^5 - (m - c)\, x^4 + (n - cm + d)\, x^3 +$
   $(r + cn - dm)\, x^2 + (cr + dn)\, x + dr$ by $x^3 - mx^2 + nx + r$.
4. $x^4 + (5 + a)\, x^3 - (4 - 5a + b)\, x^2 - (4a + 5b)\, x + 4b$ by $x^2 + 5x - 4$.
5. $x^4 - (a + b + c + d)\, x^3 + (ab + ac + ad + bc + bd + cd)\, x^2$
   $- (abc + abd + acd + bcd)\, x + abcd$ by $x^2 - (a + c)\, x + ac$.

DIVISION.

83. The following Examples in Division are of great importance.

DIVISOR.	DIVIDEND.	QUOTIENT.
$x+y$	x^2-y^2	$x-y$
$x-y$	x^2-y^2	$x+y$
$x+y$	x^3+y^3	x^2-xy+y^2
$x-y$	x^3-y^3	x^2+xy+y^2

84. Again, if we arrange two series of binomials consisting respectively of the sum and the difference of ascending powers of x and y, thus

$x+y$, x^2+y^2, x^3+y^3, x^4+y^4, x^5+y^5, x^6+y^6, and so on,

$x-y$, x^2-y^2, x^3-y^3, x^4-y^4, x^5-y^5, x^6-y^6, and so on,

$x+y$ will divide the *odd* terms in the upper line,
             and the *even* ...... in the lower ......

$x-y$ will divide *all* the terms in the lower,
             but *none* .... in the upper.

Or we may put it thus:

If n stand for any whole number,

x^n+y^n is divisible by $x+y$ when n is odd,
             by $x-y$ never;

x^n-y^n is divisible by $x+y$ when n is even,
             by $x-y$ always.

Also, it is to be observed that when the divisor is $x-y$ all the terms of the quotient are *positive*, and when the divisor is $x+y$, the terms of the quotient are alternately positive and negative.

Thus $\dfrac{x^4-y^4}{x-y} = x^3+x^2y+xy^2+y^3$,

$\dfrac{x^7+y^7}{x+y} = x^6-x^5y+x^4y^2-x^3y^3+x^2y^4-xy^5+y^6$,

$\dfrac{x^6-y^6}{x+y} = x^5-x^4y+x^3y^2-x^2y^3+xy^4-y^5$.

85. These properties may be easily remembered by taking the four simplest cases, thus, $x+y$, $x-y$, x^2+y^2, x^2-y^2, of which

the first is divisible by $x+y$,
second .... $x-y$,
third .... neither,
fourth .... both.

Again, since these properties are true for all values of x and y, suppose $y=1$, then we shall have

$$\frac{x^2-1}{x+1}=x-1, \qquad \frac{x^2-1}{x-1}=x+1,$$

$$\frac{x^3+1}{x+1}=x^2-x+1, \qquad \frac{x^3-1}{x-1}=x^2+x+1.$$

Also

$$\frac{x^5+1}{x+1}=x^4-x^3+x^2-x+1,$$

$$\frac{x^6-1}{x-1}=x^5+x^4+x^3+x^2+x+1.$$

EXAMPLES.—XVI.

Without going through the process of Division write down the quotients in the following cases:

1. When the divisor is $m+n$, and the dividends are respectively
$$m^2-n^2, \ m^3+n^3, \ m^5+n^5, \ m^6-n^6, \ m^9+n^9.$$

2. When the divisor is $m-n$, and the dividends are respectively
$$m^2-n^2, \ m^3-n^3, \ m^4-n^4, \ m^6-n^6, \ m^7-n^7.$$

3. When the divisor is $a+1$, and the dividends are respectively
$$a^2-1, \ a^3+1, \ a^5+1, \ a^7+1, \ a^8-1.$$

4. When the divisor is $y-1$, and the dividends are respectively
$$y^2-1, \ y^3-1, \ y^5-1, \ y^7-1, \ y^9-1.$$

V. ON THE RESOLUTION OF EXPRESSIONS INTO FACTORS.

86. We shall discuss in this Chapter an operation which is the opposite of that which we call Multiplication. In Multiplication we determine the product of two given factors: in the operation of which we have now to treat *the product is given and the factors have to be found.*

87. For the *resolution*, as it is called, of a product into its component factors no rule can be given which shall be applicable to all cases, but it is not difficult to explain the process in certain simple cases. We shall take these cases separately.

88. **Case I.** The simplest case for resolution is that in which all the terms of an expression have one common factor. This factor can be seen by inspection in most cases, and therefore the other factor may be at once determined.

Thus
$$a^2 + ab = a(a+b),$$
$$2a^3 + 4a^2 + 8a = 2a(a^2 + 2a + 4),$$
$$9x^3y - 18x^2y^2 + 54xy = 9xy(x^2 - 2xy + 6).$$

EXAMPLES.—xvii.

Resolve into factors:

1. $5x^2 - 15x$.
2. $3x^3 + 18x^2 - 6x$.
3. $49y^2 - 14y + 7$.
4. $4x^3y - 12x^2y^2 + 8xy^3$.
5. $x^4 - ax^3 + bx^2 + cx$.
6. $3x^5y^3 - 21x^4y^2 + 27x^3y^4$.
7. $54a^3b^6 + 108a^6b^3 - 243a^8b^2$.
8. $45x^7y^{10} - 90x^5y^7 - 360x^4y^8$.

89. Case II. The next case in point of simplicity is that in which four terms can be so arranged, that the first two have a common factor and the last two have a common factor.

Thus
$$x^2 + ax + bx + ab = (x^2 + ax) + (bx + ab)$$
$$= x(x+a) + b(x+a)$$
$$= (x+b)(x+a).$$

Again
$$ac - ad - bc + bd = (ac - ad) - (bc - bd)$$
$$= a(c-d) - b(c-d)$$
$$= (a-b)(c-d).$$

EXAMPLES.—xviii.

Resolve into factors:

1. $x^2 - ax - bx + ab.$
2. $ab + ax - bx - x^2.$
3. $bc + by - cy - y^2.$
4. $bm + mn + ab + an.$
5. $abx^2 - axy + bxy - y^2.$
6. $abx - aby + cdx - cdy.$
7. $cdx^2 + dmxy - cnxy - mny^2.$
8. $abc - b^2dx - acdy + bd^2y.$

90. Before reading the Articles that follow the student is advised to turn back to Art. 56, and to observe the manner in which the operation of multiplying a binomial by a binomial produces a *trinomial* in the Examples there given. He will then be prepared to expect that in certain cases a *trinomial can be resolved into two binomial factors*, examples of which we shall now give.

91. Case III. To find the factors of
$$x^2 + 7x + 12.$$

Our object is to find two numbers whose product is 12, and whose sum is 7.

These will evidently be 4 and 3,
$$\therefore x^2 + 7x + 12 = (x+4)(x+3).$$

Again, to find the factors of
$$x^2 + 5bx + 6b^2.$$

Our object is to find two numbers whose product is $6b^2$, and whose sum is $5b$.

These will clearly be $3b$ and $2b$,
$$\therefore x^2 + 5bx + 6b^2 = (x+3b)(x+2b).$$

RESOLUTION INTO FACTORS.

EXAMPLES.—XIX.

Resolve into factors:

1. $x^2 + 11x + 30$.
2. $x^2 + 17x + 60$.
3. $y^2 + 13y + 12$.
4. $y^2 + 21y + 110$.
5. $m^2 + 35m + 300$.
6. $m^2 + 23m + 102$.
7. $a^2 + 9ab + 8b^2$.
8. $x^2 + 13mx + 36m^2$.
9. $y^2 + 19ny + 48n^2$.
10. $x^2 + 29px + 100p^2$.
11. $x^4 + 5x^2 + 6$.
12. $x^6 + 4x^3 + 3$.
13. $x^2y^2 + 18xy + 32$.
14. $x^8y^4 + 7x^4y^2 + 12$.
15. $m^{10} + 10m^5 + 16$.
16. $n^2 + 27nq + 140q^2$.

93. **Case IV.** To find the factors of
$$x^2 - 9x + 20.$$

Our object is to find two *negative* terms whose product is 20, and whose sum is -9.

These will clearly be -5 and -4,
$$\therefore x^2 - 9x + 20 = (x - 5)(x - 4).$$

EXAMPLES.—XX.

Resolve into factors:

1. $x^2 - 7x + 10$.
2. $x^2 - 29x + 190$.
3. $y^2 - 23y + 132$.
4. $y^2 - 30y + 200$.
5. $n^2 - 43n + 460$.
6. $n^2 - 57n + 56$.
7. $x^6 - 7x^3 + 12$.
8. $a^2b^2 - 27ab + 26$.
9. $b^4c^6 - 11b^2c^3 + 30$.
10. $x^2y^2z^2 - 13xyz + 22$.

92. **Case V.** To find the factors of
$$x^2 + 5x - 84.$$

Our object is to find two terms, one positive and one negative, whose product is -84, and whose sum is 5.

These are clearly 12 and -7,
$$\therefore x^2 + 5x - 84 = (x + 12)(x - 7).$$

EXAMPLES.—xxi.

Resolve into factors:

1. $x^2 + 7x - 60$.
2. $x^2 + 12x - 45$.
3. $a^2 + 11a - 12$.
4. $a^2 + 13a - 140$.
5. $b^2 + 13b - 300$.
6. $b^2 + 25b - 150$.
7. $x^8 + 3x^4 - 4$.
8. $x^2y^2 + 3xy - 154$.
9. $m^{10} + 15m^5 - 100$.
10. $n^2 + 17n - 390$.

94. **Case VI.** To find the factors of
$$x^2 - 3x - 28.$$

Our object is to find two terms, one positive and one negative, whose product is -28, and whose sum is -3.

These will clearly be 4 and -7,
$$\therefore x^2 - 3x - 28 = (x + 4)(x - 7).$$

EXAMPLES.—xxii.

Resolve into factors:

1. $x^2 - 5x - 66$.
2. $x^2 - 7x - 18$.
3. $m^2 - 9m - 36$.
4. $n^2 - 11n - 60$.
5. $y^2 - 13y - 14$.
6. $z^2 - 15z - 100$.
7. $x^{10} - 9x^5 - 10$.
8. $c^2d^2 - 24cd - 180$.
9. $m^6n^2 - m^3n - 2$.
10. $p^8q^4 - 5p^4q^2 - 84$.

95. The results of the four preceding articles may be thus stated in general terms: a trinomial of one of the forms
$$x^2 + ax + b,\ x^2 - ax + b,\ x^2 + ax - b,\ x^2 - ax - b,$$
may be resolved into two simple factors, when b can be resolved into two factors, such that their sum, in the first two forms, or their difference, in the last two forms, is equal to a.

96. We shall now give a set of Miscellaneous Examples on the resolution into factors of expressions which come under one or other of the cases already explained.

Examples.—XXIII.

Resolve into factors:

1. $x^2 - 15x + 36$.
2. $x^2 + 4x - 45$.
3. $a^2b^2 - 16ab - 36$.
4. $x^8 - 3mx^4 - 10m^2$.
5. $y^6 + y^3 - 90$.
6. $x^4 - x^2 - 110$.
7. $x^2 + 3ax^2 + 4a^2x$.
8. $x^2 + mx + nx + mn$.
9. $y^6 - 4y^3 + 3$.
10. $x^2y - abx - cxy + abc$.
11. $x^2 + (a - b)x - ab$.
12. $x^2 - (c - d)x - cd$.
13. $ab^2 - bd + cd - abc$.
14. $4x^2 - 28xy + 48y^2$.

97. We have said, Art. 45, that when a number is multiplied by itself the result is called *the Square* of the number, and that the figure 2 placed over a number on the right hand indicates that the number is multiplied by itself.

Thus a^2 is called the square of a,

$(x - y)^2$ is called the square of $x - y$.

The **Square Root** of a given number is that number whose square is equal to the given number.

Thus the square root of 49 is 7, because the square of 7 is 49.

So also the square root of a^2 is a, because the square of a is a^2; and the square root of $(x - y)^2$ is $x - y$, because the square of $x - y$ is $(x - y)^2$.

The symbol $\sqrt{}$ placed before a number denotes that the square root of that number is to be taken: thus $\sqrt{25}$ is read "*the square root of* 25."

Note. The square root of a positive quantity may be either positive or negative. For

since a multiplied by a gives as a result a^2,

and $-a$ multiplied by $-a$ gives as a result a^2,

it follows, from our definition of a Square Root, that either a or $-a$ may be regarded as the square root of a^2.

But throughout this chapter we shall take only the *positive* value of the square root.

98. We may now take the case of Trinomials which are *perfect squares*, which are really included in the cases discussed in Arts. 91, 92, but which, from the importance they assume in a later part of our subject, demand a separate consideration.

99. **Case VII.** To find the factors of
$$x^2 + 12x + 36.$$

Seeking for the factors according to the hints given in Art. 91, we find them to be $x+6$ and $x+6$.

That is $x^2 + 12x + 36 = (x+6)^2$.

EXAMPLES.—xxiv.

Resolve into factors:

1. $x^2 + 18x + 81$.
2. $x^2 + 26x + 169$.
3. $x^2 + 34x + 289$.
4. $y^2 + 2y + 1$.
5. $z^2 + 200z + 10000$.
6. $x^4 + 14x^2 + 49$.
7. $x^2 + 10xy + 25y^2$.
8. $m^4 + 16m^2n^2 + 64n^4$.
9. $x^6 + 24x^3 + 144$.
10. $x^2y^2 + 162xy + 6561$.

100. **Case VIII.** To find the factors of
$$x^2 - 12x + 36.$$

Seeking for the factors according to the hints given in Art. 92, we find them to be $x-6$ and $x-6$.

That is, $x^2 - 12x + 36 = (x-6)^2$.

EXAMPLES.—xxv.

Resolve into factors:

1. $x^2 - 8x + 16$.
2. $x^2 - 28x + 196$.
3. $x^2 - 36x + 324$.
4. $y^2 - 40y + 400$.
5. $z^2 - 100z + 2500$.
6. $x^4 - 22x^2 + 121$.
7. $x^2 - 30xy + 225y^2$.
8. $m^4 - 32m^2n^2 + 256n^4$.
9. $x^6 - 38x^3 + 361$.

RESOLUTION INTO FACTORS. 49

101. **Case IX.** We now proceed to the most important case of Resolution into Factors, namely, that in which the expression to be resolved can be put in the form of *two squares with a negative sign between them.*

Since
$$m^2 - n^2 = (m+n)(m-n),$$
we can express the difference between the squares of two quantities by the product of two factors, determined by the following method:

> Take the square root of the first quantity, and the square root of the second quantity.
> The sum of the results will form the first factor.
> The difference of the results will form the second factor.

For example, let $a^2 - b^2$ be the given expression.
> The square root of a^2 is a.
> The square root of b^2 is b.
> The sum of the results is $a + b$.
> The difference of the results is $a - b$.

The factors will therefore be $a+b$ and $a-b$, that is, $$a^2 - b^2 = (a+b)(a-b).$$

102. The same method holds good with respect to compound quantities.

Thus, let $a^2 - (b-c)^2$ be the given expression.
> The square root of the first term is a.
> The square root of the second term is $b - c$.
> The sum of the results is $a + b - c$.
> The difference of the results is $a - b + c$.
> $$\therefore a^2 - (b-c)^2 = (a+b-c)(a-b+c).$$

Again, let $(a-b)^2 - (c-d)^2$ be the given expression.
> The square root of the first term is $a - b$.
> The square root of the second term is $c - d$.
> The sum of the results is $a - b + c - d$.
> The difference of the results is $a - b - c + d$.
> $$\therefore (a-b)^2 - (c-d)^2 = (a-b+c-d)(a-b-c+d).$$

103. The terms of an expression may often be arranged so as to form two squares with the negative sign between them, and then the expression can be resolved into factors.

Thus
$$a^2 + b^2 - c^2 - d^2 + 2ab + 2cd$$
$$= a^2 + 2ab + b^2 - c^2 + 2cd - d^2$$
$$= (a^2 + 2ab + b^2) - (c^2 - 2cd + d^2)$$
$$= (a+b)^2 - (c-d)^2$$
$$= (a+b+c-d)(a+b-c+d).$$

EXAMPLES.—XXVI.

Resolve into two or more factors:

1. $x^2 - y^2$.
2. $x^2 - 9$.
3. $4x^2 - 25$.
4. $a^4 - x^4$.
5. $x^2 - 1$.
6. $x^6 - 1$.
7. $x^8 - 1$.
8. $m^4 - 16$.
9. $36y^2 - 49z^2$.
10. $81x^2y^2 - 121a^2b^2$.
11. $(a-b)^2 - c^2$.
12. $x^2 - (m-n)^2$.
13. $(a+b)^2 - (c+d)^2$.
14. $(x+y)^2 - (x-y)^2$.
15. $x^2 - 2xy + y^2 - z^2$.
16. $(a-b)^2 - (m+n)^2$.
17. $a^2 - 2ac + c^2 - b^2 - 2bd - d^2$.
18. $2bc - b^2 - c^2 + a^2$.
19. $2xy + x^2 + y^2 - z^2$.
20. $2mn - m^2 - n^2 + a^2 + b^2 - 2ab$.
21. $(ax+by)^2 - 1$.
22. $(ax+by)^2 - (ax-by)^2$.
23. $1 - a^2 - b^2 + 2ab$.
24. $2xy - x^2 - y^2 + 1$.
25. $x^2 - 2yz - y^2 - z^2$.
26. $a^2 - 4b^2 - 9c^2 + 12bc$.
27. $a^4 - 16b^2$.
28. $1 - 49c^2$.
29. $a^2 + b^2 - c^2 - d^2 - 2ab - 2cd$.
30. $a^2 - b^2 + c^2 - d^2 - 2ac + 2bd$.
31. $3a^3x^3 - 27ax$.
32. $a^4b^6 - c^8$.
33. $(5x-2)^2 - (x-4)^2$.
34. $(7x+4y)^2 - (2x+3y)^2$.
35. $(753)^2 - (247)^2$.

104. Case X. Since

$$\frac{x^3 + a^3}{x+a} = x^2 - ax + a^2, \text{ and } \frac{x^3 - a^3}{x-a} = x^2 + ax + a^2 \quad \text{(Art. 83)}.$$

we know the following important facts:

(1) The *sum* of the *cubes* of two numbers is divisible by the *sum* of the numbers:

(2) The *difference* between the *cubes* of two numbers is divisible by the *difference* between the numbers.

Hence we may resolve into factors expressions in the form of the sum or difference of the cubes of two numbers.

Thus $\quad x^3 + 27 = x^3 + 3^3 = (x+3)(x^2 - 3x + 9)$
$\qquad\quad y^3 - 64 = y^3 - 4^3 = (y-4)(y^2 + 4y + 16).$

EXAMPLES.—XXVii.

Express in factors the following expressions:

1. $a^3 + b^3$. 2. $a^3 - b^3$. 3. $a^3 - 8$. 4. $x^3 + 343$.
5. $b^3 - 125$. 6. $x^3 + 64y^3$. 7. $a^3 - 216$. 8. $8x^3 + 27y^3$.
9. $64a^3 - 1000b^3$. 10. $729x^3 + 512y^3$.

Express in *four* factors each of the following expressions:

11. $x^6 - y^6$. 12. $x^6 - 1$. 13. $a^6 - 64$. 14. $729 - y^6$.

---

105. Before we proceed to describe other processes in Algebra, we shall give a series of examples in illustration of the principles already laid down.

The student will find it of advantage to work every example in the following series, and to accustom himself to read and to explain with facility those examples, in which illustrations are given of what may be called *the short-hand method* of expressing Arithmetical calculations by the symbols of Algebra.

EXAMPLES.—XXViii.

1. Express the sum of a and b.
2. Interpret the expression $a - b + c$.
3. How do you express the double of x?
4. By how much is a greater than 5?
5. If x be a whole number, what is the number next above it?
6. Write five numbers in order of magnitude, so that x shall be the third of the five.

7. If a be multiplied into zero, what is the result?

8. If zero be divided by x, what is the result?

9. What is the sum of $a + a + a \ldots$ written d times?

10. If the product be ac and the multiplier a, what is the multiplicand?

11. What number taken from x gives y as a remainder?

12. A is x years old, and B is y years old; how old was A when B was born?

13. A man works every day on week-days for x weeks in the year, and during the remaining weeks in the year he does not work at all. During how many days does he rest?

14. There are x boats in a race. Five are bumped. How many row over the course?

15. A merchant begins trading with a capital of x pounds. He gains a pounds each year. How much capital has he at the end of 5 years?

16. A and B sit down to play at cards. A has x shillings and B y shillings at first. A wins 5 shillings. How much has each when they cease to play?

17. There are 5 brothers in a family. The age of the eldest is x years. Each brother is 2 years younger than the one next above him in age. How old is the youngest?

18. I travel x hours at the rate of y miles an hour. How many miles do I travel?

19. From a rod 12 inches long I cut off x inches, and then I cut off y inches of the remainder. How many inches are left?

20. If n men can dig a piece of ground in q hours, how many hours will one man take to dig it?

21. By how much does 25 exceed x?

22. By how much does y exceed 25?

23. If a product has $2m$ repeated 8 times as a factor, how do you express the product?

24. By how much does $a + 2b$ exceed $a - 2b$?

25. A girl is x years of age, how old was she 5 years since?

RESOLUTION INTO FACTORS. 53

26. A boy is y years of age, how old will he be 7 years hence?

27. Express the difference between the squares of two numbers.

28. Express the product arising from the multiplication of the sum of two numbers into the difference between the same numbers.

29. What value of x will make $8x$ equal to 16?

30. What value of x will make $28x$ equal to 56?

31. What value of x will make $\dfrac{x}{7}$ equal to 4?

32. What value of x will make $x+2$ equal to 9?

33. What value of x will make $x-7$ equal to 16?

34. What value of x will make x^2+9 equal to 34?

35. What value of x will make x^2-8 equal to 92?

EXAMPLES.—XXIX.

Explain the operations symbolized in the following expressions:

1. $a+b$. 2. a^2-b^2. 3. $4a^2+b^3$. 4. $4(a^2+b^2)$.

5. $a^2-2b+3c$. 6. $a+m \times b-c$. 7. $(a+m)(b-c)$. 8. $\sqrt{x^3}$.

9. $\sqrt{x^2+y^2}$. 10. $a+2(3-c)$. 11. $(a+2)(3-c)$.

12. $\dfrac{a^2+b^2}{4ab}$. 13. $\dfrac{\sqrt{x^2-y^2}}{x-y}$. 14. $\dfrac{\sqrt{x^2+y^2}}{\sqrt{x+y}}$.

EXAMPLES.—XXX.

If a stands for 6, b for 5, x for 4, and y for 3, find the value of the following expressions:

1. $a+x-b-y$. 2. $a+y-b-x$. 3. $3a+4y-b-2x$.

4. $3(a+b)-2(x-y)$. 5. $(a+x)(b-y)$. 6. $2a+3(x-y)$.

7. $(2a+3)(x+y)$. 8. $2a+3x+y$. 9. $\dfrac{b^2+y}{a-x}$.

10. abx. 11. $ab(x+y)$. 12. $ay(b+x)^2$.

13. $ab(x-y)^2$. 14. $\sqrt{5b}$. 15. $\sqrt{y^2}$.
16. $(\sqrt{x})^2$. 17. $(\sqrt{x}+b)^2$. 18. $\sqrt{5bx}$.
19. $\sqrt{2axy}$. 20. $\dfrac{a^2+b^2+y}{x+y^2+3}$. 21. $3a+(2x-y)^2$.
22. $\{a-(b-y)\}\{a-(x-y)\}$. 24. $3(a+b-y)^3+4(a+x)^1$.
23. $(a-b-y)^2+(a-x+y)^2$. 25. $3(a-b)^2+(4x-y^2)^2$.

EXAMPLES.—XXXi.

1. Find the value of
 $3abc - a^3 + b^3 + c^3$, when $a=3$, $b=2$, $c=1$.

2. Find the value of
 $x^3 + y^3 - z^3 + 3xyz$, when $x=3$, $y=2$, $z=5$.

3. Subtract $a^2 + c^2$ from $(a+c)^2$.

4. Subtract $(x-y)^2$ from $x^2 + y^2$.

5. Find the coefficient of x in the expression
 $(a+b)^2 x - (a+bx)^2$.

6. Find the continued product of
 $2x - m$, $2x + n$, $x + 2m$, $x - 2n$.

7. Divide
 $acr^3 + (bc+ad)r^2 + (bd+ae)r + be$ by $ar+b$;
and test your result by putting
 $a=b=c=d=e=1$, and $r=10$.

8. Obtain the product of the four factors
 $(a+b+c)$, $(b+c-a)$, $(c+a-b)$, $(a+b-c)$.
What does this become when c is zero; when $b+c=a$; when $a=b=c$?

9. Find the value of
 $(a+b)(b+c) - (c+d)(d+a) - (a+c)(b-d)$,
where b is equal to d.

10. Find the value of
 $3a + (2b-c^2) + \{c^2 - (2a+3b)\} + \{3c - (2a+3b)\}^2$.
when $a=0$, $b=2$, $c=-4$.

RESOLUTION INTO FACTORS.

11. If $a=1$, $b=2$, $c=3$, $d=4$, shew that the numerical values are equal of

$$\{d-(c-b+a)\}\{(d+c)-(b+a)\},$$

and of

$$d^2-(c^2+b^2)+a^2+2(bc-ad).$$

12. Bracket together the different powers of x in the following expressions:

(α) $ax^2 + bx^2 + cx + dx$.

(β) $ax^3 - bx^3 - cx^2 - dx^2 + 2x^2$.

(γ) $4x^3 - ax^3 - 3x^2 - bx^2 - 5x - cx$.

(δ) $(a+x)^2 - (b-x)^2$.

(ε) $(mx^2 + qx + 1)^2 - (nx^2 + qx + 1)^2$.

13. Multiply the three factors $x-a$, $x-b$, $x-c$ together, and arrange the product according to *descending* powers of x.

14. Find the continued product of $(x+a)(x+b)(x+c)$.

15. Find the cube of $a+b+c$; thence without further multiplication the cubes of $a+b-c$; $b+c-a$; $c+a-b$; and subtract the sum of these three cubes from the first.

16. Find the product of $(3a+2b)(3a+2c-3b)$, and test the result by making $a=1$, $b=c=3$.

17. Find the continued product of

$$a-x,\ a+x,\ a^2+x^2,\ a^4+x^4,\ a^8+x^8.$$

18. Subtract $(b-a)(c-d)$ from $(a-b)(c-d)$. What is the value of the result when $a=2b$ and $d=2c$?

19. Add together $(b+y)(a+x)$, $x-y$, $ax-by$, and $a(x+y)$.

20. What value of x will make the difference between $(x+1)(x+2)$ and $(x-1)(x-2)$ equal to 54?

21. Add together $ax-by$, $x-y$, $x(x-y)$, and $(a-x)(b-y)$.

22. What value of x will make the difference between $(2x+4)(3x+4)$ and $(3x-2)(2x-8)$ equal to 96?

23. Add together

$2mx - 3ny$, $x+y$, $4(m+n)(x-y)$, and $mx+ny$.

24. Prove that

$$(x+y+z)^2 + x^2 + y^2 + z^2 = (x+y)^2 + (y+z)^2 + (x+z)^2.$$

25. Find the product of $(2a+3b)(2a+3c-2v)$, and test the result by making $a=1$, $b=4$, $c=2$.

26. If $a, b, c, d, e \ldots$ denote 9, 7, 5, 3, 1, find the values of
$\dfrac{ab-cd}{cd+e}$; $(bc-ad)(bd-ce)$; $\dfrac{b^2-c^2}{c+d}$; and d^a-c^d.

27. Find the value of
$3abc-a^3+b^3+c^3$ when $a=0$, $b=2$, $c=1$.

28. Find the value of
$3a^2+\dfrac{2ab^2}{c}-\dfrac{c^3}{b^2}$ when $a=4$, $b=1$, $c=2$.

29. Find the value of
$(a-b-c)^2+(b-a-c)^2+(c-a-b)^2$ when $a=1$, $b=2$, $c=3$.

30. Find the value of
$(a+b-c)^2+(a-b+c)^2+(b+c-a)^2$ when $a=1$, $b=2$, $c=4$.

31. Find the value of
$(a+b)^2+(b+c)^2+(c+a)^2$ when $a=-1$, $b=2$, $c=-3$.

32. Shew that if the sum of any two numbers divide the difference of their squares, the quotient is equal to the difference of the two numbers.

33. Shew that the product of the sum and difference of any two numbers is equal to the difference of their squares.

34. Shew that the square of the sum of any two consecutive integers is always greater by one than four times their product.

35. Shew that the square of the sum of any two consecutive even whole numbers is four times the square of the odd number between them.

36. If the number 2 be divided into any two parts, the difference of their squares will always be equal to twice the difference of the parts.

37. If the number 50 be divided into any two parts, the difference of their squares will always be equal to 50 times the difference of the parts.

38. If a number n be divided into any two parts, the difference of their squares will always be equal to n times the difference of the parts.

39. If two numbers differ by a unit, their product, together with the sum of their squares, is equal to the difference of the cubes of the numbers.

40. Shew that the sum of the cubes of any three consecutive whole numbers is divisible by three times the middle number.

VI. ON SIMPLE EQUATIONS.

106. An **Equation** is a statement that two expressions are equal.

107. An IDENTICAL EQUATION is a statement that two expressions are equal for all numerical values that can be given to the letters involved in them, provided that the same value be given to the same letter in every part of the equation.

Thus, $(x+a)^2 = x^2 + 2ax + a^2$
is an Identical Equation.

108. An EQUATION OF CONDITION is a statement that two expressions are equal for some *particular* numerical value or values that can be given to the letters involved.

Thus, $x + 1 = 6$
is an Equation of Condition, the only number which x can represent consistently with this equation being 5.

It is of such equations that we have to treat.

109. The ROOT of an Equation is that number which, when put in the place of the unknown quantity, makes both sides of the equation identical.

110. The SOLUTION of an Equation is the process of finding what number an unknown letter must stand for that the equation may be true: in other words, it is the method of finding the Root.

The letters that stand for *unknown* numbers are usually x, y, z, but the student must observe that *any* letter may stand for an unknown number.

111. A **Simple Equation** is one which contains the *first power only* of an unknown quantity. This is also called an Equation of the *First Degree*.

112. The following Axioms form the groundwork of the solution of all equations.

Ax. I. If equal quantities be added to equal quantities, the sums will be equal.

Thus, if
$$a = b,$$
$$a + c = b + c.$$

Ax. II. If equal quantities be taken from equal quantities, the remainders will be equal.

Thus, if
$$x = y,$$
$$x - z = y - z.$$

Ax. III. If equal quantities be multiplied by equal quantities, the products will be equal.

Thus, if
$$a = b,$$
$$ma = mb.$$

Ax. IV. If equal quantities be divided by equal quantities, the quotients will be equal.

Thus, if
$$xy = xz,$$
$$y = z.$$

113. On Axioms I. and II. is founded a process of great utility in the solution of equations, called THE TRANSPOSITION OF TERMS from one side of the equation to the other, which may be thus stated:

"Any term of an equation may be transferred from one side of the equation to the other *if its sign be changed.*"

For let
$$x - a = b.$$

Then, by Ax. I., if we add a to both sides, the sides remain equal:

therefore
$$x - a + a = b + a,$$
that is,
$$x = b + a.$$

Again, let
$$x + c = d.$$

Then, by Ax. II., if we subtract c from each side, the sides remain equal:

therefore
$$x + c - c = d - c,$$
that is,
$$x = d - c.$$

114. We may change all the signs of each side of an equation without altering the equality.

Thus, if
$$a - x = b - c,$$
$$x - a = c - b.$$

115. We may change the position of the two sides of the equation, leaving the signs unchanged.

Thus the equation $a - b = x - c$, may be written thus,
$$x - c = a - b.$$

116. We may now proceed to our first rule for the solution of a Simple Equation.

Rule I. Transpose the known terms to the right hand side of the equation and the unknown terms to the other, and combine all the terms on each side as far as possible.

Then divide both sides of the equation by the coefficient of the unknown quantity.

This rule we shall now illustrate by examples, in which x stands for the unknown quantity.

Ex. 1. *To solve the equation,*
$$5x - 6 = 3x + 2.$$
Transposing the terms, we get
$$5x - 3x = 2 + 6.$$
Combining like terms, we get
$$2x = 8.$$
Dividing both sides of this equation by 2, we get
$$x = 4,$$
and the value of x is determined.

Ex. 2. *To solve the equation,*
$$7x + 4 = 25x - 32$$
Transposing the terms, we get
$$7x - 25x = -32 - 4.$$
Combining like terms, we get
$$-18x = -36.$$
Changing the signs on each side, we get
$$18x = 36.$$
Dividing both sides by 18, we get
$$x = 2,$$
and the value of x is determined.

Ex. 3. *To solve the equation,*
$$2x - 3x + 120 = 4x - 6x + 132.$$
that is, $\quad 2x - 3x - 4x + 6x = 132 - 120,$
or, $\quad 8x - 7x = 12,$
therefore, $\quad x = 12.$

Ex. 4. *To solve the equation,*
$$3x + 5 - 8(13 - x) = 0,$$
that is, $\quad 3x + 5 - 104 + 8x = 0,$
or, $\quad 3x + 8x = 104 - 5,$
or, $\quad 11x = 99,$
therefore, $\quad x = 9.$

Ex. 5. *To solve the equation,*
$$6x - 2(4 - 3x) = 7 - 3(17 - x),$$
that is, $\quad 6x - 8 + 6x = 7 - 51 + 3x,$
or, $\quad 6x + 6x - 3x = 7 - 51 + 8,$
or, $\quad 12x - 3x = 15 - 51,$
or, $\quad 9x = -36,$
therefore, $\quad x = -4.$

EXAMPLES.—XXXII.

1. $7x + 5 = 5x + 11.$
2. $12x + 7 = 8x + 15.$
3. $236x + 425 = 97x + 564.$
4. $5x - 7 = 3x + 7.$
5. $12x - 9 = 8x - 1.$
6. $124x + 19 = 112x + 43.$
7. $18 - 2x = 27 - 5x.$
8. $125 - 7x = 145 - 12x.$
9. $26 - 8x = 80 - 14x.$
10. $133 - 3x = x - 83.$
11. $13 - 3x = 5x - 3.$
12. $127 + 9x = 12x + 100.$
13. $15 - 5x = 6 - 4x.$
14. $3x - 22 = 7x + 6.$
15. $8 + 4x = 12x - 16.$
16. $5x - (3x - 7) = 4x - (6x - 35).$
17. $6x - 2(9 - 4x) + 3(5x - 7) = 10x - (4 + 16x) + 35.$
18. $9x - 3(5x - 6) + 30 = 0.$
19. $12x - 5(9x + 3) + 6(7 - 8x) + 783 = 0.$
20. $x - 7(x - 11) = 14(x - 5) - 19(8 - x) - 6.$
21. $(x + 7)(x - 3) = (x - 5)(x - 15).$

22. $(x-8)(x+12) = (x+1)(x-6)$.
23. $(x-2)(7-x) + (x-5)(x+3) - 2(x-1) + 12 = 0$.
24. $(2x-7)(x+5) = (9-2x)(4-x) + 229$.
25. $(7-6x)(3-2x) = (4x-3)(3x-2)$.
26. $14 - x - 5(x-3)(x+2) + (5-x)(4-5x) = 45x - 76$.
27. $(x+5)^2 - (4-x)^2 = 21x$.
28. $5(x-2)^2 + 7(x-3)^2 = (3x-7)(4x-19) + 42$.
29. $(3x-17)^2 + (4x-25)^2 - (5x-29)^2 = 1$.
30. $(x+5)(x-9) + (x+10)(x-8) = (2x+3)(x-7) - 113$.

VII. PROBLEMS LEADING TO SIMPLE EQUATIONS.

117. WHEN we have a question to resolve by means of Algebra, we represent the number sought by an unknown symbol, and then consider in what manner the conditions of the question enable us to assert *that two expressions are equal.* Thus we obtain an equation, and by resolving it we determine the value of the number sought.

The whole difficulty connected with the solution of Algebraical Problems lies in the determination from the conditions of the question *of two different expressions having the same numerical value.*

To explain this let us take the following Problem:

Find a number such that if 15 be added to it, twice the sum will be equal to 44.

Let x represent the number.

Then $x + 15$ will represent the number increased by 15, and $2(x+15)$ will represent twice the sum.

But 44 will represent twice the sum,
therefore $\qquad 2(x+15) = 44$.

Hence $\qquad 2x + 30 = 44$,
that is, $\qquad 2x = 14$,
or, $\qquad x = 7$,
and therefore the number sought is 7.

118. We shall now give a series of Easy Problems, in which the conditions by which an equality between two expressions can be asserted may be readily seen. The student should be thoroughly familiar with the Examples in set xxviii, the use of which he will now find.

We shall insert some notes to explain the method of representing quantities by algebraic symbols in cases where some difficulty may arise.

EXAMPLES.—XXXiii.

1. To the double of a certain number I add 14 and obtain as a result 154. What is the number?

2. To four times a certain number I add 16 and obtain as a result 188. What is the number?

3. By adding 46 to a certain number I obtain as a result a number three times as large as the original number. Find the original number.

4. One number is three times as large as another. If I take the smaller from 16 and the greater from 30, the remainders are equal. What are the numbers?

5. Divide the number 92 into four parts, such that the first is greater than the second by 10, greater than the third by 18, and greater than the fourth by 24.

6. The sum of two numbers is 20, and if three times the smaller number be added to five times the greater, the sum is 84. What are the numbers?

7. The joint ages of a father and his son are 80 years. If the age of the son were doubled he would be 10 years older than his father. What is the age of each?

8. A man has six sons, each 4 years older than the one next to him. The eldest is three times as old as the youngest. What is the age of each?

9. Add £24 to a certain sum, and the amount will be as much above £80 as the sum is below £80. What is the sum?

10. Thirty yards of cloth and forty yards of silk together cost £66, and the silk is twice as valuable as the cloth. Find the cost of a yard of each.

11. Find the number, the double of which being added to 24 the result is as much above 80 as the number itself is below 100.

12. The sum of £500 is divided between A, B, C and D. A and B have together £280, A and C £260, A and D £220. How much does each receive?

13. In a company of 266 persons, composed of men, women, and children, there are twice as many men as there are women, and twice as many women as there are children. How many are there of each?

14. Divide £1520 between A, B and C, so that A has £100 less than B, and B £270 less than C.

15. Find two numbers, differing by 8, such that four times the less may exceed twice the greater by 10.

16. A and B began to play with equal sums. A won £5, and then three times A's money was equal to eleven times B's money. What had each at first?

17. A is 58 years older than B, and A's age is as much above 60 as B's age is below 50. Find the age of each.

18. A is 34 years older than B, and A is as much above 50 as B is below 40. Find the age of each.

19. A man leaves his property, amounting to £7500, to be divided between his wife, his two sons and his three daughters, as follows: a son is to have twice as much as a daughter, and the wife £500 more than all the five children together. How much did each get?

20. A vessel containing some water was filled up by pouring in 42 gallons, and there was then in the vessel 7 times as much as at first. How many gallons did the vessel hold?

21. Three persons, A, B, C, have £76. B has £10 more than A, and C has as much as A and B together. How much has each?

22. What two numbers are those whose difference is 14, and their sum 48?

23. A and B play at cards. A has £72 and B has £52 when they begin. When they cease playing, A has three times as much as B. How much did A win?

NOTE I. If we have to express algebraically two parts into which a given number, suppose 50, is divided, and we represent one of the parts by x, the other will be represented by $50 - x$.

Ex. Divide 50 into two such parts that the double of one part may be three times as great as the other part.

Let x represent one of the parts.

Then $50 - x$ will represent the other part.

Now the double of the first part will be represented by $2x$, and three times the second part will be represented by $3(50-x)$.

Hence $\quad\quad\quad 2x = 3(50-x)$,
or, $\quad\quad\quad\quad\quad 2x = 150 - 3x$,
or, $\quad\quad\quad\quad\quad 5x = 150$;
$\quad\quad\quad\quad\quad\quad \therefore x = 30$.

Hence the parts are 30 and 20.

24. Divide 84 into two such parts that three times one part may be equal to four times the other.

25. Divide 90 into two such parts that four times one part may be equal to five times the other.

26. Divide 60 into two such parts that one part is greater than the other by 24.

27. Divide 84 into two such parts that one part is less than the other by 36.

28. Divide 20 into two such parts that if three times one part be added to five times the other part the sum may be 84.

NOTE II. When we have to compare the ages of two persons at one time and also some years after or before, we must be careful to remember that *both* will be so many years older or younger.

Thus if x be the age of A at the present time, and $2x$ be the age of B at the present time,

The age of A 5 years hence will be $x + 5$,
and the age of B 5 years hence will be $2x + 5$.

Ex. A is 5 times as old as B, and 5 years hence A will only be three times as old as B. What are the ages of A and B at the present time?

Let x represent the age of B.

Then $5x$ will represent the age of A.

Now $x+5$ will represent B's age 5 years hence, and $5x+5$ will represent A's age 5 years hence.

Hence $5x+5 = 3(x+5)$,
or $5x+5 = 3x+15$,
or $2x = 10$;
$\therefore x = 5$.

Hence A is 25 and B is 5 years old.

29. A is twice as old as B, and 22 years ago he was three times as old as B. What is A's age?

30. A father is 30; his son is 6 years old. In how many years will the age of the father be just twice that of the son?

31. A is twice as old as B, and 20 years since he was three times as old. What is B's age?

32. A is three times as old as B, and 19 years hence he will be only twice as old as B. What is the age of each?

33. A man has three nephews. His age is 50, and the joint ages of the nephews are 42. How long will it be before the joint ages of the nephews will be equal to the age of the uncle?

NOTE III. In problems involving weights and measures, after assuming a symbol to represent one of the unknown quantities, we must be careful to express the other quantities *in the same terms*. Thus, if x represent a number of *pence*, all the sums involved in the problem *must be reduced to pence*.

Ex. A sum of money consists of fourpenny pieces and sixpences, and it amounts to £1. 16s. 8d. The number of coins is 78. How many are there of each sort?

[S.A.]   E

Let x be the number of fourpenny pieces.

Then $4x$ is their value *in pence*.

Also $78-x$ is the number of sixpences.

And $6(78-x)$ is their value *in pence*.

Also £1. 16s. 8d. is equivalent to 440 *pence*.

Hence $\qquad 4x + 6(78-x) = 440,$
$\qquad\qquad$ or $4x + 468 - 6x = 440,$
from which we find $x = 14$.

Hence there are 14 fourpenny pieces,
$\qquad\qquad$ and 64 sixpences.

34. A bill of £100 was paid with guineas and half-crowns, and 48 more half-crowns than guineas were used. How many of each were paid?

35. A person paid a bill of £3. 14s. with shillings and half-crowns, and gave 41 pieces of money altogether. How many of each were paid?

36. A man has a sum of money amounting to £11. 13s. 4d., consisting only of shillings and fourpenny pieces. He has in all 300 pieces of money. How many has he of each sort?

37. A bill of £50 is paid with sovereigns and moidores of 27 shillings each, and 3 more sovereigns than moidores are given. How many of each are used?

38. A sum of money amounting to £42. 8s. is made up of shillings and half-crowns, and there are six times as many half-crowns as there are shillings. How many are there of each sort?

39. I have £5. 11s. 3d. in sovereigns, shillings and pence. I have twice as many shillings and three times as many pence as I have sovereigns. How many have I of each sort?

VIII. ON THE METHOD OF FINDING THE HIGHEST COMMON FACTOR.

119. An expression is said to be a *Factor* of another expression when the latter is divisible by the former.

Thus $3a$ is a factor of $12a$,
$5xy$ .... of $15x^2y^2$.

120. An expression is said to be a *Common Factor* of two or more other expressions, when *each* of the latter is divisible by the former.

Thus $3a$ is a common factor of $12a$ and $15a$,
$3xy$........................ of $15x^2y^2$ and $21x^3y^3$,
$4z$ .... of $8z$, $12z^2$ and $16z^3$.

121. The *Highest Common Factor* of two or more expressions is the expression of *highest dimensions* by which each of the former is divisible.

Thus $6a^2$ is the Highest Common Factor of $12a^2$ and $18a^3$,
$5x^2y$ .... of $10x^3y$, $15x^2y^2$ and $25x^4y^3$.

NOTE. That which we call the Highest Common Factor is named by others the *Greatest Common Measure* or the *Highest Common Divisor*. Our reasons for rejecting these names will be given at the end of the chapter.

122. The words Highest Common Factor are abbreviated thus, H.C.F.

123. To take a simple example in Arithmetic, it will readily be admitted that the highest number which will divide 12, 18, and 30 is 6.

Now,
$12 = 2 \times 3 \times 2$,
$18 = 2 \times 3 \times 3$,
$30 = 2 \times 3 \times 5$.

Having thus reduced the numbers to their *simplest* factors, it appears that we may determine the Highest Common Factor in the following way.

Set down the factors of one of the numbers in any order.

Place beneath them the factors of the second number, in such order that factors *like any of those of the first number shall stand under those factors.*

Do the same for the third number.

Then the number of vertical columns in which the numbers are alike will be the number of factors in the H.C.F., and if we multiply the figures at the head of those columns together the result will be the H.C.F. required.

Thus in the example given above two vertical columns are alike, and therefore there are two factors in the H.C.F.

And the numbers 2 and 3 which stand at the heads of those columns being multiplied together will give the H.C.F. of 12, 18, and 30.

124. Ex. 1. To find the H.C.F. of a^3b^2x and $a^2b^3x^2$.

$$a^3b^2x = aaa \cdot bb \cdot x,$$
$$a^2b^3x^2 = aa \cdot bbb \cdot xx\,;$$
$$\therefore \text{H.C.F.} = aabbx$$
$$= a^2b^2x.$$

Ex. 2. To find the H.C.F. of $34a^2b^6c^4$ and $51a^3b^4c^2$.

$$34a^2b^6c^4 = 2 \times 17 \times aa \cdot bbbbbb \cdot cccc,$$
$$51a^3b^4c^2 = 3 \times 17 \times aaa \cdot bbbb \cdot cc\,;$$
$$\therefore \text{H.C.F.} = 17aabbbbcc$$
$$= 17a^2b^4c^2.$$

EXAMPLES.—XXXIV.

Find the Highest Common Factor of

1. a^4b and a^2b^3.
2. x^3y^2z and $x^2y^2z^2$.
3. $14x^2y^2$ and $24x^3y$.
4. $45m^2n^2p$ and $60m^3np^2$.

5. $18ab^2c^2d$ and $36a^2bcd^2$.
6. a^2b^2, a^2b^3 and a^1b^1.
7. $4ab$, $10ac$ and $30bc$.
8. $17pq^2$, $34p^2q$ and $51p^3q^3$.
9. $8x^2y^3z^4$, $12x^3y^2z^3$ and $20x^4y^3z^2$.
10. $30x^4y^5$, $90x^2y^3$ and $120x^3y^4$.

125. The student must be urged to commit to memory the following Table of forms which can or cannot be resolved into factors. Where a blank occurs after the sign = it signifies that the form on the left hand cannot be resolved into simpler factors.

$x^2 - y^2 = (x+y)(x-y)$ $x^2 - 1 = (x+1)(x-1)$
$x^2 + y^2 =$ $x^2 + 1 =$
$x^3 - y^3 = (x-y)(x^2 + xy + y^2)$ $x^3 - 1 = (x-1)(x^2 + x + 1)$
$x^3 + y^3 = (x+y)(x^2 - xy + y^2)$ $x^3 + 1 = (x+1)(x^2 - x + 1)$
$x^4 - y^4 = (x^2 + y^2)(x^2 - y^2)$ $x^4 - 1 = (x^2 + 1)(x^2 - 1)$
$x^4 + y^4 =$ $x^4 + 1 =$
$x^2 + 2xy + y^2 = (x+y)^2$ $x^2 + 2x + 1 = (x+1)^2$
$x^2 - 2xy + y^2 = (x-y)^2$ $x^2 - 2x + 1 = (x-1)^2$
$x^3 + 3x^2y + 3xy^2 + y^3 = (x+y)^3$ $x^3 + 3x^2 + 3x + 1 = (x+1)^3$
$x^3 - 3x^2y + 3xy^2 - y^3 = (x-y)^3$ $x^3 - 3x^2 + 3x - 1 = (x-1)^3$

The left-hand side of the table gives the *general* forms, the right-hand side the *particular* cases in which $y = 1$.

126. **Ex.** To find the H.C.F. of $x^2 - 1$, $x^2 - 2x + 1$, and $x^2 + 2x - 3$.

$$x^2 - 1 = (x-1)(x+1),$$
$$x^2 - 2x + 1 = (x-1)(x-1),$$
$$x^2 + 2x - 3 = (x-1)(x+3),$$
$$\therefore \text{H.C.F.} = x - 1.$$

EXAMPLES.—XXXV.

1. $a^2 - b^2$ and $a^3 - b^3$.
2. $a^2 - b^2$ and $a^4 - b^4$.
3. $a^2 - x^2$ and $(a-x)^2$.
4. $a^3 + x^3$ and $(a+x)^3$.
5. $9x^2 - 1$ and $(3x + 1)^2$.
6. $1 - 25a^2$ and $(1 - 5a)^2$.
7. $x^2 - y^2$, $(x+y)^2$ and $x^2 + 3xy + 2y^2$.
8. $x^2 - y^2$, $x^3 - y^3$ and $x^2 - 7xy + 6y^2$.
9. $x^2 - 1$, $x^3 - 1$ and $x^2 + x - 2$.
10. $1 - a^2$, $1 + a^3$ and $a^2 + 5a + 4$.

127. In large numbers the factors cannot often be determined by inspection, and if we have to find the H.C.F. of two such numbers we have recourse to the following Arithmetical Rule:

"Divide the greater of the two numbers by the less, and the divisor by the remainder, repeating the process until no remainder is left: the last divisor is the H.C.F. required."

Thus, to find the H.C.F. of 689 and 1573.

```
    689 ) 1573 ( 2
          1378
          ‾‾‾‾
          195 ) 689 ( 3
                585
                ‾‾‾
                104 ) 195 ( 1
                      104
                      ‾‾‾
                       91 ) 104 ( 1
                            91
                            ‾‾
                            13 ) 91 ( 7
                                 91
```

∴ 13 is the H.C.F. of 689 and 1573.

EXAMPLES.—XXXVI.

Find the H.C.F. of

1. 6906 and 10359.
2. 1908 and 2736.
3. 49608 and 169416.
4. 126025 and 40115.
5. 1581227 and 16758766.
6. 35175 and 236845.

128. The Arithmetical Rule is founded on the following operation in Algebra, which is called the Proof of the Rule for finding the Highest Common Factor of two expressions.

Let a and b be two expressions, arranged according to descending powers of some common letter, of which a is not of lower dimensions than b.

Let b divide a with p as quotient and remainder c,
c .... b ...... q .... d.
d .... c ...... r .... with no remainder.

The form of the operation may be shewn thus :

$$\begin{array}{r}b)\ a\ (p \\ pb \\ \hline c)\ b\ (q \\ qc \\ \hline d)\ c\ (r \\ rd \end{array}$$

Then we can shew

I. That d is a common factor of a and b.

II. That any other common factor of a and b is a factor of d, and that therefore d is the Highest Common Factor of a and b.

For (I.) to shew that d is a factor of a and b :

$$\begin{aligned} b &= qc + d \\ &= qrd + d \\ &= (qr+1)\,d, \text{ and } \therefore d \text{ is a factor of } b\,; \end{aligned}$$

and
$$\begin{aligned} a &= pb + c \\ &= p(qc+d)+c \\ &= pqc + pd + c \\ &= pqrd + pd + rd \\ &= (pqr+p+r)\,d, \text{ and } \therefore d \text{ is a factor of } a. \end{aligned}$$

And (II.) to shew that any common factor of a and b is a factor of d.

Let δ be any common factor of a and b, such that

$$a = m\delta \text{ and } b = n\delta.$$

Then we can shew that δ is a factor of d.

For
$$\begin{aligned} d &= b - qc \\ &= b - q(a - pb) \\ &= b - qa + pqb \\ &= n\delta - qm\delta + pqn\delta \\ &= (n - qm + pqn)\,\delta, \text{ and } \therefore \delta \text{ is a factor of } d. \end{aligned}$$

Now no expression higher than d can be a factor of d ;

$\therefore d$ is the Highest Common Factor of a and b.

129. **Ex.** To find the H.C.F. of $x^2 + 2x + 1$ and
$$x^3 + 2x^2 + 2x + 1.$$

$$
\begin{array}{r}
x^2+2x+1 \overline{\smash{\big)}\, x^3+2x^2+2x+1} \,(x \\
\underline{x^3+2x^2+x} \\
x+1 \overline{\smash{\big)}\, x^2+2x+1} \,(x+1 \\
\underline{x^2+x} \\
x+1 \\
x+1
\end{array}
$$

Hence $x + 1$ being the last divisor is the H.C.F. required.

130. In the algebraical process four devices are frequently useful. These we shall now state, and exemplify each in the next Article.

   I. If the sign of the first term of a remainder be *negative*, we may change the signs of all the terms.

   II. If a remainder contain a factor which is clearly not a common factor of the given expressions it may be removed.

   III. We may multiply or divide either of the given expressions by any number which does not introduce or remove a *common* factor.

   IV. If the given expressions have a common factor which can be seen by inspection, we may remove it from both, and find the Highest Common Factor of the parts which remain. *If we multiply this result by the ejected factor*, we shall obtain the Highest Common Factor of the given expressions.

131. **Ex. I.** To find the H.C.F. of $2x^2 - x - 1$ and
$$6x^2 - 4x - 2.$$

$$
\begin{array}{r}
2x^2-x-1 \overline{\smash{\big)}\, 6x^2-4x-2} \,(3 \\
\underline{6x^2-3x-3} \\
-x+1
\end{array}
$$

Change the signs of the remainder, and it becomes $x-1$.

$$x-1 \overline{\smash{\big)}\, 2x^2 - x - 1} \, (2x+1$$
$$\underline{2x^2 - 2x}$$
$$x-1$$
$$x-1$$

The H.C.F. required is $x-1$.

Ex. II. To find the H.C.F. of $x^2 + 3x + 2$ and $x^2 + 5x + 6$.

$$x^2 + 3x + 2 \overline{\smash{\big)}\, x^2 + 5x + 6} \, (1$$
$$\underline{x^2 + 3x + 2}$$
$$2x + 4$$

Divide the remainder by 2, and it becomes $x + 2$.

$$x+2 \overline{\smash{\big)}\, x^2 + 3x + 2} \, (x+1$$
$$\underline{x^2 + 2x}$$
$$x + 2$$
$$x + 2$$

The H.C.F. required is $x + 2$.

Ex. III. To find the H.C.F. of $12x^2 + x - 1$ and $15x^2 + 8x + 1$.

$$\begin{array}{r} \text{Multiply} \quad 15x^2 + 8x + 1 \\ \text{by} \quad 4 \end{array}$$

$$12x^2 + x - 1 \overline{\smash{\big)}\, 60x^2 + 32x + 4} \, (5$$
$$\underline{60x^2 + 5x - 5}$$
$$27x + 9$$

Divide the remainder by 9, and the result is $3x + 1$.

$$3x+1 \overline{\smash{\big)}\, 12x^2 + x - 1} \, (4x - 1$$
$$\underline{12x^2 + 4x}$$
$$-3x - 1$$
$$-3x - 1$$

The H.C.F. is therefore $3x + 1$.

Ex. IV. To find the H.C.F. of $x^3 - 5x^2 + 6x$ and $x^3 - 10x^2 + 21x$.

Remove and reserve the factor x, which is common to both expressions.

Then we have remaining $x^2 - 5x + 6$ and $x^2 - 10x + 21$.

The H.C.F. of these expressions is $x - 3$.

The H.C.F. of the original expressions is therefore $x^2 - 3x$.

EXAMPLES.—XXXVii.

Find the H.C.F. of the following expressions:

1. $x^2 + 7x + 12$ and $x^2 + 9x + 20$.
2. $x^2 + 12x + 20$ and $x^2 + 14x + 40$.
3. $x^2 - 17x + 70$ and $x^2 - 13x + 42$.
4. $x^2 + 5x - 84$ and $x^2 + 21x + 108$.
5. $x^2 + x - 12$ and $x^2 - 2x - 3$.
6. $x^2 + 5xy + 6y^2$ and $x^2 + 6xy + 9y^2$.
7. $x^2 - 6xy + 8y^2$ and $x^2 - 8xy + 16y^2$.
8. $x^2 - 13xy - 30y^2$ and $x^2 - 18xy + 45y^2$.
9. $x^3 - y^3$ and $x^2 - 2xy + y^2$.
10. $x^3 + y^3$ and $x^3 + 3x^2y + 3xy^2 + y^3$.
11. $x^4 - y^4$ and $x^2 - 2xy + y^2$.
12. $x^5 + y^5$ and $x^3 + y^3$.
13. $x^4 - y^4$ and $x^2 + 2xy + y^2$.
14. $a^2 - b^2 + 2bc - c^2$ and $a^2 + 2ab + b^2 - 2ac - 2bc + c^2$.
15. $12x^2 + 7xy + y^2$ and $28x^2 + 3xy - y^2$.
16. $6x^2 + xy - y^2$ and $39x^2 - 22xy + 3y^2$.
17. $15x^2 - 8xy + y^2$ and $40x^2 - 3xy - y^2$.
18. $x^5 - 5x^3 + 5x^2 - 1$ and $x^4 + x^3 - 4x^2 + x + 1$.
19. $x^4 + 4x^2 + 16$ and $x^5 + x^4 - 2x^3 + 17x^2 - 10x + 20$.
20. $x^4 + x^2y^2 + y^4$ and $x^4 + 2x^3y + 3x^2y^2 + 2xy^3 + y^4$.
21. $x^6 - 6x^4 + 9x^2 - 4$ and $x^6 + x^5 - 2x^4 + 3x^2 - x - 2$.

22. $15a^4 + 10a^3b + 4a^2b^2 + 6ab^3 - 3b^4$ and $6a^3 + 19a^2b + 8ab^2 - 5b^3$.
23. $15x^3 - 14x^2y + 24xy^2 - 7y^3$ and $27x^3 + 33x^2y - 20xy^2 + 2y^3$.
24. $21x^2 - 83xy - 27x + 22y^2 + 99y$ and $12x^2 - 35xy - 6x - 33y^2 + 22y$.
25. $3a^3 - 12a^2 - a^2b + 10ab - 2b^2$ and $6a^3 - 17a^2b + 8ab^2 - b^3$.
26. $18a^3 - 18a^2x + 6ax^2 - 6x^3$ and $60a^2 - 75ax + 15x^2$.
27. $21x^3 - 26x^2 + 8x$ and $6x^2 - x - 2$.
28. $6x^4 + 29a^2x^2 + 9a^4$ and $3x^3 - 15ax^2 + a^2x - 5a^3$.
29. $x^8 + x^6y^2 + x^2y + y^3$ and $x^4 - y^4$.
30. $2x^3 + 10x^2 + 14x + 6$ and $x^3 + x^2 + 7x + 39$.
31. $45a^3x + 3a^2x^2 - 9ax^3 + 6x^4$ and $18a^2x - 8x^3$.

132. It is sometimes easier to find the H.C.F. by *reversing* the order in which the expressions are given.

Thus to find the H.C.F. of $21x^2 + 38x + 5$ and $129x^2 + 221x + 10$ the easier course is to reverse the expressions, so that they stand thus, $5 + 38x + 21x^2$ and $10 + 221x + 129x^2$, and then to proceed by the ordinary process. The H.C.F. is $3x + 5$. Other examples are

(1) $187x^3 - 84x^2 + 31x - 6$ and $253x^3 - 14x^2 + 29x - 12$,
(2) $371y^3 + 26y^2 - 50y + 3$ and $469y^3 + 75y^2 - 103y - 21$,

of which the H.C.F. are respectively $11x - 3$ and $7y + 3$.

133. If the Highest Common Factor of *three* expressions a, b, c be required, find first the H.C.F. of a and b. If d be the H.C.F. of a and b, then the H.C.F. of d and c will be the H.C.F. of a, b, c.

134. Ex. To find the H.C.F. of
$x^3 + 7x^2 - x - 7$, $x^3 + 5x^2 - x - 5$, and $x^2 - 2x + 1$.

The H.C.F. of $x^3 + 7x^2 - x - 7$ and $x^3 + 5x^2 - x - 5$ will be found to be $x^2 - 1$.

The H.C.F. of $x^2 - 1$ and $x^2 - 2x + 1$ will be found to be $x - 1$.

Hence $x - 1$ is the H.C.F. of the three expressions.

EXAMPLES.—XXXVIII.

Find the Highest Common Factor of

1. $x^2 + 5x + 6$, $x^2 + 7x + 10$, and $x^2 + 12x + ?$.
2. $x^3 + 4x^2 - 5$, $x^3 - 3x + 2$, and $x^3 + 4x^2 - 8x + 3$.
3. $2x^2 + x - 1$, $x^2 + 5x + 4$, and $x^3 + 1$.
4. $y^3 - y^2 - y + 1$, $3y^2 - 2y - 1$, and $y^3 - y^2 + y - 1$.
5. $x^3 - 4x^2 + 9x - 10$, $x^3 + 2x^2 - 3x + 20$, and $x^3 + 5x^2 - 9x + 35$.
6. $x^3 - 7x^2 + 16x - 12$, $3x^3 - 14x^2 + 16x$, and $5x^3 - 10x^2 + 7x - 14$.
7. $y^3 - 5y^2 + 11y - 15$, $y^3 - y^2 + 3y + 5$, and $2y^3 - 7y^2 + 16y - 15$.

NOTE. We use the name Highest Common Factor instead of *Greatest Common Measure* or *Highest Common Divisor* for the following reasons:

(1) We have used the word "*Measure*" in Art. 35 in a different sense, that is, to denote the number of times any quantity contains the *unit* of measurement.

(2) *Divisor* does not necessarily imply a quantity which is contained in another an *exact* number of times. Thus in performing the operation of dividing 333 by 13, we call 13 *divisor*, but we do not mean that 333 contains 13 an exact number of times.

IX. FRACTIONS

135. A QUANTITY a is called an EXACT DIVISOR of a quantity b, when b contains a an exact number of times.

A quantity a is called a MULTIPLE of a quantity b, when a contains b an exact number of times.

FRACTIONS. 77

136. Hitherto we have treated of quantities which contain the unit of measurement in each case an exact number of times.

We have now to treat of quantities *which contain some exact divisor of a primary unit* an exact number of times.

137. We **must** first explain what we mean by *a primary unit*.

We said in Art. 33 that to *measure* any quantity we take a known standard or unit of the same kind. Our choice as to the quantity to be taken as the unit is at first unrestricted, but when once made we must adhere to it, or at least we must give distinct notice of any change which we make with respect to it. To such a unit we give the name of PRIMARY UNIT.

138. Next, to explain what we mean by *an exact divisor of a primary unit*.

Keeping our Primary Unit as our main standard of measurement, we may conceive it to be divided into a number of parts of equal magnitude, any one of which we may take as a *Subordinate Unit*.

Thus we may take a pound as the unit by which we measure sums of money, and retaining this steadily as the *primary* unit, we may still conceive it to be subdivided into 20 equal parts. We call each of the subordinate units in this case a shilling, and we say that one of these equal *subordinate* units is *one-twentieth part* of the primary unit, that is, of a pound.

These subordinate units, then, are *exact divisors of the primary unit*.

139. Keeping the primary unit still clearly in view, we represent *one* of the subordinate units by the following notation.

We agree to represent the words one-third, one-fifth, and one-twentieth by the symbols $\frac{1}{3}$, $\frac{1}{5}$, $\frac{1}{20}$, and we say that if the Primary Unit be divided into *three* equal parts, $\frac{1}{3}$ will represent *one* of these parts.

If we have to represent *two* of these subordinate units, we do so by the symbol $\frac{2}{3}$; if *three*, by the symbol $\frac{3}{3}$; if *four*, by the symbol $\frac{4}{3}$, and so on. And, generally, if the Primary Unit be divided into b equal parts, we represent a of those parts by the symbol $\frac{a}{b}$.

140. The symbol $\frac{a}{b}$ we call the Fraction Symbol, or, more briefly, a **Fraction**. The number *below* the line is called the DENOMINATOR, because it denominates the number of equal parts into which the Primary Unit is divided. The number *above* the line is called the NUMERATOR, because it enumerates how many of these equal parts, or Subordinate Units, are taken.

141. The term *number* may be correctly applied to Fractions, since they are measured by units, but we must be careful to observe the following distinction:

An Integer or Whole Number is a multiple of the Primary Unit.

A Fractional Number is a multiple of the Subordinate Unit.

142. The Denominator of a Fraction shews what multiple the Primary Unit is of the Subordinate Unit.

The Numerator of a Fraction shews what multiple the Fraction is of the Subordinate Unit.

143. The Numerator and Denominator of a fraction are called the Terms of the Fraction.

144. Having thus explained the nature of Fractions, we next proceed to treat of the operations to which they are subjected in Algebra.

145. DEF. If the quantity x be divided into b equal parts, and a of those parts be taken, the result is said to be the fraction $\frac{a}{b}$ of x.

If x be the unit, this is called the fraction $\frac{a}{b}$.

FRACTIONS.

146. If the unit be divided into b equal parts,

$\dfrac{1}{b}$ will represent one of the parts.

$\dfrac{2}{b}$ .... two ....

$\dfrac{3}{b}$ .... three ....

And generally,

$\dfrac{a}{b}$ will represent a of the parts.

147. Next let us suppose that each of the b parts is subdivided into c equal parts: then the unit has been divided into bc equal parts, and

$\dfrac{1}{bc}$ will represent one of the subdivisions.

$\dfrac{2}{bc}$ .... two ....

And generally,

$\dfrac{a}{bc}$ .... a ....

148. To shew that $\dfrac{ac}{bc} = \dfrac{a}{b}$.

Let the unit be divided into b equal parts.

Then $\dfrac{a}{b}$ will represent *a of these parts.* .... (1).

Next let each of the b parts be subdivided into c equal parts.

Then the primary unit has been divided into bc equal parts, and $\dfrac{ac}{bc}$ will represent *ac of these subdivisions.* .... (2).

Now one of the parts in (1) is equal to c of the subdivisions in (2),

∴ a parts are equal to ac subdivisions;

$$\therefore \dfrac{a}{b} = \dfrac{ac}{bc}.$$

Cor. We draw from this proof two inferences:

I. If the numerator and denominator of a fraction be *multiplied* by the same number, the value of the fraction is not altered.

II. If the numerator and denominator of a fraction be *divided* by the same number, the value of the fraction is not altered.

149. To make the important Theorem established in the preceding Article more clear, we shall give the following proof that $\frac{4}{5} = \frac{16}{20}$, by taking a straight line as the unit of length.

Let the line AC be divided into 5 equal parts.

Then, if B be the point of division nearest to C,

$$AB \text{ is } \frac{4}{5} \text{ of } AC. \quad\ldots\ldots\ldots\ldots\ldots (1)$$

Next, let each of the parts be subdivided into 4 equal parts.

Then AC contains 20 of these subdivisions,
and AB .... 16 ....

$$\therefore AB \text{ is } \frac{16}{20} \text{ of } AC. \quad\ldots\ldots\ldots\ldots\ldots (2).$$

Comparing (1) and (2), we conclude that

$$\frac{4}{5} = \frac{16}{20}.$$

150. From the Theorem established in Art. 148 we derive the following rule for reducing a fraction to its lowest terms:

Find the Highest Common Factor of the numerator and denominator and divide both by it. The resulting fraction will be one equivalent to the original fraction expressed in the simplest terms.

FRACTIONS.

151. When the numerator and denominator each consist of a single term the H.C.F. may be determined by inspection, or we may proceed as in the following Example:

To reduce the fraction $\dfrac{10a^3b^2c^4}{12a^2b^3c^2}$ to its lowest terms,

$$\frac{10a^3b^2c^4}{12a^2b^3c^2} = \frac{2 \times 5 \times aaabbcccc}{2 \times 6 \times aabbbcc};$$

We may then remove factors common to the numerator and denominator, and we shall have remaining $\dfrac{5 \times acc}{6 \times b}$;

$\therefore$ the required result will be $\dfrac{5ac^2}{6b}$.

152. Two cases are especially to be noticed.

(1) If *every one* of the factors of the numerator be removed, the number 1 (being always a factor of every algebraical expression) will still remain to form a numerator.

Thus $\quad \dfrac{3a^2c}{12a^3c^2} = \dfrac{3aac}{3 \times 4 \times aaacc} = \dfrac{1}{4ac}$.

(2) If every one of the factors of the denominator be removed, the result will be a whole number.

Thus $\quad \dfrac{12a^3c^2}{3a^2c} = \dfrac{3 \times 4 \times aaacc}{3 \times aac} = 4ac$.

This is, in fact, a case of exact division, such as we have explained in Art. 74.

EXAMPLES.—XXXIX.

Reduce to equivalent fractions in their simplest terms the following fractions:

1. $\dfrac{4a^2}{12a^3}$
2. $\dfrac{8x^3}{36x^2}$
3. $\dfrac{10a^2b^3}{24a^3b^2}$
4. $\dfrac{18x^5y^2z^3}{45x^3y^2z^4}$
5. $\dfrac{7a^5b^7c^8}{21a^3b^2c^5}$
6. $\dfrac{4axy}{3abc}$
7. $\dfrac{51ay^2z}{34a^2yz^2}$ [S.A.]
8. $\dfrac{15ab^4c^3}{12a^3b^2c^2}$
9. $\dfrac{8x^3y^2z^3}{6x^5y^8z^3}$

10. $\dfrac{210m^3n^2p}{42m^2n^2p^2}$.   11. $\dfrac{a^2}{a^2+ab}$.   12. $\dfrac{14m^2x}{21m^3p}-\dfrac{}{7mx}$.

13. $\dfrac{xy}{3xy^2-5x^2yz}$.   14. $\dfrac{4ax+2x^2}{8ax^3-2x^2}$.   15. $\dfrac{ay+y^2}{abc+bcy}$.

16. $\dfrac{4a^2x+6a^2y}{8x^2-18y^2}$.   17. $\dfrac{12ab^2-6ab}{8b^2c-2c}$.   18. $\dfrac{c^2-4a^2}{c^2+4ac+4a^2}$.

19. $\dfrac{3x^4+3c^2y^2}{5x^4+5x^2y^2}$.   24. $\dfrac{7ab^3x^8-7ab^3y^2}{14a^3bcx^8-14a^3bcy^2}$.

20. $\dfrac{10x-10y}{4x^2-8xy+4y^2}$.   25. $\dfrac{5x^9+45dx^2}{10cx^9+90cdx^2}$.

21. $\dfrac{ax+by}{7a^2x^2-7b^2y^2}$.   26. $\dfrac{10a^2+20ab+10b^2}{5a^3+5a^2b}$.

22. $\dfrac{6ab+8cd}{27a^2b^2x-48c^2d^2x}$.   27. $\dfrac{4x^2-8xy+4y^2}{48(x-y)^2}$.

23. $\dfrac{xy-xyz}{2az-2az^2}$.   28. $\dfrac{3mx+5nx^2}{3my+5nxy}$.

153. We shall now give a set of Examples, some of which may be worked by Resolution into Factors. In others the H.C.F. of the numerator and denominator must be found by the usual process. As an example of the latter sort let us take the following:

To reduce the fraction $\dfrac{x^3-4x^2-19x-14}{2x^3-9x^2-38x+21}$ to its lowest terms.

Proceeding by the usual rule for finding the H.C.F. of the numerator and denominator we find it to be $x-7$.

Now if we divide $x^3-4x^2-19x-14$ by $x-7$, the result is x^2+3x+2, and if we divide $2x^3-9x^2-38x+21$ by $x-7$, the result is $2x^2+5x-3$.

Hence the fraction $\dfrac{x^2+3x+2}{2x^2+5x-3}$ is equivalent to the proposed fraction and is in its lowest terms.

EXAMPLES.—xl.

1. $\dfrac{a^2+7a+10}{a^2+5a+6}$.   2. $\dfrac{x^2-9x+20}{x^2-7x+12}$.   3. $\dfrac{x^2-2x-3}{x^2-10x+21}$.

6. $\dfrac{x^2 - 18xy + 45y^2}{x^2 - 8xy - 105y^2}$

5. $\dfrac{x^4 + x^2 + 1}{x^2 + x + 1}$.

6. $\dfrac{x^6 + 2x^3y^3 + y^6}{x^6 - y^6}$.

7. $\dfrac{x^3 - 4x^2 + 9x - 10}{x^3 + 2x^2 - 3x + 20}$.

14. $\dfrac{m^3 + 3m^2 - 4m}{m^3 - 7m + 6}$.

8. $\dfrac{x^3 - 5x^2 + 11x - 15}{x^3 - x^2 + 3x + 5}$.

15. $\dfrac{a^3 + 1}{a^3 + 2a^2 + 2a + 1}$.

9. $\dfrac{x^3 - 8x^2 + 21x - 18}{3x^3 - 16x^2 + 21x}$.

16. $\dfrac{3ax^2 - 13ax + 14a}{7x^3 - 17x^2 + 6x}$.

10. $\dfrac{x^3 - 7x^2 + 16x - 12}{3x^3 - 14x^2 + 16x}$.

17. $\dfrac{14x^2 - 34x + 12}{9ax^2 - 39ax + 42a}$.

11. $\dfrac{x^4 + x^3y + xy^3 - y^4}{x^4 - x^3y - xy^3 - y^4}$.

18. $\dfrac{10a - 24a^2 + 14a^3}{15 - 24a + 3a^2 + 6a^3}$.

12. $\dfrac{a^3 + 4a^2 - 5}{a^3 - 3a + 2}$.

19. $\dfrac{2ab^3 + ab^2 - 8ab + 5a}{7b^3 - 12b^2 + 5b}$.

13. $\dfrac{b^3 + 4b^2 - 5b}{b^3 - 6b + 5}$.

20. $\dfrac{a^3 - 3a^2 + 3a - 2}{a^3 - 4a^2 + 6a - 4}$.

21. $\dfrac{3x^2 + 2x - 1}{x^3 + x^2 - x - 1}$. 22. $\dfrac{a^2 - a - 20}{a^2 + a - 12}$. 23. $\dfrac{x^3 - 3x^2 + 4x - 2}{x^3 - x^2 - 2x + 2}$.

24. $\dfrac{(x+y+z)^2 + (z-y)^2 + (x-z)^2 + (y-x)^2}{x^2 + y^2 + z^2}$.

25. $\dfrac{2x^4 - x^3 - 9x^2 + 13x - 5}{7x^3 - 19x^2 + 17x - 5}$.

33. $\dfrac{15a^2 + ab - 2b^2}{9a^2 + 3ab - 2b^2}$.

26. $\dfrac{16x^4 - 53x^2 + 45x + 6}{8x^4 - 30x^3 + 31x^2 - 12}$.

34. $\dfrac{x^3 - 7x + 10}{2x^2 - x - 6}$.

27. $\dfrac{4x^2 - 12ax + 9a^2}{8x^3 - 27a^3}$.

35. $\dfrac{x^3 + 3x^2 + 4x + 12}{x^3 + 4x^2 + 4x + 3}$.

28. $\dfrac{6x^3 - 23x^2 + 16x - 3}{6x^3 - 17x^2 + 11x - 2}$.

36. $\dfrac{x^4 - x^2 - 2x + 2}{2x^3 - x - 1}$.

29. $\dfrac{x^3 - 6x^2 + 11x - 6}{x^3 - 2x^2 - x + 2}$.

37. $\dfrac{x^3 - 2x^2 - 15x + 36}{3x^2 - 4x - 15}$.

30. $\dfrac{m^3 + m^2 + m - 3}{m^3 + 3m^2 + 5m + 3}$.

38. $\dfrac{3x^3 + x^2 - 5x + 21}{6x^3 + 29x^2 + 26x - 21}$.

31. $\dfrac{x^5 + 5x^4 - x^2 - 5x}{x^4 + 3x^3 - x - 3}$.

39. $\dfrac{x^4 - x^3 - 4x^2 - x + 1}{4x^3 - 3x^2 - 8x - 1}$.

32. $\dfrac{a^2 - b^2 - 2bc - c^2}{a^2 + 2ab + b^2 - c^2}$.

40. $\dfrac{a^3 - 7a^2 + 16a - 12}{3a^3 - 14a^2 + 16a}$.

154. The fraction $\frac{a}{b}$ is said to be a *proper* fraction, when a is less than b.

The fraction $\frac{a}{b}$ is said to be an *improper* fraction, when a is greater than b.

155. A whole number x may be written as a fractional number by writing 1 beneath it as a denominator, thus $\frac{x}{1}$.

156. To prove that $\frac{a}{b}$ of $\frac{c}{d} = \frac{ac}{bd}$.

Divide the unit into bd parts.

Then $\frac{a}{b}$ of $\frac{c}{d} = \frac{a}{b}$ of $\frac{bc}{bd}$ (Art. 148)

$= \frac{a}{b}$ of bc of these parts   (Art. 147)

$= \frac{ac}{bc}$ of bc of these parts   (Art. 148)

$= ac$ of these parts   (Art. 147).

But $\frac{ac}{bd} = ac$ of these parts;

$\therefore \frac{a}{b}$ of $\frac{c}{d} = \frac{ac}{bd}$.

This is an important Theorem, for from it is derived the Rule for what is called MULTIPLICATION OF FRACTIONS. We extend the meaning of the sign $\times$ and define $\frac{a}{b} \times \frac{c}{d}$ (which according to our definition in Art. 36 has no meaning) to mean $\frac{a}{b}$ of $\frac{c}{d}$, and we conclude that $\frac{a}{b} \times \frac{c}{d} = \frac{ac}{bd}$, which in words gives us this rule—"Take the product of the numerators to form the numerator of the resulting fraction, and the product of the denominators to form the denominator."

The same rule holds good for the multiplication of three or more fractions.

157. To shew that $\dfrac{a}{b} \div \dfrac{c}{d} = \dfrac{ad}{bc}$.

The quotient, x, of $\dfrac{a}{b}$ divided by $\dfrac{c}{d}$ is such a number that x multiplied by the divisor $\dfrac{c}{d}$ will give as a result the dividend $\dfrac{a}{b}$.

$$\therefore \dfrac{xc}{d} = \dfrac{a}{b};$$

$$\therefore \dfrac{d}{c} \text{ of } \dfrac{xc}{d} = \dfrac{d}{c} \text{ of } \dfrac{a}{b};$$

$$\therefore \dfrac{xcd}{cd} = \dfrac{ad}{bc};$$

$$\therefore x = \dfrac{ad}{bc}.$$

Hence we obtain a rule for what is called DIVISION OF FRACTIONS.

Since $\dfrac{a}{b} \div \dfrac{c}{d} = \dfrac{ad}{bc}$,

$\dfrac{a}{b} \div \dfrac{c}{d} = \dfrac{a}{b} \times \dfrac{d}{c}.$

Hence we reduce the process of division to that of multiplication by inverting the divisor.

158. The following are examples of the Multiplication and Division of Fractions.

1. $\dfrac{2x}{3a^2} \times 3a = \dfrac{2x}{3a^2} \times \dfrac{3a}{1} = \dfrac{6ax}{3a^2} = \dfrac{2x}{a}.$

2. $\dfrac{3x}{2b} \div 3a = \dfrac{3x}{2b} \div \dfrac{3a}{1} = \dfrac{3x}{2b} \times \dfrac{1}{3a} = \dfrac{3x}{6ab} = \dfrac{x}{2ab}.$

3. $\dfrac{4a^2}{9c^2} \times \dfrac{3c}{2a} = \dfrac{3 \times 4 \times a^2 c}{2 \times 9 \times ac^2} = \dfrac{2a}{3c}.$

4. $\dfrac{14x^2}{27y^2} \div \dfrac{7x}{9y} = \dfrac{14x^2}{27y^2} \times \dfrac{9y}{7x} = \dfrac{9 \times 14 \times x^2 y}{7 \times 27 \times xy^2} = \dfrac{2x}{3y}.$

5. $\dfrac{2a}{3b} \times \dfrac{9b}{10c} \times \dfrac{5c}{4a} = \dfrac{2a \times 9b \times 5c}{3b \times 10c \times 4a} = \dfrac{3}{4}.$

6. $\dfrac{x^2-4x}{x^3+7x^2} \times \dfrac{x^2+7x}{x-4} = \dfrac{x(x-4)}{x^2(x+7)} \times \dfrac{x(x+7)}{x-4}$

$= \dfrac{x(x-4)x(x+7)}{x^2(x+7)(x-4)} = 1.$

7. $\dfrac{a^2-b^2}{a^2+2ab+b^2} \div \dfrac{4(a^2-ab)}{a^2+ab} = \dfrac{a^2-b^2}{a^2+2ab+b^2} \times \dfrac{a^2+ab}{4(a^2-ab)}$

$= \dfrac{(a+b)(a-b)}{(a+b)(a+b)} \times \dfrac{a(a+b)}{4a(a-b)}$

$= \dfrac{(a+b)(a-b)a(a+b)}{(a+b)(a+b)4a(a-b)} = \dfrac{1}{4}.$

EXAMPLES.—xli.

Simplify the following expressions:

1. $\dfrac{3x}{4y} \times \dfrac{7x}{9y}.$
2. $\dfrac{3a}{4b} \times \dfrac{2b}{3a}.$
3. $\dfrac{4x^2}{9y^2} \times \dfrac{3x}{2y}.$

4. $\dfrac{8a^2b^3}{45x^2y} \times \dfrac{15xy^2}{24a^3b^2}.$
5. $\dfrac{9x^2y^2z}{10a^2b^2c} \times \dfrac{20a^3b^2c}{18xy^2z}.$
6. $\dfrac{2a}{5b} \times \dfrac{4b}{3c} \times \dfrac{5c}{6a}.$

7. $\dfrac{3x^2y}{4xz^3} \times \dfrac{5y^2z}{6xy} \times \dfrac{12xz}{20xy^2}.$
8. $\dfrac{7a^5b^4}{5c^2d^3} \times \dfrac{20c^3d^2}{42a^4b^3} \times \dfrac{4ac}{3bd}.$

9. $\dfrac{9m^2n^2}{8p^3q^3} \times \dfrac{5p^2q}{2xy} \times \dfrac{24x^2y^2}{90mn}.$
10. $\dfrac{25k^3m^2}{14n^2q^2} \times \dfrac{70n^3q}{75p^2m} \times \dfrac{3pm}{4k^2n}.$

EXAMPLES.—xlii.

Reduce to simple fractions in their lowest terms:

1. $\dfrac{a-b}{a^2+ab} \times \dfrac{a^2-b^2}{a^2-ab}.$
4. $\dfrac{x^2+x-2}{x^2-7x} \times \dfrac{x^2-13x+42}{x^2+2x}.$

2. $\dfrac{x^2+4x}{x^2-3x} \times \dfrac{4x^2-12x}{3x^2+12x}.$
5. $\dfrac{x^2-11x+30}{x^2-6x+9} \times \dfrac{x^2-3x}{x^2-5x}.$

3. $\dfrac{x^2+3x+2}{x^2-5x+6} \times \dfrac{x^2-7x+12}{x^2+x}.$
6. $\dfrac{x^2-4}{x^2+5x} \times \dfrac{x^2-25}{x^2+2x}.$

7. $\dfrac{a^2-4a+3}{a^2-5a+4} \times \dfrac{a^2-9a+20}{a^2-10a+21} \times \dfrac{a^2-7a}{a^2-5a}.$

8. $\dfrac{b^2-7b+6}{b^2+3b-4} \times \dfrac{b^2+10b+24}{b^2-14b+48} \times \dfrac{b^3-8b^2}{b^2+6b}.$

9. $\dfrac{x^2 - y^2}{x^2 - 3xy + 2y^2} \times \dfrac{xy - 2y^2}{x^2 + xy} \times \dfrac{x^2 - xy}{(x-y)^2}.$

10. $\dfrac{(a+b)^2 - c^2}{a^2 - (b-c)^2} \times \dfrac{c^2 - (a-b)^2}{c^2 - (a+b)^2}.$

11. $\dfrac{(x-m)^2 - n^2}{(x-n)^2 - m^2} \times \dfrac{x^2 - (n-m)^2}{x^2 - (m-n)^2}.$

12. $\dfrac{(a+b)^2 - (c+d)^2}{(a+c)^2 - (b+d)^2} \times \dfrac{(a-b)^2 - (d-c)^2}{(a-c)^2 - (d-b)^2}.$

13. $\dfrac{x^2 - 2xy + y^2 - z^2}{x^2 + 2xy + y^2 - z^2} \times \dfrac{x + y - z}{x - y + z}.$

EXAMPLES.—xliii.

Simplify the following expressions:

1. $\dfrac{2a}{x} \div \dfrac{3b}{5c}.$

2. $\dfrac{15y}{14z} \div \dfrac{5y^2}{7z}.$

3. $\dfrac{8x^4y}{15ab^3} \div \dfrac{2x^3}{30ab^2}.$

4. $\dfrac{4a}{nx} \div 3ab.$

5. $\dfrac{3p}{2p-2} \div \dfrac{2p}{p-1}.$

6. $1 \div \dfrac{4a}{5x}.$

7. $\dfrac{5x}{7} \div 2.$

8. $\dfrac{1}{x^2 - 3x + 2} \div \dfrac{1}{x - 1}.$

9. $\dfrac{1}{x^2 - 17x + 30} \div \dfrac{1}{x - 15}.$

158. We are now able to justify the use of the Fraction Symbol as one of the Division Symbols in Art. 73, that is, we can shew that $\dfrac{a}{b}$ is a proper representation of the quotient resulting from the division of a by b.

For let x be this quotient.

Then, by the definition of a quotient, Art. 72,
$$b \times x = a.$$

But, from the nature of fractions,
$$b \times \dfrac{a}{b} = a;$$

$$\therefore \dfrac{a}{b} = x.$$

159. Here we may state an important theorem, which we shall require in the next chapter.

If $ad = bc$, to shew that $\dfrac{a}{b} = \dfrac{c}{d}$

Since $ad = bc$,

$$\dfrac{ad}{bd} = \dfrac{bc}{bd}$$

$$\therefore \dfrac{a}{b} = \dfrac{c}{d}$$

X. THE LOWEST COMMON MULTIPLE.

160. An expression is a COMMON MULTIPLE of two or more other expressions when the former is exactly divisible by each of the latter.

Thus $24x^3$ is a common multiple of 6, $8x^2$ and $12x^3$.

161. The LOWEST COMMON MULTIPLE of two or more expressions is the expression *of lowest dimensions* which is exactly divisible by each of them.

Thus $18x^4$ is the Lowest Common Multiple of $6x^4$, $9x^2$, and $3x$.

The words Lowest Common Multiple are abbreviated into L.C.M.

162. Two numbers are said to be *prime* to each other which have no common factor but unity.

Thus 2 and 3 are prime to each other.

163. If a and b be prime to each other the fraction $\dfrac{a}{b}$ is in its lowest terms.

Hence if a and b be prime to each other, and $\dfrac{a}{b} = \dfrac{c}{d}$, and if m be the H.C.F. of c and d,

$$a = \dfrac{c}{m} \text{ and } b = \dfrac{d}{m}.$$

164. In finding the Lowest Common Multiple of two or more expressions, each consisting of a single term, we may proceed as in Arithmetic, thus:

(1) To find the L.C.M. of $4a^3x$ and $18ax^3$,

$$\begin{array}{c|cc} 2 & 4a^3x, & 18ax^3 \\ \hline a & 2a^3x, & 9ax^3 \\ \hline x & 2a^2x, & 9x^3 \\ \hline & 2a^2, & 9x^2 \end{array}$$

L.C.M. $= 2 \times a \times x \times 2a^2 \times 9x^2 = 36a^3x^3$.

(2) To find the L.C.M. of ab, ac, bc,

$$\begin{array}{c|ccc} a & ab, & ac, & bc \\ \hline b & b, & c, & bc \\ \hline c & 1, & c, & c \\ \hline & 1, & 1, & 1 \end{array}$$

L.C.M. $= a \times b \times c = abc$.

(3) To find the L.C.M. of $12a^2c$, $14bc^2$ and $36ab^2$,

$$\begin{array}{c|ccc} 2 & 12a^2c, & 14bc^2, & 36ab^2 \\ \hline 6 & 6a^2c, & 7bc^2, & 18ab^2 \\ \hline a & a^2c, & 7bc^2, & 3ab^2 \\ \hline b & ac, & 7bc^2, & 3b^2 \\ \hline c & ac, & 7c^2, & 3b \\ \hline & a, & 7c, & 3b \end{array}$$

L.C.M. $= 2 \times 6 \times a \times b \times c \times a \times 7c \times 3b = 252a^2b^2c^2$.

EXAMPLES.—xliv.

Find the L.C.M. of

1. $4a^3x$ and $6a^2x^2$.
2. $3x^2y$ and $12xy^2$.
3. $4a^3b$ and $8a^2b^2$.
4. ax, a^2x and a^2x^2.
5. $2ax$, $4ax^2$ and x^3.
6. ab, a^2c and b^2c^3.
7. a^2x, a^3y and x^2y^2.
8. $51a^2x^2$, $34ax^3$ and ax^4.
9. $5p^2q$, $10q^2r$ and $20pqr$.
10. $18ax^2$, $72ay^2$ and $12xy$.

165. The method of finding the L.C.M., given in the preceding article, may be extended to the case of compound expressions, when one or more of their factors can be readily determined. Thus we may take the following Examples:

(1) To find the L.C.M. of $a-x$, a^2-x^2, and a^2+ax,

$$\begin{array}{c|ccc} a-x & a-x, & a^2-x^2, & a^2+ax \\ a+x & 1, & a+x, & a^2+ax \\ \hline & 1, & 1, & a \end{array}$$

L.C.M. $= (a-x)(a+x)a = (a^2-x^2)a = a^3-ax^2$.

(2) To find the L.C.M. of x^2-1, x^4-1, and $4x^6-4x^4$,

$$\begin{array}{c|ccc} x^2-1 & x^2-1, & x^4-1, & 4x^6-4x^4 \\ \hline & 1, & x^2+1, & 4x^4 \end{array}$$

L.C.M. $= (x^2-1)(x^2+1)4x^4 = (x^4-1)4x^4 = 4x^8-4x^4$.

166. The student who is familiar with the methods of resolving simple expressions into factors, especially those given in Art. 125, may obtain the L.C.M. of such expressions by a process which may be best explained by the following Examples:

Ex. 1. To find the L.C.M. of a^2-x^2 and a^3-x^3.

$a^2-x^2 = (a-x)(a+x)$,
$a^3-x^3 = (a-x)(a^2+ax+x^2)$.

Now the L.C.M. must contain in itself each of the factors in each of these products, and no others.

∴ L.C.M. is $(a-x)(a+x)(a^2+ax+x^2)$,

the factor $a-x$ occurring *once* in each product, and therefore *once* only in the L.C.M.

Ex. 2. To find the L.C.M. of
a^2-b^2, $a^2-2ab+b^2$, and $a^2+2ab+b^2$.

$a^2-b^2 = (a+b)(a-b)$,
$a^2-2ab+b^2 = (a-b)(a-b)$,
$a^2+2ab+b^2 = (a+b)(a+b)$;

L.C.M. is $(a+b)(a-b)(a-b)(a+b)$,

the factor $a-b$ occurring *twice* in one of the products, and $a+b$ occurring *twice* in another of the products, and therefore each of these factors must occur *twice* in the L.C.M.

EXAMPLES.—xlv.

Find the L.C.M. of the following expressions:

1. x^2 and $ax + x^2$.
2. $x^2 - 1$ and $x^2 - x$.
3. $a^2 - b^2$ and $a^2 + ab$.
4. $2x - 1$ and $4x^2 - 1$.
5. $a + b$ and $a^3 + b^3$.
6. $x + 1, x - 1$ and $x^2 - 1$.
7. $x + 1, x^3 - 1$ and $x^2 + x + 1$.
8. $x + 1, x^2 + 1$ and $x^3 + 1$.
9. $x - 1, x^2 - 1$ and $x^3 - 1$.
10. $x^2 - 1, x^2 + 1$ and $x^4 - 1$.
11. $x^2 - x, x^3 - 1$ and $x^3 + 1$.
12. $x^2 - 1, x^2 - x$ and $x^3 - 1$.
13. $2a + 1, 4a^2 - 1$ and $8a^3 + 1$.
14. $x + y$ and $2x^2 + 2xy$.
15. $(a + b)^2$ and $a^2 - b^2$.
16. $a + b, a - b$ and $a^2 - b^2$.
17. $4(1 + x), 4(1 - x)$ and $2(1 - x^2)$.
18. $x - 1, x^2 + x + 1$ and $x^3 - 1$.
19. $(a - b)(a - c)$ and $(a - c)(b - c)$.
20. $(x + 1)(x + 2), (x + 2)(x + 3)$ and $(x + 1)(x + 3)$.
21. $x^2 - y^2, (x + y)^2$ and $(x - y)^2$.
22. $(a + 3)(a + 1), (a + 3)(a - 1)$ and $a^2 - 1$.
23. $x^2(x - y), x(x^2 - y^2)$ and $x + y$.
24. $(x + 1)(x + 3), (x + 2)(x + 3)(x + 4)$ and $(x + 1)(x + 2)$
25. $x^2 - y^2, 3(x - y)^2$ and $12(x^3 + y^3)$.
26. $3(x^2 + xy), 8(xy - y^2)$ and $10(x^2 - y^2)$.

167. The chief use of the rule for finding the L.C.M. is for the reduction of fractions to common denominators, and in the simple examples, which we shall have to put before the student in a subsequent chapter, the rules which we have already given will be found generally sufficient. But as we may have to find the L.C.M. of two or more expressions in which the elementary factors cannot be determined by inspection, we must now proceed to discuss a Rule for finding the L.C.M. of two expressions which is applicable to every case.

168. The rule for finding the L.C.M. of two expressions a and b is this.

Find d the highest common factor of a and b.

Then the L.C.M. of a and $b = \dfrac{a}{d} \times b$,

or, $= \dfrac{b}{d} \times a$.

In words, the L.C.M. of two expressions is found by the following process :

Divide one of the expressions by the H.C.F. *and multiply the quotient by the other expression. The result is the* L.C.M.

The *proof* of this rule we shall now give.

169. To find the L.C.M. of two algebraical expressions.

Let a and b be the two algebraical expressions.

Let d be their H.C.F.,

x the required L.C.M.

Now since x is a multiple of a and b, we may say that

$$x = ma, \quad x = nb \; ;$$
$$\therefore ma = nb \; ;$$
$$\therefore \dfrac{m}{n} = \dfrac{b}{a} \quad \text{(Art. 159)}.$$

Now since x is the *Lowest* Common Multiple of a and b, m and n can have no common factor ;

$\therefore$ the fraction $\dfrac{m}{n}$ must be in its lowest terms ;

$\therefore m = \dfrac{b}{d}$ and $n = \dfrac{a}{d}$ (Art. 163).

Hence, since $x = ma$,

$$x = \dfrac{b}{d} \times a.$$

Also, since $x = nb$,

$$x = \dfrac{a}{d} \times b.$$

THE LOWEST COMMON MULTIPLE.

170. Ex. Find the L.C.M. of $x^2 - 13x + 42$ and $x^2 - 19x + 84$.

First we find the H.C.F. of the two expressions to be $x - 7$.

Then L.C.M. $= \dfrac{(x^2 - 13x + 42) \times (x^2 - 19x + 84)}{x - 7}$.

Now each of the factors composing the numerator is divisible by $x - 7$.

Divide $x^2 - 13x + 42$ by $x - 7$, and the quotient is $x - 6$.

Hence L.C.M. $= (x - 6)(x^2 - 19x + 84) = x^3 - 25x^2 + 198x - 504$.

EXAMPLES.—xlvi.

Find the L.C.M. of the following expressions:

1. $x^2 + 5x + 6$ and $x^2 + 6x + 8$.
2. $a^2 - a - 20$ and $a^2 + a - 12$.
3. $x^2 + 3x + 2$ and $x^2 + 4x + 3$.
4. $x^2 + 11x + 30$ and $x^2 + 12x + 35$.
5. $x^2 - 9x - 22$ and $x^2 - 13x + 22$.
6. $2x^2 + 3x + 1$ and $x^2 - x - 2$.
7. $x^3 + x^2y + xy + y^2$ and $x^4 - y^4$.
8. $x^2 - 8x + 15$ and $x^2 + 2x - 15$.
9. $21x^2 - 26x + 8$ and $7x^3 - 4x^2 - 21x + 12$.
10. $x^3 + x^2y + xy^2 + y^3$ and $x^3 - x^2y + xy^2 - y^3$.
11. $a^3 + 2a^2b - ab^2 - 2b^3$ and $a^3 - 2a^2b - ab^2 + 2b^3$.

171. To find the L.C.M. of three expressions, denoted by a, b, c, we find m the L.C.M. of a and b, and then find M the L.C.M. of m and c. M is the L.C.M. of a, b and c.

The proof of this rule may be thus stated:

Every common multiple of a and b is a multiple of m,
and every multiple of m is a multiple of a and b,
therefore every common multiple of m and c is a common multiple of a, b and c,
and every common multiple of a, b and c is a common multiple of m and c,
and therefore the L.C.M. of m and c is the L.C.M. of a, b and c.

EXAMPLES.—xlvii.

Find the L.C.M. of the following expressions:

1. $x^2 - 3x + 2$, $x^2 - 4x + 3$ and $x^2 - 5x + 4$.
2. $x^2 + 5x + 4$, $x^2 + 4x + 3$ and $x^2 + 7x + 12$.
3. $x^2 - 9x + 20$, $x^2 - 12x + 35$ and $x^2 - 11x + 28$.
4. $6x^2 - x - 2$, $21x^2 - 17x + 2$ and $14x^2 + 5x - 1$.
5. $x^2 - 1$, $x^2 + 2x - 3$ and $6x^2 - x - 2$.
6. $x^3 - 27$, $x^2 - 15x + 36$ and $x^3 - 3x^2 - 2x + 6$.

XI. ON ADDITION AND SUBTRACTION OF FRACTIONS.

172. Having established the Rules for finding the Lowest Common Multiple of given expressions, we may now proceed to treat of the method by which Fractions are combined by the processes of ADDITION and SUBTRACTION.

173. Two Fractions may be replaced by two equivalent fractions with a Common Denominator by the following rule:

Find the L.C.M. of the denominators of the given fractions.

Divide the L.C.M. by the Denominator of each fraction.

Multiply the first Numerator by the first Quotient.

Multiply the second Numerator by the second Quotient.

The two Products will be the Numerators of the equivalent fractions whose common denominator is the L.C.M. of the original denominators.

The same rule holds for three, four, or more fractions.

174. Ex. 1. Reduce to equivalent fractions with the lowest common denominator,

$$\frac{2x+5}{3} \text{ and } \frac{4x-7}{4}.$$

Denominators 3, 4.
Lowest Common Multiple 12.
Quotients 4, 3.
New Numerators $8x + 20$, $12x - 21$.
Equivalent Fractions $\dfrac{8x+20}{12}$, $\dfrac{12x-21}{12}$.

Ex. 2. Reduce to equivalent fractions with the lowest common denominator,
$$\frac{5b+4c}{ab}, \frac{6a-2c}{ac}, \frac{3a-5b}{bc}.$$

Denominators ab, ac, bc.
Lowest Common Multiple abc.
Quotients c, b, a.
New Numerators $5bc + 4c^2$, $6ab - 2bc$, $3a^2 - 5ab$.
Equivalent Fractions $\dfrac{5bc+4c^2}{abc}$, $\dfrac{6ab-2bc}{abc}$, $\dfrac{3a^2-5ab}{abc}$.

EXAMPLES.—xlviii.

Reduce to equivalent fractions with the lowest common denominator:

1. $\dfrac{3x}{4}$ and $\dfrac{4x}{5}$.

2. $\dfrac{3x-7}{6}$ and $\dfrac{4x-9}{18}$.

3. $\dfrac{2x-4y}{5x^2}$ and $\dfrac{3x-8y}{10x}$.

4. $\dfrac{4a+5b}{2a^2}$ and $\dfrac{3a-4b}{5a}$.

5. $\dfrac{4a-5c}{5ac}$ and $\dfrac{3a-2c}{12a^2c}$.

6. $\dfrac{a-b}{a^3b}$ and $\dfrac{a^2-ab}{ab^2}$.

7. $\dfrac{3}{1+x}$ and $\dfrac{3}{1-x}$.

8. $\dfrac{2}{1-y^2}$ and $\dfrac{2}{1+y^2}$.

9. $\dfrac{5}{1-x}$ and $\dfrac{6}{1-x^2}$.

10. $\dfrac{a}{c}$ and $\dfrac{b}{c(b+x)}$.

11. $\dfrac{1}{(a-b)(b-c)}$ and $\dfrac{1}{(a-b)(a-c)}$.

12. $\dfrac{1}{ab(a-b)(a-c)}$ and $\dfrac{1}{ac(a-c)(b-c)}$.

175. To shew that $\dfrac{a}{b} + \dfrac{c}{d} = \dfrac{ad+bc}{bd}$.

Suppose the unit to be divided into bd equal parts.

Then $\dfrac{ad}{bd}$ will represent ad of these parts,

and $\dfrac{bc}{bd}$ will represent bc of these parts.

Now $\dfrac{a}{b} = \dfrac{ad}{bd}$, by Art. 148,

and $\dfrac{c}{d} = \dfrac{bc}{bd}$.

Hence $\dfrac{a}{b} + \dfrac{c}{d}$ will represent $ad+bc$ of the parts.

But $\dfrac{ad+bc}{bd}$ will represent $ad+bc$ of the parts.

Therefore $\dfrac{a}{b} + \dfrac{c}{d} = \dfrac{ad+bc}{bd}$.

By a similar process it may be shewn that
$$\dfrac{a}{b} - \dfrac{c}{d} = \dfrac{ad-bc}{bd}.$$

176. Since $\dfrac{a}{b} + \dfrac{c}{d} = \dfrac{ad+bc}{bd}$,

our Rule for *Addition* of Fractions will run thus:

"Reduce the fractions to equivalent fractions having the Lowest Common Denominator. Then *add* the Numerators of the equivalent fractions and place the result as the Numerator of a fraction, whose Denominator is the Common Denominator of the equivalent fractions.

The fraction will be equal to the sum of the original fractions."

The beginner should, however, generally take *two* fractions at a time, and then combine a *third* with the resulting fraction, as will be shewn in subsequent Examples.

Also, since $\dfrac{a}{b} - \dfrac{c}{d} = \dfrac{ad-bc}{bd}$,

the Rule for *Subtracting* one fraction from another will be:

"Reduce the fractions to equivalent fractions having the Lowest Common Denominator. Then *subtract* the Numerator of the second of the equivalent fractions from the Numerator of the first of the equivalent fractions, and place the result as the Numerator of a fraction, whose Denominator is the Common Denominator of the equivalent fractions. This fraction will be equal to the difference of the original fractions."

These rules we shall illustrate by examples of various degrees of difficulty.

Note. When a negative sign precedes a fraction, it is best to place the numerator of that fraction in a bracket, before combining it with the numerators of other fractions.

177. **Ex. 1.** To simplify

$$\frac{4x-3y}{7} + \frac{3x+7y}{14} - \frac{5x-2y}{21} + \frac{9x+2y}{42}.$$

Lowest Common Multiple of denominators is 42.

Multiplying the numerators by 6, 3, 2, 1 respectively,

$$\frac{24x-18y}{42} + \frac{9x+21y}{42} - \frac{10x-4y}{42} + \frac{9x+2y}{42}$$

$$= \frac{24x-18y+9x+21y-(10x-4y)+9x+2y}{42}$$

$$= \frac{24x-18y+9x+21y-10x+4y+9x+2y}{42}$$

$$= \frac{32x+9y}{42}.$$

Ex. 2. To simplify $\dfrac{2x+1}{3x} - \dfrac{4x+2}{5x} + \dfrac{1}{7}.$

Lowest Common Multiple of denominators is $105x$.

Multiplying the numerators by 35, 21, 15x, respectively.

$$\frac{70x+35}{105x} - \frac{84x+42}{105x} + \frac{15x}{105x}$$

$$= \frac{70x+35-(84x+42)+15x}{105x}$$

$$= \frac{70x+35-84x-42+15x}{105x} = \frac{x-7}{105x}.$$

[S.A.]

EXAMPLES.—xlix.

1. $\dfrac{4x+7}{5} + \dfrac{3x-4}{15}.$

2. $\dfrac{3a-4b}{7} - \dfrac{2a-b+c}{3} + \dfrac{13a-4c}{12}.$

3. $\dfrac{4x-3y}{7} + \dfrac{3x+7y}{14} - \dfrac{5x-2y}{21} + \dfrac{9x+2y}{42}.$

4. $\dfrac{3x-2y}{5x} + \dfrac{5x-7y}{10x} + \dfrac{8x+2y}{25}.$

5. $\dfrac{4x^2-7y^2}{3x^2} + \dfrac{3x-8y}{6x} + \dfrac{5-2y}{12}.$

6. $\dfrac{4a^2+5b^2}{2b^2} + \dfrac{3a+2b}{5b} + \dfrac{7-2a}{9}.$

7. $\dfrac{4x+5}{3} - \dfrac{3x-7}{5x} + \dfrac{9}{12x^2}.$

8. $\dfrac{5a+2b}{3c} - \dfrac{4c-3b}{2a} + \dfrac{6ab-7bc}{14ac}.$

9. $\dfrac{2a+5c}{a^2c} + \dfrac{4ac-3c^2}{ac^2} - \dfrac{5ac-2c^2}{a^2c^2}.$

10. $\dfrac{3xy-4}{x^2y^2} - \dfrac{5y^2+7}{xy^3} - \dfrac{6x^2-11}{x^3y}.$

11. $\dfrac{a-b}{a^3b} + \dfrac{4a-5b}{a^2bc} + \dfrac{3a-7b}{b^2c^2}.$

178. **Ex.** To simplify
$$\dfrac{a-b}{a+b} + \dfrac{a+b}{a-b}.$$

L.C.M. of denominators is a^2-b^2.

Multiplying the numerators by $a-b$ and $a+b$ respectively, we get

$$\dfrac{a^2-2ab+b^2}{a^2-b^2} + \dfrac{a^2+2ab+b^2}{a^2-b^2}$$
$$= \dfrac{a^2-2ab+b^2+a^2+2ab+b^2}{a^2-b^2}$$
$$= \dfrac{2a^2+2b^2}{a^2-b^2}.$$

EXAMPLES.—1.

1. $\dfrac{1}{x-6} + \dfrac{1}{x+5}$.
2. $\dfrac{1}{x-7} - \dfrac{1}{x-3}$.
3. $\dfrac{1}{1+x} + \dfrac{1}{1-x}$.
4. $\dfrac{x+y}{x-y} - \dfrac{x-y}{x+y}$.
5. $\dfrac{1}{1-x} - \dfrac{2}{1-x^2}$.
6. $\dfrac{a}{c} - \dfrac{(ad-bc)x}{c(c+dx)}$.
7. $\dfrac{x}{x+y} + \dfrac{x}{x-y}$.
8. $\dfrac{1}{x-y} + \dfrac{x}{(x-y)^2}$.
9. $\dfrac{2}{x+a} + \dfrac{3a}{(x+a)^2}$.
10. $\dfrac{1}{2a(a+x)} + \dfrac{1}{2a(a-x)}$.

179. Ex. 1. To simplify
$$\frac{3}{1+y} + \frac{5}{1-y} - \frac{6}{1+y^2}.$$

Taking the first two fractions
$$\frac{3}{1+y} + \frac{5}{1-y}$$
$$= \frac{3-3y}{1-y^2} + \frac{5+5y}{1-y^2}$$
$$= \frac{8+2y}{1-y^2};$$

we can now combine with this result the *third* of the original fractions, and we have

$$\frac{3}{1+y} + \frac{5}{1-y} - \frac{6}{1+y^2}$$
$$= \frac{8+2y}{1-y^2} - \frac{6}{1+y^2}$$
$$= \frac{8+2y+8y^2+2y^3}{1-y^4} - \frac{6-6y^2}{1-y^4}$$
$$= \frac{8+2y+8y^2+2y^3-6+6y^2}{1-y^4}$$
$$= \frac{2y^3+14y^2+2y+2}{1-y^4}.$$

Ex. 2. To simplify

$$\frac{2}{(a-b)(b-c)} + \frac{2}{(a-b)(c-a)} + \frac{2}{(b-c)(c-a)},$$

L.C.M. of first two denominators being $(a-b)(b-c)(c-a)$

$$= \frac{2c-2a}{(a-b)(b-c)(c-a)} + \frac{2b-2c}{(a-b)(b-c)(c-a)} + \frac{2}{(b-c)(c-a)}$$

$$= \frac{2b-2a}{(a-b)(b-c)(c-a)} + \frac{2}{(b-c)(c-a)}.$$

L.C.M. of the two denominators being $(a-b)(b-c)(c-a)$

$$= \frac{2b-2a+2a-2b}{(a-b)(b-c)(c-a)} = \frac{0}{(a-b)(b-c)(c-a)} = 0.$$

EXAMPLES.—li.

1. $\dfrac{1}{1+a} + \dfrac{1}{1-a} + \dfrac{2a}{1-a^2}.$ 4. $\dfrac{1}{a-b} - \dfrac{1}{a+b} - \dfrac{2b}{a^2+b^2} - \dfrac{4b^3}{a^4+b^4}.$

2. $\dfrac{1}{1-x} - \dfrac{1}{1+x} + \dfrac{2x}{1+x^2}.$ 5. $\dfrac{x}{y} + \dfrac{y}{x+y} + \dfrac{x^2}{x^2+xy}.$

3. $\dfrac{x}{1-x} - \dfrac{x^2}{1-x^2} + \dfrac{x}{1+x^2}.$ 6. $\dfrac{x+3}{x+4} + \dfrac{x-4}{x-3} + \dfrac{x+5}{x+7}.$

7. $\dfrac{x-1}{x-2} + \dfrac{x-2}{x-3} + \dfrac{x-3}{x-4}.$

8. $\dfrac{3}{x-a} + \dfrac{4a}{(x-a)^2} - \dfrac{5a^2}{(x-a)^3}.$

9. $\dfrac{1}{x-1} - \dfrac{1}{x+2} - \dfrac{3}{(x+1)(x+2)}.$

10. $\dfrac{1}{(x+1)(x+2)} - \dfrac{3}{(x+1)(x+2)(x+3)}.$

11. $\dfrac{x^2}{x^2-1} + \dfrac{x}{x-1} + \dfrac{x}{x+1}.$

12. $\dfrac{1}{(a+c)(a+d)} - \dfrac{1}{(a+c)(a+e)}.$

13. $\dfrac{a-b}{(b+c)(c+a)} + \dfrac{b-c}{(c+a)(a+b)} + \dfrac{c-a}{(a+b)(b+c)}.$

14. $\dfrac{x-a}{x-b}+\dfrac{x-b}{x-a}-\dfrac{(a-b)^2}{(x-a)(x-b)}.$

15. $\dfrac{x+y}{y}-\dfrac{2x}{x+y}+\dfrac{x^2y-x^3}{y(x^2-y^2)}.$

16. $\dfrac{a+b}{(b-c)(c-a)}+\dfrac{b+c}{(c-a)(a-b)}+\dfrac{c+a}{(a-b)(b-c)}.$

17. $\dfrac{x}{x^2+xy+y^2}+\dfrac{2xy}{x^3-y^3}.$

18. $\dfrac{2}{a-b}+\dfrac{2}{b-c}+\dfrac{2}{c-a}+\dfrac{(a-b)^2+(b-c)^2+(c-a)^2}{(a-b)(b-c)(c-a)}.$

19. $\dfrac{a+b}{b}-\dfrac{2a}{a+b}+\dfrac{a^2b-a^3}{a^2b-b^3}.$

20. $\dfrac{1}{(n+1)(n+2)}-\dfrac{1}{(n+1)(n+2)(n+3)}-\dfrac{1}{(n+1)(n+3)}.$

21. $\dfrac{a^2-bc}{(a+b)(a+c)}+\dfrac{b^2-ac}{(b+a)(b+c)}+\dfrac{c^2-ab}{(c+b)(c+a)}.$

180. Since $\dfrac{ab}{b}=a,$ and $\dfrac{-ab}{-b}=a,$ Art. 77,

$$\dfrac{ab}{b}=\dfrac{-ab}{-b}.$$

From this we learn that we may change the sign of the denominator of a fraction if we also change the sign of the numerator.

Hence if the numerator or denominator, or both, be expressions with more than one term, we may change the sign of every term in the denominator if we also change the sign of every term in the numerator

For
$$\dfrac{a-b}{c-d}=\dfrac{-(a-b)}{-(c-d)}$$
$$=\dfrac{-a+b}{-c+d};$$

or, writing the terms of the new fraction so that the positive terms may stand first,

$$=\dfrac{b-a}{d-c}.$$

181. Ex. To simplify $\dfrac{x(a+x)}{a-x} - \dfrac{5ax - x^2}{x-a}$.

Changing the signs of the numerator and denominator of the second fraction,

$$\dfrac{x(a+x)}{a-x} - \dfrac{-5ax + x^2}{a-x}$$

$$= \dfrac{ax + x^2 - (-5ax + x^2)}{a-x} = \dfrac{ax + x^2 + 5ax - x^2}{a-x} = \dfrac{6ax}{a-x}.$$

182. Again, since $-ab =$ the product of $-a$ and b,
and $ab =$ the product of $+a$ and b,

the sign of a product will be changed by changing the signs of one of the factors composing the product.

Hence $(a-b)(b-c)$ will give a set of terms,

and $(b-a)(b-c)$ will give *the same set of terms with different signs*.

This may be seen by actual multiplication:

$(a-b)(b-c) = ab - ac - b^2 + bc$,
$(b-a)(b-c) = -ab + ac + b^2 - bc$.

Consequently if we have a fraction

$$\dfrac{1}{(a-b)(b-c)},$$

and we change the factor $a-b$ into $b-a$, we shall in effect *change the sign of every term* of the expression which would result from the multiplication of $(a-b)$ into $(b-c)$.

Now we may change the signs of the denominator if we also change the signs of the numerator (Art. 180);

$$\therefore \dfrac{1}{(a-b)(b-c)} = \dfrac{-1}{(b-a)(b-c)}.$$

If we change the signs of *two* factors in a denominator, the sign of the numerator will remain unaltered, thus

$$\dfrac{1}{(a-b)(b-c)} = \dfrac{1}{(b-a)(c-b)}.$$

183. Ex. Simplify
$$\frac{1}{(a-b)(b-c)} + \frac{1}{(b-a)(a-c)} - \frac{1}{(c-a)(c-b)}.$$

First change the signs of the factor $(b-a)$ in the second fraction, changing also the sign of the numerator; and change the signs of the factor $(c-a)$ in the third fraction, changing also the sign of the numerator,

the result is $\dfrac{1}{(a-b)(b-c)} + \dfrac{-1}{(a-b)(a-c)} - \dfrac{-1}{(a-c)(c-b)}.$

Next, change the signs of the factor $(c-b)$ in the third, changing also the sign of the numerator,

the result is $\dfrac{1}{(a-b)(b-c)} + \dfrac{-1}{(a-b)(a-c)} - \dfrac{1}{(a-c)(b-c)}.$

L.C.M. of the three denominators is $(a-b)(b-c)(a-c)$,

$$= \frac{a-c}{(a-b)(b-c)(a-c)} + \frac{-b+c}{(a-b)(a-c)(b-c)} - \frac{a-b}{(a-b)(a-c)(b-c)}$$

$$= \frac{a-c-b+c-(a-b)}{(a-b)(b-c)(a-c)} = \frac{0}{(a-b)(b-c)(a-c)} = 0.$$

EXAMPLES.—lii.

1. $\dfrac{x}{x-y} + \dfrac{x-y}{y-x}.$ 2. $\dfrac{3+2x}{2-x} - \dfrac{2-3x}{2+x} + \dfrac{16x-x^2}{x^2-4}.$

3. $\dfrac{x}{x+1} - \dfrac{x}{1-x} + \dfrac{x^2}{x^2-1}.$ 4. $\dfrac{1}{6y+6} - \dfrac{1}{2y-2} + \dfrac{4}{3-3y^2}.$

5. $\dfrac{1}{(m-2)(m-3)} + \dfrac{2}{(m-1)(3-m)} + \dfrac{1}{(m-1)(m-2)}.$

6. $\dfrac{1}{(a-b)(x+b)} + \dfrac{1}{(b-a)(x+a)}.$ 7. $\dfrac{a^2+b^2}{a^2-b^2} - \dfrac{2ab^2}{a^3-b^3} + \dfrac{2a^2b}{a^3+b^3}.$

8. $\dfrac{1}{4(1+x)} - \dfrac{1}{4(x-1)} + \dfrac{1}{2(1+x^2)}.$

9. $\dfrac{1}{(x-y)(y-z)} + \dfrac{1}{(y-x)(x-z)} + \dfrac{1}{(z-x)(z-y)}.$

10. $\dfrac{1}{a(a-b)(a-c)} + \dfrac{1}{b(b-a)(b-c)} + \dfrac{1}{c(c-a)(c-b)}.$

184. Ex. To simplify

$$\frac{1}{x^2-11x+30}+\frac{1}{x^2-12x+35}.$$

Here the denominators may be expressed in factors, and we have

$$\frac{1}{(x-5)(x-6)}+\frac{1}{(x-5)(x-7)}.$$

The L.C.M. of the denominators is $(x-5)(x-6)(x-7)$, and we have

$$\frac{x-7}{(x-5)(x-6)(x-7)}+\frac{x-6}{(x-5)(x-6)(x-7)}$$

$$=\frac{2x-13}{(x-5)(x-6)(x-7)}.$$

EXAMPLES.—liii.

1. $\dfrac{1}{x^2+9x+20}+\dfrac{1}{x^2+12x+35}.$

2. $\dfrac{1}{x^2-13x+42}+\dfrac{1}{x^2-15x+54}.$

3. $\dfrac{1}{x^2+7x-44}+\dfrac{1}{x^2-2x-143}.$

4. $\dfrac{1}{x^2+3x+2}+\dfrac{2x}{x^2+4x+3}+\dfrac{1}{x^2+5x+6}.$

5. $\dfrac{m}{n}+\dfrac{2m}{m+n}-\dfrac{2mn}{(m+n)^2}.$

6. $\dfrac{1+x}{1+x+x^2}+\dfrac{1-x}{1-x+x^2}-\dfrac{2}{1+x^2+x^4}.$

7. $\dfrac{5}{3(1-x)}-\dfrac{2}{1+x}+\dfrac{7x}{3x^2+3}-\dfrac{7x}{3x^2-3}.$

8. $\dfrac{1}{8(x-1)}+\dfrac{1}{4(3-x)}+\dfrac{1}{8(x-5)}+\dfrac{1}{(1-x)(x-3)(x-5)}.$

9. $1-x+x^2-x^3+\dfrac{x^4}{1+x}.$

XII. ON FRACTIONAL EQUATIONS.

185. We shall explain in this Chapter the method of solving, first, Equations in which fractional terms occur, and secondly, Problems leading to such Equations.

186. An Equation involving fractional terms may be reduced to an equivalent Equation without fractions *by multiplying every term of the equation by the Lowest Common Multiple of the denominators of the fractional terms.*

This process is in accordance with the principle laid down in Ax. III. page 58; for if both sides of an equation be multiplied by the same expression, the resulting products will, by that Axiom, be equal to each other.

187. The following examples will illustrate the process of clearing an Equation of Fractions.

Ex. 1. $\dfrac{x}{2} + \dfrac{x}{6} = 8$.

The L.C.M. of the denominators is 6.

Multiplying both sides by 6, we get

$$\frac{6x}{2} + \frac{6x}{6} = 48,$$

or, $\quad 3x + x = 48,$

or, $\quad 4x = 48;$

$\therefore x = 12.$

Ex. 2. $\dfrac{x}{2} + \dfrac{x+1}{7} = x - 2$.

The L.C.M. of the denominators is 14.

Multiplying both sides by 14, we get

$$\frac{14x}{2} + \frac{14x + 14}{7} = 14x - 28,$$

or, $\qquad 7x + 2x + 2 = 14x - 28,$

or, $\qquad 7x + 2x - 14x = -28 - 2,$

or, $\qquad -5x = -30.$

Changing the signs of both sides, we get
$$5x = 30;$$
$$\therefore x = 6.$$

188. The process may be shortened from the following considerations. If we have to multiply a fraction by a multiple of its denominator, we may first *divide the multiplier by the denominator, and then multiply the numerator by the quotient.* The result will be a whole number.

Thus, $\qquad \dfrac{x}{3} \times 12 = x \times 4 = 4x,$

$$\dfrac{x-1}{7} \times 56 = (x-1) \times 8 = 8x - 8.$$

Ex. 1. $\dfrac{x}{2} + \dfrac{x}{3} + \dfrac{x}{4} = 39.$

The L.C.M. of the denominators being 12, if we multiply the numerators of the fractions by 6, 4, and 3 respectively, and the other side of the equation by 12, we get
$$6x + 4x + 3x = 468,$$
or, $\qquad 13x = 468;$
$$\therefore x = 36.$$

Ex. 2. $\dfrac{8}{x} - \dfrac{15}{2x} + \dfrac{7}{3x} = \dfrac{17}{12}.$

The L.C.M. of the denominators is $12x$. Hence, if we multiply the numerators by 12, 6, 4, and x respectively, we get
$$96 - 90 + 28 = 17x,$$
or, $\qquad 34 = 17x,$

or, $\qquad 17x = 34;$
$$\therefore x = 2.$$

EXAMPLES.—liv.

1. $\dfrac{x}{2} = 8.$

2. $\dfrac{3x}{4} = 9.$

3. $\dfrac{x}{3} + \dfrac{x}{5} = 8.$

4. $\dfrac{x}{4} - \dfrac{x}{7} = 3.$

5. $36 - \dfrac{4x}{9} = 8.$

6. $\dfrac{2x}{3} = \dfrac{176 - 4x}{5}.$

7. $\dfrac{2x}{3} + 4 = \dfrac{7x}{12} + 9.$

17. $\dfrac{x+2}{5} + \dfrac{x-1}{7} = \dfrac{x-2}{2}.$

8. $\dfrac{2x}{3} + 12 = \dfrac{4x}{5} + 6.$

18. $\dfrac{x}{2} + \dfrac{x}{3} - 9\dfrac{3}{4} = \dfrac{x}{4}.$

9. $\dfrac{3x}{4} + 5 = \dfrac{5x}{6} + 2.$

19. $\dfrac{x+9}{4} + \dfrac{2x}{7} = \dfrac{3x-6}{5} + 3.$

10. $\dfrac{7x}{8} - 5 = \dfrac{9x}{10} - 8.$

20. $\dfrac{17 - 3x}{5} = \dfrac{29 - 11x}{3} + \dfrac{28x + 14}{21}.$

11. $\dfrac{5x}{9} - 8 = 74 - \dfrac{7x}{12}.$

21. $\dfrac{2x - 10}{7} = 0.$

12. $\dfrac{x}{6} - 4 = 24 - \dfrac{x}{8}.$

22. $\dfrac{3x + 4}{7} + \dfrac{4x - 51}{47} = 0.$

13. $56 - \dfrac{3x}{4} = 48 - \dfrac{5x}{8}.$

23. $\dfrac{3}{x} - 3 = \dfrac{1}{x} - 1.$

14. $\dfrac{3x}{4} + \dfrac{180 - 5x}{6} = 29.$

24. $\dfrac{12 + x}{x} - 5 = \dfrac{6}{x}.$

15. $\dfrac{3x}{4} - 11 = \dfrac{x - 8}{2}.$

25. $\dfrac{1}{4}x + \dfrac{1}{10}x + \dfrac{1}{20}x = 40.$

16. $\dfrac{x}{2} + \dfrac{x}{3} + \dfrac{x}{4} = \dfrac{13}{12}.$

26. $2\dfrac{1}{4}x + \dfrac{3-x}{2} = 3\dfrac{5}{8}x - 43\dfrac{1}{2}.$

27. $2\dfrac{3}{4} - \dfrac{3}{x} = \dfrac{1}{x} - \dfrac{325}{100}.$

28. $2\dfrac{1}{2} + \dfrac{18 - x}{3} = 1\dfrac{1}{9}x + \dfrac{1}{3} + \dfrac{3 - 2x}{10} + \dfrac{2}{5}.$

29. $\dfrac{x}{3} + \dfrac{x}{4} - \dfrac{5x}{6} - 12 = 1\dfrac{2}{3}x - 58.$

30. $\dfrac{7x + 2}{10} - 12 - \dfrac{3x}{4} = \dfrac{3x + 13}{5} - \dfrac{17x}{4}.$

189. It must next be observed that in clearing an equation of fractions, whenever a fraction is preceded by a negative sign, we must place the result obtained by multiplying that numerator *in a bracket*, after the removal of the denominator.

For example, we ought to proceed thus :—

Ex. 1. $\dfrac{x+2}{5} = \dfrac{x-2}{2} - \dfrac{x-1}{7}.$

Multiply by 70, the L.C.M. of the denominators, and we get
$$14x + 28 = 35x - 70 - (10x - 10),$$
or $\quad 14x + 28 = 35x - 70 - 10x + 10,$

from which we shall find $x = 8$.

Ex. 2. $\dfrac{17 - 2x}{5x} - \dfrac{4x + 2}{3x} = 1.$

Multiplying by $15x$, the L.C.M. of the denominators, we get
$$51 - 6x - (20x + 10) = 15x,$$
or $\quad 51 - 6x - 20x - 10 = 15x,$

from which we shall find $x = 1$.

Note. It is from want of attention to this way of treating fractions preceded by a negative sign that beginners make so many mistakes in the solution of equations.

EXAMPLES.—lv.

1. $5x - \dfrac{x+2}{2} = 71.$

2. $x - \dfrac{3-x}{3} = 5\dfrac{2}{3}.$

3. $\dfrac{5-2x}{4} + 2 = x - \dfrac{6x-8}{2}.$

4. $\dfrac{5x}{2} - \dfrac{5x}{4} = \dfrac{9}{4} - \dfrac{3-x}{2}.$

5. $2x - \dfrac{5x-4}{6} = 7 - \dfrac{1-2x}{5}.$

6. $\dfrac{x+2}{2} = \dfrac{14}{9} - \dfrac{3+5x}{4}.$

7. $\dfrac{5x+3}{8} - \dfrac{3-4x}{3} + \dfrac{x}{2} = \dfrac{31}{2} - \dfrac{9-5x}{6}.$

8. $\dfrac{x+5}{7} - \dfrac{x-2}{5} = \dfrac{x+9}{11}.$

9. $\dfrac{x+1}{3} - \dfrac{x-4}{7} = \dfrac{x+4}{5}.$

10. $x - 3 - \dfrac{x+2}{8} = \dfrac{x}{3}.$

11. $\dfrac{x+5}{7} = \dfrac{x+2}{4} - \dfrac{x-2}{3}.$

12. $\dfrac{x}{3} - \dfrac{x-1}{11} = x - 9.$

13. $\dfrac{x+2}{5} = \dfrac{x-2}{2} - \dfrac{x-1}{7}.$

14. $\dfrac{x+9}{4} - \dfrac{3x-6}{5} = 3 - \dfrac{2x}{7}.$

15. $\dfrac{x+1}{2} - \dfrac{x-3}{3} = \dfrac{x+30}{13}.$

16. $\dfrac{2x}{7} - \dfrac{x+3}{5} = 3x - 21.$

17. $\dfrac{2x+7}{7} - \dfrac{9x-8}{11} = \dfrac{x-11}{2}.$

18. $\dfrac{7x-31}{4} - \dfrac{8+15x}{26} = \dfrac{7x-8}{22}.$

19. $\dfrac{8x-15}{3} - \dfrac{11x-1}{7} = \dfrac{7x+2}{13}.$

20. $\dfrac{7x+9}{8} - \dfrac{3x+1}{7} = \dfrac{9x-13}{4} - \dfrac{249-9x}{14}.$

21. $\dfrac{x}{10} + 10x = \dfrac{x}{2} + \dfrac{x}{5} + \dfrac{x}{40} - \dfrac{10-x}{7} + 93\dfrac{3}{4}.$

190. Literal equations are those in which known quantities are represented by *letters*, usually the first in the alphabet. The following are examples:—

Ex. 1. *To solve the equation*
$$ax + bc = bx + ac,$$
that is, $\quad ax - bx = ac - bc,$
or, $\quad (a-b)x = (a-b)c,$
therefore, $\quad x = c.$

Ex. 2. *To solve the equation*
$$a^2x + bx - c = b^2x + cx - d,$$
that is, $\quad a^2x + bx - b^2x - cx = c - d,$
or, $\quad (a^2 + b - b^2 - c)x = c - d,$
therefore, $\quad x = \dfrac{c-d}{a^2 + b - b^2 - c}.$

EXAMPLES.—lvi.

1. $ax + bx = c.$
2. $2a - cx = 3c - 5bx.$
3. $bc + ax - d = a^2b - fx.$
4. $dm - 5x = bc - ax.$
5. $abc - a^2x = ax - a^2b.$
6. $3acx - 6bcd = 12cdx + abc.$

7. $k^2 + 3ackx + 3k = kx + 3abk - k^2 - ackx$.
8. $-ac^2 + b^2c + abcx = abc + cmx - ac^2x + b^2c - mc$.
9. $(a+x+b)(a+b-x) = (a+x)(b-x) - ab$.
10. $(a-x)(a+x) = 2a^2 + 2ax - x^2$.
11. $(a^2+x)^2 = x^2 + 4a^2 + a^4$.
12. $(a^2-x)(a^2+x) = a^4 + 2ax - x^2$.
13. $\dfrac{ax-b}{c} + a = \dfrac{x+ac}{c}$.
14. $ax - \dfrac{3a-bx}{2} = \dfrac{1}{2}$.
15. $6a - \dfrac{4ax-2b}{3} = x$.
16. $ax - \dfrac{bx+1}{x} = \dfrac{a(x^2-1)}{x}$.
17. $\dfrac{m(p^2x+x^3)}{px} = mqx + \dfrac{mx^2}{p}$.
18. $\dfrac{x}{a} - b = \dfrac{c}{d} - x$.
19. $\dfrac{x^2-a}{bc} - \dfrac{a-x}{b} = \dfrac{2x}{b} - \dfrac{a}{x}$.
20. $\dfrac{3}{c} - \dfrac{ab-x^2}{bx} = \dfrac{4x-ac}{cx}$.
21. $\dfrac{ab+x}{b^2} - \dfrac{b^2-x}{a^2b} = \dfrac{x-b}{a^2} - \dfrac{ab-x}{b^2}$.
22. $\dfrac{3ax-2b}{3b} - \dfrac{ax-a}{2b} = \dfrac{ax}{b} - \dfrac{2}{3}$.
23. $am - b - \dfrac{ax}{b} + \dfrac{x}{m} = 0$.
24. $\dfrac{2a^2b^3}{(a+b)} - \dfrac{b^2x}{a(a+b)} + \dfrac{3a^2c}{a+b} = \dfrac{3acx}{b} - \dfrac{b^3 - 2ab^2x}{(a+b)}$.
25. $\dfrac{ax^2}{b-cx} + a + \dfrac{ax}{c} = 0$.
26. $\dfrac{a(d^2+x^2)}{dx} = ac + \dfrac{ax}{d}$.
27. $\dfrac{ab}{x} = bc + d + \dfrac{1}{x}$.
28. $c = a + \dfrac{m(a-x)}{3a+x}$.
29. $(a+x)(b+x) - a(b+c) = \dfrac{a^2c}{b} + x^2$.
30. $\dfrac{ace}{d} - \dfrac{(a+b)^2 \cdot x}{a} - bx = ae - 3bx$.

191. In the examples already given the L.C.M. of the denominators can generally be determined by inspection. When *compound* expressions appear in the denominators, it is sometimes desirable to collect the fractions into *two*, one

on each side of the equation. When this has been done, we can clear the equation of fractions by multiplying the numerator on the *left* by the denominator on the *right*, and the numerator on the *right* by the denominator on the *left*, and making the products equal.

For, if $\dfrac{a}{b} = \dfrac{c}{d}$, it is evident that $ad = bc$.

Ex. $\dfrac{4x+5}{10} - \dfrac{13x-6}{7x+4} = \dfrac{2x-3}{5}$;

$\therefore \dfrac{4x+5}{10} - \dfrac{2x-3}{5} = \dfrac{13x-6}{7x+4}$;

$\therefore \dfrac{4x+5-(4x-6)}{10} = \dfrac{13x-6}{7x+4}$;

$\therefore \dfrac{11}{10} = \dfrac{13x-6}{7x+4}$;

$\therefore 11(7x+4) = 10(13x-6)$;

whence we find $x = \dfrac{104}{53}$.

EXAMPLES.—lvii.

1. $\dfrac{3x+7}{4x+5} = \dfrac{3x+5}{4x+3}$.

2. $\dfrac{x+6}{2x+5} = \dfrac{x}{2x-5}$.

3. $\dfrac{2x+7}{x+2} = \dfrac{4x-1}{2x-1}$.

4. $\dfrac{5x-1}{2x+3} = \dfrac{5x-3}{2x-3}$.

5. $\dfrac{1}{3x-2} + \dfrac{2}{4x-3} = 0$.

6. $\dfrac{2}{1-5x} - \dfrac{5}{1-2x} = 0$.

7. $\dfrac{1}{x-1} + \dfrac{1}{x+1} = \dfrac{3}{x^2-1}$.

8. $\dfrac{4x+3}{9} = \dfrac{8x+19}{18} - \dfrac{7x-29}{5x-12}$.

9. $\dfrac{x}{3} - \dfrac{x^2-5x}{3x-7} = \dfrac{2}{3}$.

10. $\dfrac{3x+2}{x-1} + \dfrac{2x-4}{x+2} = 5$.

11. $\dfrac{1}{6}(x+3) - \dfrac{1}{7}(11-x) = \dfrac{2}{5}(x-4) - \dfrac{1}{21}(x-3)$.

12. $\dfrac{(x+1)(2x+2)}{(x-3)(x+6)} - 2 = 0$.

13. $\dfrac{(2x+3)x}{2x+1} + \dfrac{1}{3x} = x+1$.

14. $\dfrac{3}{x+1} - \dfrac{x+1}{x-1} = \dfrac{x^2}{1-x^2}$.

15. $\dfrac{2}{1-x} + \dfrac{8}{1+x} = \dfrac{45}{1-x^2}$.

16. $\dfrac{4}{x-8} + \dfrac{3}{2x-16} - 1\dfrac{5}{24} = \dfrac{2}{3x-24}.$

17. $\dfrac{x^4 - (4x^2 - 20x + 24)}{x^2 - 2x + 4} = x^2 + 2x - 4.$

18. $\dfrac{2x^4 + 2x^3 - 23x^2 + 31x}{x^2 + 3x - 4} = 2x^2 - 4x - 3.$

19. $\dfrac{1}{4}x - 1 = \dfrac{1}{16x}\left(4x^2 - 3x - 1\dfrac{5}{8}\right).$

20. $5 - x\left(3\dfrac{1}{2} - \dfrac{2}{x}\right) = \dfrac{1}{2}x - \dfrac{3x - (4 - 5x)}{4}.$

192. Equations into which Decimal Fractions enter do not present any serious difficulty, as may be seen from the following Examples:—

Ex. 1. *To solve the equation*
$$\cdot 5x = \cdot 03x + 1\cdot 41.$$

Turning the decimals into the form of Vulgar Fractions, we get
$$\dfrac{5x}{10} = \dfrac{3x}{100} + \dfrac{141}{100}.$$

Then multiplying both sides by 100, we get
$$50x = 3x + 141;$$
therefore $\quad 47x = 141;$
therefore $\quad x = 3.$

Ex. 2. $1\cdot 2x - \dfrac{\cdot 18x - \cdot 05}{\cdot 5} = \cdot 4x + 8\cdot 9.$

First clear *the fraction* of decimals by multiplying its numerator and denominator by 100, and we get
$$1\cdot 2x - \dfrac{18x - 5}{50} = \cdot 4x + 8\cdot 9;$$
therefore $\quad \dfrac{12x}{10} - \dfrac{18x - 5}{50} = \dfrac{4x}{10} + \dfrac{89}{10};$
therefore $\quad 60x - 18x + 5 = 20x + 445;$
therefore $\quad 22x = 440;$
therefore $\quad x = 20.$

EXAMPLES.—lviii.

1. $\cdot 5x - 2 = \cdot 25x + \cdot 2x - 1.$
2. $3 \cdot 25x - 5 \cdot 1 + x - \cdot 75x = 3 \cdot 9 + \cdot 5x.$
3. $\cdot 125x + \cdot 01x = 13 - \cdot 2x + \cdot 4.$
4. $\cdot 3x + 1 \cdot 305x + \cdot 5x = 22 \cdot 95 - \cdot 195x.$
5. $\cdot 2x - \cdot 01x + \cdot 005x = 11 \cdot 7.$
6. $2 \cdot 4x - \dfrac{\cdot 36x - \cdot 05}{\cdot 5} = \cdot 8x + 8 \cdot 9.$
7. $2 \cdot 4x - 10 \cdot 75 = \cdot 25x.$
8. $\cdot 5x + 2 - \cdot 75x = \cdot 4x - 11.$
9. $\dfrac{4 \cdot 05}{9x} + 3 \cdot 875 = 4 \cdot 025.$
10. $2 \cdot 5x - \dfrac{2+x}{7}\left(\dfrac{1}{4} - 2\right) = \cdot 5 - \dfrac{5x+3}{8}.$
11. $\dfrac{8 \cdot 5}{2} - \dfrac{\cdot 2}{x} = 4\dfrac{1}{4} - \dfrac{1 - \cdot 1x}{x}.$
12. $\dfrac{\cdot 48x}{6} - \dfrac{3 - 4x}{\cdot 2} = 1993.$
13. $\dfrac{2 - 3x}{1 \cdot 5} + \dfrac{5x}{1 \cdot 25} - \dfrac{2x - 3}{9} = \dfrac{x - 2}{1 \cdot 8} + 2\dfrac{7}{9}.$
14. $\dfrac{24 \cdot 08}{x} + \dfrac{1}{x} \cdot \cdot 04(x + \cdot 9) = 241 \cdot 2.$
15. $\cdot 5x + \dfrac{\cdot 45x - \cdot 75}{\cdot 6} = \dfrac{1 \cdot 2}{\cdot 2} - \dfrac{\cdot 3x - \cdot 6}{\cdot 9}.$
16. $\cdot 5 - \dfrac{3 \cdot 5x}{x - 2} - \dfrac{24 - 3x}{8} = \cdot 375x.$
17. $\cdot 15x + \dfrac{\cdot 135x - \cdot 225}{\cdot 6} = \dfrac{\cdot 36}{\cdot 2} - \dfrac{\cdot 09x - \cdot 18}{\cdot 9}.$

193. *To shew that a simple equation can only have one root.*

Let $x = a$ be the equation, a form to which all equations of the first degree may be reduced.

Now suppose a and β to be two roots of the equation. Then, by Art. 109,
$$a = a,$$
and
$$\beta = a,$$
therefore
$$a = \beta;$$
in other words, the two supposed roots are identical.

[S.A.]

XIII. PROBLEMS IN FRACTIONAL EQUATIONS.

194. We shall now give a series of Easy Problems resulting for the most part in FRACTIONAL EQUATIONS.

Take the following as an example of the *form* in which such Problems should be set out by a beginner.

"Find a number such that the sum of its third and fourth parts shall be equal to 7."

Suppose x to represent the number.

Then $\dfrac{x}{3}$ will represent the third part of the number,

and $\dfrac{x}{4}$ will represent the fourth part of the number.

Hence $\dfrac{x}{3} + \dfrac{x}{4}$ will represent the sum of the two parts.

But 7 will represent the sum of the two parts.

Therefore $\qquad \dfrac{x}{3} + \dfrac{x}{4} = 7.$

Hence $\qquad\qquad 4x + 3x = 84,$
that is, $\qquad\qquad 7x = 84,$
that is, $\qquad\qquad\ x = 12,$
and therefore the number sought is 12.

EXAMPLES.—lix.

1. What is the number of which the half, the fourth, and the fifth parts added together give as a result 95?

2. What is the number of which the twelfth, twentieth, and fortieth parts added together give as a result 38?

3. What is the number of which the fourth part exceeds the fifth part by 4?

PROBLEMS IN FRACTIONAL EQUATIONS. 115

4. What is the number of which the twenty-fifth part exceeds the thirty-fifth part by 8?

5. Divide 60 into two such parts that a seventh part of one may be equal to an eighth part of the other.

6. Divide 50 into two such parts that one-fourth of one part being added to five-sixths of the other part the sum may be 40.

7. Divide 100 into two such parts that if a third part of the one be subtracted from a fourth part of the other the remainder may be 11.

8. What is the number which is greater than the sum of its third, tenth, and twelfth parts by 58?

9. When I have taken away from 33 the fourth, fifth, and tenth parts of a certain number, the remainder is zero. What is the number?

10. What is the number of which the fourth, fifth, and sixth parts added together exceed the half of the number by 112?

11. If to the sum of the half, the third, the fourth, and the twelfth parts of a certain number I add 30, the sum is twice as large as the original number. Find the number.

12. The difference between two numbers is 8, and the quotient resulting from the division of the greater by the less is 3. What are the numbers?

13. The seventh part of a man's property is equal to his whole property diminished by £1626. What is his property?

14. The difference between two numbers is 504, and the quotient resulting from the division of the greater by the less is 15. What are the numbers?

15. The sum of two numbers is 5760, and their difference is equal to one-third of the greater. What are the numbers?

16. To a certain number I add its half, and the result is as much above 60 as the number itself is below 65. Find the number.

17. The difference between two numbers is 20, and one-seventh of the one is equal to one-third of the other. What are the numbers?

18. The sum of two numbers is 31207. On dividing one by the other the quotient is found to be 15 and the remainder 1335. What are the numbers?

19. The ages of two brothers amount to 27 years. On dividing the age of the elder by that of the younger the quotient is $3\frac{1}{2}$. What is the age of each?

20. Divide 237 into two such parts that one is four-fifths of the other.

21. Divide £1800 between A and B, so that B's share may be two-sevenths of A's share.

22. Divide 46 into two such parts that the sum of the quotients obtained by dividing one part by 7 and the other by 3 may be equal to 10.

23. Divide the number a into two such parts that the sum of the quotients obtained by dividing one part by m and the other by n may be equal to b.

24. The sum of two numbers is a, and their difference is b. Find the numbers.

25. On multiplying a certain number by 4 and dividing the product by 3, I obtain 24. What is the number?

26. Divide £864 between A, B, and C, so that A gets $\frac{5}{11}$ of what B gets, and C's share is equal to the sum of the shares of A and B.

27. A man leaves the half of his property to his wife, a sixth part to each of his two children, a twelfth part to his brother, and the rest, amounting to £600, to charitable uses. What was the amount of his property?

28. Find two numbers, of which the sum is 70, such that the first divided by the second gives 2 as a quotient and 1 as a remainder.

29. Find two numbers of which the difference is 25, such that the second divided by the first gives 4 as a quotient and 4 as a remainder.

30. Divide the number 208 into two parts such that the sum of the fourth of the greater and the third of the less is less by 4 than four times the difference between the two parts.

31. There are thirteen days between division of term and the end of the first two-thirds of the term. How many days are there in the term?

32. Out of a cask of wine of which a fifth part had leaked away 10 gallons were drawn, and then the cask was two-thirds full. How much did it hold?

33. The sum of the ages of a father and son is half what it will be in 25 years : the difference is one-third what the sum will be in 20 years. Find the respective ages.

34. A mother is 70 years old, her daughter is exactly half that age. How many years have passed since the mother was $3\frac{1}{3}$ times the age of the daughter?

35. A is 72, and B is two-thirds of that age. How long is it since A was 5 times as old as B?

NOTE I. If a man can do a piece of work in x hours, the part of the work which he can do in one hour will be represented by $\frac{1}{x}$.

Thus if A can reap a field in 12 hours, he will reap in one hour $\frac{1}{12}$ of the field.

EX. A can do a piece of work in 5 days, and B can do it in 12 days. How long will A and B working together take to do the work?

Let x represent the number of days A and B will take.

Then $\frac{1}{x}$ will represent the part of the work they do daily.

Now $\frac{1}{5}$ represents the part A does daily,

and $\frac{1}{12}$ represents the part B does daily.

Hence $\frac{1}{5} + \frac{1}{12}$ will represent the part A and B do daily.

Consequently $\frac{1}{5} + \frac{1}{12} = \frac{1}{x}.$

Hence $\qquad 12x + 5x = 60,$

or $\qquad 17x = 60;$

$$\therefore x = \frac{60}{17}.$$

That is, they will do the work in $3\frac{9}{17}$ days.

36. A can do a piece of work in 2 days. B can do it in 3 days. In what time will they do it if they work together?

37. A can do a piece of work in 50 days, B in 60 days, and C in 75 days. In what time will they do it all working together?

38. A and B together finish a work in 12 days; A and C in 15 days; B and C in 20 days. In what time will they finish it all working together?

39. A and B can do a piece of work in 4 hours; A and C in $3\frac{3}{5}$ hours; B and C in $5\frac{1}{7}$ hours. In what time can A do it alone?

40. A can do a piece of work in $2\frac{1}{2}$ days, B in $3\frac{1}{3}$ days, and C in $3\frac{3}{4}$ days. In what time will they do it all working together?

41. A does $\frac{3}{5}$ of a piece of work in 10 days. He then calls in B, and they finish the work in 3 days. How long would B take to do one-third of the work by himself?

NOTE II. If a tap can fill a vessel in x hours, the part of the vessel filled by it in one hour will be represented by $\frac{1}{x}.$

EX. Three taps running separately will fill a vessel in 20, 30, and 40 minutes respectively. In what time will they fill it when they all run at the same time?

Let x represent the number of minutes they will take.

Then $\dfrac{1}{x}$ will represent the part of the vessel filled in 1 minute.

Now $\dfrac{1}{20}$ represents the part filled by the first tap in 1 minute,

$\dfrac{1}{30}$ .... second ....

$\dfrac{1}{40}$ .... third ....

Hence $\qquad \dfrac{1}{20} + \dfrac{1}{30} + \dfrac{1}{40} = \dfrac{1}{x}$,

or, multiplying both sides by $120x$,

$$6x + 4x + 3x = 120,$$

that is, $\qquad 13x = 120 ;$

$$\therefore x = \dfrac{120}{13}.$$

Hence they will take $9\dfrac{3}{13}$ minutes to fill the vessel.

42. A vessel can be filled by two pipes, running separately, in 3 hours and 4 hours respectively. In what time will it be filled when both run at the same time?

43. A vessel may be filled by three different pipes: by the first in $1\dfrac{1}{3}$ hours, by the second in $3\dfrac{1}{3}$ hours, and by the third in 5 hours. In what time will the vessel be filled when all three pipes are opened at once?

44. A bath is filled by a pipe in 40 minutes. It is emptied by a waste-pipe in an hour. In what time will the bath be full if both pipes are opened at once?

45. If three pipes fill a vessel in a, b, c minutes running separately, in what time will the vessel be filled when all three are opened at once?

46. A vessel containing $755\frac{1}{4}$ gallons can be filled by three pipes. The first lets in 12 gallons in $3\frac{1}{4}$ minutes, the second $15\frac{1}{3}$ gallons in $2\frac{1}{2}$ minutes, the third 17 gallons in 3 minutes: in what time will the vessel be filled by the three pipes all running together?

47. A vessel can be filled in 15 minutes by three pipes, one of which lets in 10 gallons more and the other 4 gallons less than the third each minute. The cistern holds 2400 gallons. How much comes through each pipe in a minute?

NOTE III. In questions involving distance travelled over in a certain time at a certain rate, it is to be observed that

$$\frac{\text{Distance}}{\text{Rate}} = \text{Time.}$$

That is, if I travel 20 miles at the rate of 5 miles an hour,

$$\text{number of hours I take} = \frac{20}{5}.$$

Ex. A and B set out, one from Newmarket and the other from Cambridge, at the same time. The distance between the towns is 13 miles. A walks 4 miles an hour, and B 3 miles an hour. Where will they meet?

Let x represent their distance from Cambridge when they meet.

Then $13-x$ will represent their distance from Newmarket.

Then $\frac{x}{3}=$ time in hours that B has been walking,

$\frac{13-x}{4}=$ .... A ....

And since both have been walking the same time,

$$\frac{x}{3} = \frac{13-x}{4},$$

or $4x = 39 - 3x$,

or $7x = 39$;

$$\therefore x = \frac{39}{7}.$$

That is, they meet at a distance of $5\frac{4}{7}$ miles from Cambridge.

48. A person starts from Ely to walk to Cambridge (which is distant 16 miles) at the rate of $4\frac{4}{9}$ miles an hour, at the same time that another person leaves Cambridge for Ely walking at the rate of a mile in 18 minutes. Where will they meet?

49. A person walked to the top of a mountain at the rate of $2\frac{1}{3}$ miles an hour, and down the same way at the rate of $3\frac{1}{2}$ miles an hour, and was out 5 hours. How far did he walk altogether?

50. A man walks a miles in b hours. Write down
   (1) The number of miles he will walk in c hours.
   (2) The number of hours he will be walking d miles.

51. A steamer which started from a certain place is followed after 2 days by another steamer on the same line. The first goes 244 miles a day, and the second 286 miles a day. In how many days will the second overtake the first?

52. A messenger who goes $31\frac{1}{2}$ miles in 5 hours is followed after 8 hours by another who goes $22\frac{1}{2}$ miles in 3 hours. When will the second overtake the first?

53. Two men set out to walk, one from Cambridge to London, the other from London to Cambridge, a distance of 60 miles. The former walks at the rate of 4 miles, the latter at the rate of $3\frac{3}{4}$ miles an hour. At what distance from Cambridge will they meet?

54. A sets out and travels at the rate of 7 miles in 5 hours. Eight hours afterwards B sets out from the same place, and travels along the same road at the rate of 5 miles in 3 hours. After what time will B overtake A?

122   PROBLEMS IN FRACTIONAL EQUATIONS.

Note IV. In problems relating to clocks the chief point to be noticed is that the minute-hand moves 12 times as fast as the hour-hand.

The following examples should be carefully studied.

Find the time between 3 and 4 o'clock when the hands of a clock are

  (1) Opposite to each other.
  (2) At right angles to each other.
  (3   Coincident.

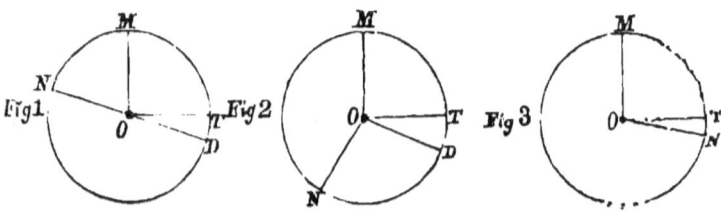

(1) Let ON represent the position of the minute-hand in Fig. I.

OD represents the position of the hour-hand in Fig. I.

M marks the 12 o'clock point.

T .... 3 o'clock ......

The lines OM, OT represent the position of the hands at 3 o'clock.

Now suppose the time to be x minutes past 3.

Then the minute-hand has since 3 o'clock moved over the arc MDN.

And the hour-hand has since 3 o'clock moved over the arc TD.

Hence arc $MDN =$ *twelve times* arc TD.

If then we represent MDN by x,

we shall represent TD by $\dfrac{x}{12}$.

Also we shall represent MT by 15,

and DN by 30.

Now $MDN = MT + TD + DN$,

that is, $x = 15 + \dfrac{x}{12} + 30$,

or $12x = 180 + x + 360$
or $11x = 540$;
$\therefore x = \dfrac{540}{11}.$

Hence the time is $49\dfrac{1}{11}$ minutes past 3.

(2) In Fig. II. the description given of the state of the clock in Fig. I. applies, except that DN will be represented by 15 instead of 30.

Now suppose the time to be x minutes past 3.
Then since
$$MDN = MT + TD + DN,$$
$$x = 15 + \dfrac{x}{12} + 15.$$
from which we get
$$x = \dfrac{360}{11},$$
that is, the time is $32\dfrac{8}{11}$ minutes past 3.

(3) In Fig. III. the hands are both in the position ON.
Now suppose the time to be x minutes past 3.
Then since
$$MN = MT + TN,$$
$$x = 15 + \dfrac{x}{12},$$
or $12x = 180 + x$,
or $x = \dfrac{180}{11}$,

that is, the time is $16\dfrac{4}{11}$ minutes past 3.

55. At what time are the hands of a watch opposite to each other,

(1) Between 1 and 2,
(2) Between 4 and 5,
(3) Between 8 and 9?

PROBLEMS IN FRACTIONAL EQUATIONS.

56. At what time are the hands of a watch at right angles to each other,
   (1) Between 2 and 3.
   (2) Between 4 and 5,
   (3) Between 7 and 8 ?

57. At what time are the hands of a watch together,
   (1) Between 3 and 4,
   (2) Between 6 and 7,
   (3) Between 9 and 10 ?

58. A person buys a certain number of apples at the rate of five for twopence. He sells half of them at two a penny, and the remaining half at three a penny, and clears a penny by the transaction. How many does he buy?

59. A man gives away half a sovereign more than half as many sovereigns as he has; and again half a sovereign more than half the sovereigns then remaining to him, and now has nothing left. How much had he at first?

60. What must be the value of n in order that $\dfrac{2a+n}{3n+69a}$ may be equal to $\dfrac{1}{33}$ when a is $\dfrac{1}{3}$?

61. A body of troops retreating before the enemy, from which it is at a certain time 25 miles distant, marches 18 miles a day. The enemy pursues it at the rate of 23 miles a day, but is first a day later in starting, then after 2 days is forced to halt for one day to repair a bridge, and this they have to do again after two days' more marching. After how many days from the beginning of the retreat will the retreating force be overtaken?

62. A person, after paying an income-tax of sixpence in the pound, gave away one-thirteenth of his remaining income, and had £540 left. What was his original income?

63. From a sum of money I take away £50 more than the half, then from the remainder £30 more than the fifth, then from the second remainder £20 more than the fourth part; and at last only £10 remains. What was the original sum?

PROBLEMS IN FRACTIONAL EQUATIONS. 125

64. I bought a certain number of eggs at 2 a penny, and the same number at 3 a penny. I sold them at 5 for twopence, and lost a penny. How many eggs did I buy?

65. A cistern, holding 1200 gallons, is filled by 3 pipes A, B, C in 24 minutes. The pipe A requires 30 minutes more than C to fill the cistern, and 10 gallons less run through C per minute than through A and B together. What time would each pipe take to fill the cistern by itself?

66. A, B, and C drink a barrel of beer in 24 days. A and B drink $\frac{4}{3}$rds of what C does, and B drinks twice as much as A. In what time would each separately drink the cask?

67. A and B shoot by turns at a target. A puts 7 bullets out of 12 into the centre, and B puts in 9 out of 12. Between them they put in 32 bullets. How many shots did each fire?

68. A farmer sold at market 100 head of stock, horses, oxen, and sheep, selling two oxen for every horse. He obtained on the sale £2, 7s. a head. If he sold the horses, oxen, and sheep at the respective prices £22, £12, 10s., and £1, 10s., how many horses, oxen, and sheep respectively did he sell?

69. In a Euclid paper A gets 160 marks, and B just passes. A gets full marks for book-work, and twice as many marks for riders as B gets altogether. Also B, sending answers to all the questions, gets no marks for riders and half marks for book-work. Supposing it necessary to get $\frac{1}{5}$ of full marks in order to pass, find the number of marks which the paper carries.

70. It is between 2 and 3 o'clock, but a person looking at the clock and mistaking the hour-hand for the minute-hand, fancies that the time of day is 55 minutes earlier than the reality. What is the true time?

71. An army in a defeat loses one-sixth of its number in killed and wounded, and 4000 prisoners. It is reinforced by 3000 men, but retreats, losing a fourth of its number in doing so. There remain 18000 men. What was the original force?

72. The national debt of a country was increased by one-fourth in a time of war. During twenty years of peace which

followed £25,000,000 was paid off, and at the end of that time the interest was reduced from 4½ to 4 per cent. It was then found that the interest was the same in amount as before the war. What was the amount of the debt before the war?

73. An artesian well supplies a brewery. The consumption of water goes on each week-day from 3 A.M. to 6 P.M. at double the rate at which the water flows into the well. If the well contained 2250 gallons when the consumption began on Monday morning, and it was just emptied when the consumption ceased in the evening of the next Thursday but one, what is the rate of the influx of water into the well in gallons per hour?

XIV. ON MISCELLANEOUS FRACTIONS.

195. IN this Chapter we shall treat of various matters connected with FRACTIONS, so as to exhibit the mode of applying the elementary rules to the simplification of expressions of a more complicated kind than those which have hitherto been discussed.

196. The attention of the student must first be directed to a point in which the notation of Algebra differs from that of Arithmetic, namely *when a whole number and a fraction stand side by side with no sign between them.*

Thus in Arithmetic $2\frac{3}{7}$ stands for the *sum* of 2 and $\frac{3}{7}$.

But in Algebra $x\frac{y}{z}$ stands for the *product* of x and $\frac{y}{z}$.

So in Algebra $3\dfrac{a+b}{c}$ stands for the product of 3 and $\dfrac{a+b}{c}$;

that is, $3\dfrac{a+b}{c} = \dfrac{3a+3b}{c}$

EXAMPLES.—lx.

Simplify the following fractions:

1. $a + x + 3\dfrac{a}{x}$.

2. $\dfrac{a^2 + ax}{x^2} - 2\dfrac{x-a}{x}$.

3. $\dfrac{x-y}{x} + 2\dfrac{y}{x-y}$.

4. $4\dfrac{a+b}{a-b} - 2\dfrac{a^2 - b^2}{a^2 + b^2}$.

197. A fraction of which the Numerator or Denominator is itself a fraction, is called a **Complex Fraction**.

Thus $\dfrac{x}{a}$, $\dfrac{\frac{y}{c}}{d}$ and $\dfrac{\frac{x}{y}}{\frac{m}{n}}$ are complex fractions.

A Fraction whose terms are whole numbers is called a **Simple Fraction**.

All Complex Fractions may be reduced to Simple Fractions by the processes already described. We may take the following Examples:

(1) $\dfrac{\frac{a}{b}}{\frac{m}{n}} = \dfrac{a}{b} \div \dfrac{m}{n} = \dfrac{a}{b} \times \dfrac{n}{m} = \dfrac{an}{bm}$.

(2) $\dfrac{\frac{a}{b} - \frac{c}{d}}{\frac{m}{n} - \frac{p}{q}} = \left(\dfrac{a}{b} - \dfrac{c}{d}\right) \div \left(\dfrac{m}{n} - \dfrac{p}{q}\right) = \dfrac{ad - bc}{bd} \div \dfrac{mq - np}{nq}$

$= \dfrac{ad - bc}{bd} \times \dfrac{nq}{mq - np} = \dfrac{nq(ad - bc)}{bd(mq - np)}$.

(3) $\dfrac{1 + x}{1 + \frac{1}{x}} = (1 + x) \div \left(1 + \dfrac{1}{x}\right) = (1 + x) \div \dfrac{x + 1}{x}$

(4) $\dfrac{\dfrac{1}{1-x}-\dfrac{1}{1+x}}{\dfrac{x}{1-x}+\dfrac{1}{1+x}} = \left(\dfrac{1}{1-x}-\dfrac{1}{1+x}\right) \div \left(\dfrac{x}{1-x}+\dfrac{1}{1+x}\right)$

$= \dfrac{1+x-1+x}{1-x^2} \div \dfrac{x+x^2+1-x}{1-x^2}$

$= \dfrac{2x}{1-x^2} \times \dfrac{1-x^2}{1+x^2} = \dfrac{2x}{1+x^2}.$

(5) $\dfrac{3}{1+\dfrac{3}{1+\dfrac{3}{1-x}}} = \dfrac{3}{1+\dfrac{3}{\dfrac{1-x+3}{1-x}}} = \dfrac{3}{1+\dfrac{3(1-x)}{1-x+3}} = \dfrac{3}{1+\dfrac{3-3x}{4-x}}$

$= \dfrac{3}{\dfrac{4-x+3-3x}{4-x}} = \dfrac{3(4-x)}{4-x+3-3x} = \dfrac{12-3x}{7-4x}.$

EXAMPLES.—lxi.

Simplify the following expressions:

1. $\dfrac{x+\dfrac{4}{5}}{7} - \dfrac{x}{3\dfrac{1}{23}}.$

2. $\dfrac{\dfrac{x}{y}-\dfrac{y}{x}}{x-y}.$

3. $\dfrac{1-x^2}{1+\dfrac{1}{x}}.$

4. $\dfrac{y\left(\dfrac{x}{y}+1\right)}{x\left(1-\dfrac{y}{x}\right)}.$

5. $\dfrac{5+x+\dfrac{1}{x^2}}{2-x+\dfrac{1}{x^2}}.$

6. $\dfrac{x+\dfrac{1}{x^2}}{1+\dfrac{1}{x}}.$

7. $\dfrac{a-\dfrac{1}{a^2}}{1-\dfrac{1}{a}}.$

8. $\dfrac{\dfrac{x}{x+a}+\dfrac{x}{x-a}}{\dfrac{2x}{x^2-a^2}}.$

9. $\dfrac{2x}{x^2+\dfrac{1}{1 \div x^2}}.$

10. $\dfrac{x}{1+\dfrac{1}{x}}+1-\dfrac{1}{x+1}.$

11. $\dfrac{\dfrac{x-y}{x+y}+\dfrac{x+y}{x-y}}{\dfrac{x-y}{x+y}-\dfrac{x+y}{x-y}}.$

12. $\dfrac{1+x+x^2}{1+\dfrac{1}{x}}.$

13. $\dfrac{\dfrac{a+b}{b}+\dfrac{b}{a+b}}{\dfrac{1}{a}+\dfrac{1}{b}}.$

14. $\dfrac{2m-3+\dfrac{1}{m}}{\dfrac{2m-1}{m}}$

15. $\dfrac{\dfrac{1}{ab}+\dfrac{1}{ac}+\dfrac{1}{bc}}{\dfrac{a^2-(b+c)^2}{ab}}.$

198. Any fraction may be split up into a number of fractions equal to the number of terms in its numerator. Thus

$$\frac{x^3+x^2+x+1}{x^4}=\frac{x^3}{x^4}+\frac{x^2}{x^4}+\frac{x}{x^4}+\frac{1}{x^4}$$
$$=\frac{1}{x}+\frac{1}{x^2}+\frac{1}{x^3}+\frac{1}{x^4}.$$

EXAMPLES.—lxii.

Split up into four fractions, each in its lowest terms, the following fractions:

1. $\dfrac{a^4+3a^3+2a^2+5a}{2a^4}.$

2. $\dfrac{a^2bc+ab^2d+abc^2+bcd^2}{abcd}.$

3. $\dfrac{x^3-3x^2y+3xy^2-y^3}{x^2y^2}.$

4. $\dfrac{9a^3-12a^2+6a-3}{108}.$

5. $\dfrac{18p^2+12q^2-36r^2+72s^2}{3pqrs}.$

6. $\dfrac{10x^3-25x^2+75x-125}{1000}.$

199. The quotient obtained by dividing the unit by any fraction of that unit is called THE RECIPROCAL of that fraction.

Thus $\dfrac{1}{\dfrac{a}{b}}$, that is, $\dfrac{b}{a}$, is the Reciprocal of $\dfrac{a}{b}.$

200. We have shewn in Art. 158, that the fraction symbol $\dfrac{a}{b}$ is a proper representative of the Division of a by b. In

[S.A.]

Chapter IV. we treated of cases of division in which the divisor is contained an exact number of times in the dividend. We now proceed to treat of cases in which the divisor is not contained exactly in the dividend, and to shew the proper method of representing the Quotient in such cases.

Suppose we have to divide 1 by $1-a$. We may at once represent the result by the fraction $\dfrac{1}{1-a}$. But we may actually perform the operation of division in the following way.

$$
\begin{array}{r}
1-a\,)\ 1\ (1+a+a^2+a^3+\ldots \\
\underline{1-a} \\
a \\
\underline{a-a^2} \\
a^2 \\
\underline{a^2-a^3} \\
a^3 \\
\underline{a^3-a^4} \\
a^4
\end{array}
$$

The Quotient in this case is *interminable*. We may carry on the operation to any extent, but an exact and terminable Quotient we shall never find. It is clear, however, that the terms of the Quotient are formed by a certain law, and such a succession of terms is called a SERIES. If, as in the case before us, the series may be indefinitely extended, it is called an INFINITE SERIES.

If we wish to express in a concise form the result of the operation, we may stop at any term of the quotient and write the result in the following way.

$$\frac{1}{1-a} = 1 + \frac{a}{1-a},$$

$$\frac{1}{1-a} = 1 + a + \frac{a^2}{1-a},$$

$$\frac{1}{1-a} = 1 + a + a^2 + \frac{a^3}{1-a},$$

$$\frac{1}{1-a} = 1 + a + a^2 + a^3 + \frac{a^4}{1-a},$$

always being careful to attach to that term of the quotient, at which we intend to stop, the *remainder* at that point of the division, placed as the numerator of a fraction of which the divisor is the denominator.

EXAMPLES.—lxiii.

Carry on each of the following divisions to 5 terms in the quotient.

1. 2 by $1+a$.
2. m by $m+2$.
3. $a-b$ by $a+b$.
4. a^2+x^2 by a^2-x^2.
5. ax by $a-x$.
6. b by $a+x$.
7. 1 by $1+2x-2x^2$.
8. $1+x$ by $1-x+x^2$.
9. $1+b$ by $1-2b$.
10. x^3-b^3 by $x+b$.
11. a^2 by $x-b$.
12. a^2 by $(a+x)^2$.

13. If the divisor be $x-a$, the quotient x^2-2ax, and the remainder $4a^3$, what is the dividend?

14. If the divisor be $m-5$, the quotient $m^3+5m^2+15m+34$, and the remainder 75, what is the dividend?

201. If we are required to *multiply* such an expression as
$$\frac{x^2}{2}+\frac{x}{3}+\frac{1}{4} \text{ by } \frac{x}{2}-\frac{1}{3},$$
we may multiply each term of the former by each term of the latter, and combine the results by the ordinary methods of addition and subtraction of fractions, thus

$$\frac{x^2}{2}+\frac{x}{3}+\frac{1}{4}$$

$$\frac{x}{2}-\frac{1}{3}$$

$$\frac{x^3}{4}+\frac{x^2}{6}+\frac{x}{8}$$

$$-\frac{x^2}{6}-\frac{x}{9}-\frac{1}{12}$$

$$\frac{x^3}{4} \quad +\frac{x}{72}-\frac{1}{12}$$

Or we may first reduce the multiplicand and the multiplier to single fractions and proceed in the following way:

$$\left(\frac{x^2}{2}+\frac{x}{3}+\frac{1}{4}\right)\times\left(\frac{x}{2}-\frac{1}{3}\right)$$

$$=\frac{6x^2+4x+3}{12}\times\frac{3x-2}{6}=\frac{18x^3+x-6}{72}$$

$$=\frac{18x^3}{72}+\frac{x}{72}-\frac{6}{72}=\frac{x^3}{4}+\frac{x}{72}-\frac{1}{12}.$$

This latter process will be found the simpler by a beginner.

EXAMPLES.—lxiv.

Multiply

1. $\dfrac{x^2}{3}+\dfrac{x}{2}+\dfrac{1}{5}$ by $\dfrac{x}{3}+\dfrac{1}{4}$.

2. $\dfrac{a^2}{5}-\dfrac{a}{6}+\dfrac{1}{3}$ by $\dfrac{a}{4}-\dfrac{1}{5}$.

3. $x^3+x+\dfrac{1}{x}+\dfrac{1}{x^3}$ by $x-\dfrac{1}{x}$.

4. $x^2-1+\dfrac{1}{x^2}$ by $x^2+1+\dfrac{1}{x^2}$.

5. $\dfrac{1}{a^2}+\dfrac{1}{b^2}$ by $\dfrac{1}{a^2}-\dfrac{1}{b^2}$.

6. $\dfrac{1}{a}-\dfrac{1}{b}+\dfrac{1}{c}$ by $\dfrac{1}{a}+\dfrac{1}{b}+\dfrac{1}{c}$.

7. $1+\dfrac{b}{a}+\dfrac{b^2}{a^2}$ by $1-\dfrac{b}{a}+\dfrac{b^2}{a^2}$.

8. $1+\dfrac{1}{2}x+\dfrac{1}{4}x^2$ by $1-\dfrac{1}{2}x+\dfrac{1}{8}x^2-\dfrac{1}{16}x^3$.

9. $\dfrac{5}{2x^2}+\dfrac{3}{x}-\dfrac{7}{3}$ by $\dfrac{2}{x^2}-\dfrac{1}{x}-\dfrac{1}{2}$.

10. $\dfrac{a^2}{b^2}+\dfrac{b^2}{a^2}+2$ by $\dfrac{a^2}{b^2}-\dfrac{b^2}{a^2}-2$.

202. If we have to *divide* such an expression as

$$x^3+3x+\frac{3}{x}+\frac{1}{x^3}$$

by $x+\dfrac{1}{x}$, we may proceed as in the division of whole numbers, carefully observing that the order of descending powers of x is

$$\ldots\ldots x^3,\ x^2,\ x,\ \frac{1}{x},\ \frac{1}{x^2},\ \frac{1}{x^3}\ldots\ldots$$

Any isolated digits, as 1, 2, 3 ... will stand between x and $\frac{1}{x}$.

Thus the expression

$$4 + x^3 + \frac{1}{x^3} + 3x^2 + \frac{3}{x^2} + 5x + \frac{5}{x},$$

arranged according to *descending* powers of x, will stand thus,

$$x^3 + 3x^2 + 5x + 4 + \frac{5}{x} + \frac{3}{x^2} + \frac{1}{x^3}.$$

The reason for this arrangement will be given in the Chapter on the Theory of Indices.

Ex.
$$x + \frac{1}{x} \Big) x^3 + 3x + \frac{3}{x} + \frac{1}{x^3} \Big(x^2 + 2 + \frac{1}{x^2}$$

$$\underline{x^3 + \ x}$$

$$2x + \frac{3}{x}$$

$$\underline{2x + \frac{2}{x}}$$

$$\frac{1}{x} + \frac{1}{x^3}$$

$$\underline{\frac{1}{x} + \frac{1}{x^3}}$$

Or we may proceed in the following way, which will be found simpler by the beginner.

$$\left(x^3 + 3x + \frac{3}{x} + \frac{1}{x^3}\right) \div \left(x + \frac{1}{x}\right)$$

$$= \frac{x^6 + 3x^4 + 3x^2 + 1}{x^3} \div \frac{x^2 + 1}{x}$$

$$= \frac{x^6 + 3x^4 + 3x^2 + 1}{x^3} \times \frac{x}{x^2 + 1}$$

$$= \frac{x^4 + 2x^2 + 1}{x^2} = \frac{x^4}{x^2} + \frac{2x^2}{x^2} + \frac{1}{x^2} = x^2 + 2 + \frac{1}{x^2}.$$

EXAMPLES.—lxv.

Divide:

1. $x^2 - \dfrac{1}{x^2}$ by $x + \dfrac{1}{x}$.

2. $a^2 - \dfrac{1}{b^2}$ by $a - \dfrac{1}{b}$.

3. $m^3 + \dfrac{1}{n^3}$ by $m + \dfrac{1}{n}$.

4. $c^6 - \dfrac{1}{d^6}$ by $c - \dfrac{1}{d}$.

5. $\dfrac{x^2}{y^2} + 2 + \dfrac{y^2}{x^2}$ by $\dfrac{x}{y} + \dfrac{y}{x}$.

6. $\dfrac{1}{a^4} + \dfrac{1}{a^2 b^2} + \dfrac{1}{b^4}$ by $\dfrac{1}{a^2} - \dfrac{1}{ab} + \dfrac{1}{b^2}$.

7. $\dfrac{x^3}{y^3} - \dfrac{y^3}{x^3} - 3\dfrac{x}{y} + 3\dfrac{y}{x}$ by $\dfrac{x}{y} - \dfrac{y}{x}$.

8. $\dfrac{3x^6}{4} - 4x^4 + \dfrac{77}{8}x^3 - \dfrac{43}{4}x^2 - \dfrac{33}{4}x + 27$ by $\dfrac{x^2}{2} - x + 3$.

9. $\dfrac{a^3}{b^3} + \dfrac{b^3}{a^3}$ by $\dfrac{a}{b} + \dfrac{b}{a}$.

10. $\dfrac{1}{a^3} + \dfrac{1}{b^3} + \dfrac{1}{c^3} - \dfrac{3}{abc}$ by $\dfrac{1}{a} + \dfrac{1}{b} + \dfrac{1}{c}$.

203. In dealing with expressions involving Decimal Fractions two methods may be adopted, as will be seen from the following example.

Multiply $\cdot 1x - \cdot 2y$ by $\cdot 03x + \cdot 4y$.

We may proceed thus, applying the Rules for Multiplication, Addition, and Subtraction of Decimals.

$$\begin{array}{r} \cdot 1x - \cdot 2y \\ \cdot 03x + \cdot 4y \\ \hline \cdot 003x^2 - \cdot 006xy \\ + \cdot 04\ xy - \cdot 08y^2 \\ \hline \cdot 003x^2 + \cdot 034xy - \cdot 08y^2 \end{array}$$

Or thus,

$$(\cdot 1x - \cdot 2y)(\cdot 03x + \cdot 4y) = \left(\dfrac{x}{10} - \dfrac{2y}{10}\right)\left(\dfrac{3x}{100} + \dfrac{4y}{10}\right)$$

$$= \dfrac{x - 2y}{10} \times \dfrac{3x + 40y}{100}$$

$$= \dfrac{3x^2 + 34xy - 80y^2}{1000}$$

$$= \cdot 003x^2 + \cdot 034xy - \cdot 08y^2.$$

The latter method will be found the simpler for a beginner.

EXAMPLES.—lxvi.

Multiply :
1. $\cdot 1x - 3$ by $\cdot 5x + \cdot 07$,
2. $\cdot 05x + 7$ by $\cdot 2x - 3$,
3. $\cdot 3x - \cdot 2y$ by $\cdot 4x + \cdot 7y$,
4. $4\cdot 3x + 5\cdot 2y$ by $\cdot 04x - \cdot 06y$.
5. Find the value of
$a^3 - b^3 + c^3 + 3abc$ when $a = \cdot 03$, $b = \cdot 1$, and $c = \cdot 07$.
6. Find the value of
$x^3 - 3ax^2 + 3a^2x - a^3$ when $x = \cdot 7$ and $a = \cdot 03$.

204. When any expression E is put in a form of which f is a factor, then $\dfrac{E}{f}$ is the other factor.

Thus
$$a + b = a\left(\dfrac{a+b}{a}\right)$$
$$= a\left(1 + \dfrac{b}{a}\right).$$

So
$$ab + ac + bc = abc \cdot \dfrac{ab + ac + bc}{abc}$$
$$= abc \cdot \left(\dfrac{1}{c} + \dfrac{1}{b} + \dfrac{1}{a}\right),$$

and
$$x^2 + 2xy + y^2 = x^2 \cdot \left(\dfrac{x^2 + 2xy + y^2}{x^2}\right)$$
$$= x^2 \cdot \left(1 + \dfrac{2y}{x} + \dfrac{y^2}{x^2}\right).$$

EXAMPLES.—lxvii.

1. Write in factors, one of which is $a_1 x$, the series
$$a_1 x + a_2 x^2 + a_3 x^3 + a_4 x^4 + \ldots$$
2. Write in factors, one of which is xyz, the expression
$$xy - xz + yz.$$
3. Write in factors, one of which is x^2, the expression
$$x^2 + xy + y^2.$$
4. Write in factors, one of which is $a + b$, the expression
$$(a+b)^3 - c(a+b)^2 - d(a+b) + e.$$

205. We shall now give two examples of a process by which, when certain fractions are known to be equal, other relations between the quantities involved in them may be determined.

This process will be found of great use in a later part of the subject, and the student is advised to pay particular attention to it.

(1) If $\dfrac{a}{b} = \dfrac{c}{d}$, shew that

$$\frac{a+b}{a-b} = \frac{c+d}{c-d}.$$

Let
$$\frac{a}{b} = \lambda.$$

Then
$$\frac{c}{d} = \lambda ;$$

$$\therefore a = \lambda b,$$
$$\text{and} \quad c = \lambda d.$$

Now
$$\frac{a+b}{a-b} = \frac{\lambda b + b}{\lambda b - b} = \frac{b(\lambda+1)}{b(\lambda-1)} = \frac{\lambda+1}{\lambda-1},$$

and
$$\frac{c+d}{c-d} = \frac{\lambda d + d}{\lambda d - d} = \frac{d(\lambda+1)}{d(\lambda-1)} = \frac{\lambda+1}{\lambda-1}.$$

Hence $\dfrac{a+b}{a-b}$ and $\dfrac{c+d}{c-d}$ being each equal to $\dfrac{\lambda+1}{\lambda-1}$ are equal to one another.

(2) If $\dfrac{m}{a-b} = \dfrac{n}{b-c} = \dfrac{r}{c-a}$, shew that $m+n+r=0$.

Let
$$\frac{m}{a-b} = \lambda,$$

$$\frac{n}{b-c} = \lambda,$$

$$\frac{r}{c-a} = \lambda,$$

then
$$m = \lambda a - \lambda b,$$
$$n = \lambda b - \lambda c,$$
$$r = \lambda c - \lambda a ;$$
$$\therefore m + n + r = \lambda a - \lambda b + \lambda b - \lambda c + \lambda c - \lambda a = 0.$$

EXAMPLES.—lxviii.

1. If $\dfrac{a}{b}=\dfrac{c}{d}$ prove the following relations:

   (1) $\dfrac{a-b}{b}=\dfrac{c-d}{d}.$

   (2) $\dfrac{a}{a+b}=\dfrac{c}{c+d}.$

   (3) $\dfrac{3a}{4a-5b}=\dfrac{3c}{4c-5d}.$

   (4) $\dfrac{a^2+b^2}{a^2-b^2}=\dfrac{c^2+d^2}{c^2-d^2}.$

   (5) $\dfrac{8a+b}{4a+7b}=\dfrac{8c+d}{4c+7d}.$

   (6) $\dfrac{a^2-b^2}{c^2-d^2}=\dfrac{ab}{cd}.$

   (7) $\dfrac{11a+b}{11c+d}=\dfrac{13a+b}{13c+d}.$

   (8) $\dfrac{a^2-ab+b^2}{a^2+ab+b^2}=\dfrac{c^2-cd+d^2}{c^2+cd+d^2}.$

2. If $\dfrac{l}{a-b}=\dfrac{m}{b-c}=\dfrac{n}{c-a}$, then $l+m+n=0$.

3. If $\dfrac{a}{b}=\dfrac{c}{d}=\dfrac{e}{f}$, prove that $\dfrac{a}{b}=\dfrac{la+mc+ne}{lb+md+nf}.$

4. If $\dfrac{a+b}{b}=\dfrac{b+c}{c}=\dfrac{c+a}{a}$, prove that $a=b=c$.

5. If $\dfrac{a_1}{b_1}=\dfrac{a_2}{b_2}=\dfrac{a_3}{b_3}$, shew that $\dfrac{a_1}{b_1}=\dfrac{2a_1+3a_2+4a_3}{2b_1+3b_2+4b_3}.$

6. If $\dfrac{a}{b}$, $\dfrac{c}{d}$, $\dfrac{e}{f}$ be in descending order of magnitude, shew that $\dfrac{a+c+e}{b+d+f}$ is less than $\dfrac{a}{b}$ and greater than $\dfrac{e}{f}.$

7. If $\dfrac{x_1}{y_1}=\dfrac{x_2}{y_2}$, shew that $\dfrac{4x_1+5y_1}{7x_1+9y_1}=\dfrac{4x_2+5y_2}{7x_2+9y_2}.$

8. If $\dfrac{a}{b}=\dfrac{c}{d}$, shew that $\dfrac{a^2+ab}{c^2+cd}=\dfrac{ab-b^2}{cd-d^2}.$

9. If $\dfrac{a}{b}=\dfrac{c}{d}$, shew that $\dfrac{7a+b}{3a+5b}=\dfrac{7c+d}{3c+5d}.$

10. If $\frac{a}{b}$ be a *proper* fraction, shew that $\frac{a+c}{b+c}$ is greater than $\frac{a}{b}$, c being a positive quantity.

11. If $\frac{a}{b}$ be an *improper* fraction, shew that $\frac{a+c}{b+c}$ is less than $\frac{a}{b}$, c being a positive quantity.

206. We shall now give a series of examples in the working of which most of the processes connected with fractions will be introduced.

EXAMPLES.—lxix.

1. Find the value of $3a^2 + \frac{2ab^2}{c} - \frac{c^3}{b^2}$ when
$$a = 4, \ b = \frac{1}{2}, \ c = 1.$$

2. Simplify $\dfrac{2x^3 + x^2 - 8x + 5}{7x^2 - 12x + 5}$ and $\dfrac{a^3 - 39a + 70}{a^2 + 4a - 45}$.

3. Simplify $\left(\dfrac{a+p}{a-p} - \dfrac{a-p}{a+p}\right) \div \left(\dfrac{a+p}{a-p} + \dfrac{a-p}{a+p}\right)$.

4. Add together
$$\frac{x^2}{4} - \frac{y^2}{6} + \frac{z^2}{8}, \ \frac{y^2}{4} - \frac{z^2}{6} + \frac{x^2}{8} \text{ and } \frac{z^2}{4} + \frac{x^2}{6} + \frac{y^2}{8},$$
and subtract $z^2 - x^2 + \dfrac{y^2}{2}$ from the result.

5. Find the value of $\dfrac{a^2 + b^2 - c^2 + 2ab}{a^2 - b^2 - c^2 + 2bc}$ when
$$a = 4, \ b = \frac{1}{2}, \ c = 1.$$

6. Multiply $\dfrac{5}{2}x^2 + 3ax - \dfrac{7}{3}a^2$ by $2x^2 - ax - \dfrac{a}{2}$.

7. Shew that $\dfrac{a^3 - b^3}{(a-b)^2} = a + 2b + \dfrac{3b^2}{a-b}$.

ON MISCELLANEOUS FRACTIONS.

8. Simplify $\dfrac{x-y}{x} + \dfrac{2y}{x-y} + \dfrac{y^3 - xy^2}{x^3 - xy^2}$

9. Shew that $\dfrac{60x^3 - 17x^2 - 4x + 1}{5x^2 + 9x - 2} = 12x - 25 + \dfrac{49}{x+2}$.

10. Simplify $\dfrac{x^4 - 9x^3 + 7x^2 + 9x - 8}{x^4 + 7x^3 - 9x^2 - 7x + 8}$.

11. Simplify $\dfrac{x^3}{x^4 - 1} + \dfrac{1}{1 - x - \dfrac{1}{1 + x - \dfrac{1}{1-x}}}$.

12. Simplify $a + ab + b^2 \left(a + ab + b^2 \dfrac{a}{1-b} \right)$.

13. Multiply together $\left(l + \dfrac{1}{l}\right)\left(l^2 + \dfrac{1}{l^2}\right)\left(l - \dfrac{1}{l}\right)$.

14. Add together $\dfrac{1}{a+1}, \dfrac{1}{b+1}, \dfrac{1}{c+1}$, and shew that if their sum be equal to 1, then $abc = a+b+c+2$.

15. Divide $\dfrac{x}{a} - 1 - \dfrac{b}{a} - \dfrac{b^2}{a^2} + \dfrac{b}{x} + \dfrac{b^2}{x^2}$ by $x - a$.

16. Simplify $\dfrac{\dfrac{a}{b} \div c + \dfrac{b}{c} \div a + \dfrac{c}{a} \div b}{\dfrac{b}{a} \div c + \dfrac{c}{b} \div a + \dfrac{a}{c} \div b}$, and shew that it is equal

to $\dfrac{s(s-a) + (s-b)(s-c)}{bc}$ if $2s = a + b + c$.

17. Shew that $\dfrac{1}{1 + \dfrac{1}{a \div x}} + \dfrac{1}{1 - \dfrac{1}{a \div x}} + \dfrac{2}{1 + \dfrac{1}{a^2 \div x^2}} = -\dfrac{4a^4}{a^4 - x^4}$.

18. Simplify $\dfrac{a+b}{a-b} + \dfrac{a-b}{a+b} - 2\dfrac{a^2 - b^2}{a^2 + b^2}$.

19. Simplify $\dfrac{b}{a+b} - \dfrac{a+b}{2a} + \dfrac{a^2 + b^2}{2a(a-b)}$.

20. Simplify $\dfrac{a^2 - ab + b^2}{a^3 - 3ab(a-b) - b^3} \times \dfrac{a^2 - b^2}{a^2 + b^2}$.

21. Simplify $\dfrac{2}{(x^2-1)^2} - \dfrac{1}{2x^2-4x+2} - \dfrac{1}{1-x^2}$.

22. Simplify $\dfrac{a^2+b^2+2ab-c^2}{c^2-a^2-b^2+2ab} \div \dfrac{a+b+c}{b+c-a}$.

23. Simplify $\left(\dfrac{x}{1+\dfrac{1}{x}} + 1 - \dfrac{1}{x+1}\right) \div \left(\dfrac{x}{1-\dfrac{1}{x}} - x - \dfrac{1}{x-1}\right)$.

24. Find the value of $\left(\dfrac{x-a}{x-b}\right)^3 - \dfrac{x-2a+b}{x+a-2b}$, when $x = \dfrac{a+b}{2}$.

25. Simplify $\dfrac{a^2-(b-c)^2}{(a+c)^2-b^2} + \dfrac{b^2-(a-c)^2}{(a+b)^2-c^2} + \dfrac{c^2-(a-b)^2}{(b+c)^2-a^2}$.

26. Simplify $\dfrac{(x^2-4x)(x^2-4)^2}{(x^2-2x)^2}$.

27. Simplify $\dfrac{(a^2-1)(a^6-1)}{(a+1)^2(a^2-a)^2}$.

28. Simplify $\dfrac{1}{x^3} + \dfrac{1}{x^2} - \dfrac{1}{x} - \dfrac{1}{(x^2+1)^2} + \dfrac{x-1}{x^2+1} - \dfrac{3}{x^2(x^2+1)^2}$.

29. Divide $\dfrac{x^3}{a^3} - \dfrac{x}{a} + \dfrac{a}{x} - \dfrac{a^3}{x^3}$ by $\dfrac{x}{a} - \dfrac{a}{x}$.

30. Simplify $\left\{\dfrac{a+b}{2(a-b)} - \dfrac{a-b}{2(a+b)} + \dfrac{2b^2}{a^2-b^2}\right\} \dfrac{a-b}{2b}$.

31. Simplify $\dfrac{(a+b+c)^2+(b-c)^2+(c-a)^2+(a-b)^2}{a^2+b^2+c^2}$.

32. Take $\dfrac{1-x-3x^2}{(3-2x-7x^2)^3}$ from $\dfrac{1+3x^2+2x^3}{(3-2x-7x^2)^4}$.

33. Simplify $\left(\dfrac{x^2+y^2}{x^2-y^2} - \dfrac{x^2-y^2}{x^2+y^2}\right) \div \left(\dfrac{x+y}{x-y} - \dfrac{x-y}{x+y}\right)$.

34. Simplify $\left(\dfrac{x^2}{y^2}-1\right)\left(\dfrac{x}{x-y}-1\right) + \left(\dfrac{x^3}{y^3}-1\right)\left(\dfrac{x^2+xy}{x^2+xy+y^2}-1\right)$.

35. Simplify

$\dfrac{a^2-ab}{a^3-b^3} \times \dfrac{a^2+ab+b^2}{a+b} + \left(\dfrac{2a^3}{a^3+b^3}-1\right)\left(1-\dfrac{2ab}{a^2+ab+b^2}\right)$.

36. Simplify
$$\frac{1}{2(x-1)^2}-\frac{1}{4(x-1)}+\frac{1}{4(x+1)}-\frac{1}{(x-1)^2(x+1)}.$$

37. Prove that
$$\frac{1}{abc}+\frac{1}{a(a-b)(x-a)}+\frac{1}{b(b-a)(x-b)}=\frac{1}{x(x-a)(x-b)}.$$

38. If $s=a+b+c+\ldots$ to n terms, shew that
$$\frac{s-a}{a}+\frac{s-b}{b}+\frac{s-c}{c}+\ldots=s\left(\frac{1}{a}+\frac{1}{b}+\frac{1}{c}+\ldots\right)-n.$$

39. Multiply $\left(\dfrac{x^2}{x^2-y^2}-\dfrac{y^2}{x^2+y^2}\right)$ by $\dfrac{(x^2-y^2)^2}{(x^2-y^2)^2+(x^2+y^2)^2}.$

40. Simplify $\dfrac{1+\dfrac{a-x}{a+x}}{1-\dfrac{a-x}{a+x}}\div\dfrac{1+\dfrac{a^2-x^2}{a^2+x^2}}{1-\dfrac{a^2-x^2}{a^2+x^2}}.$

41. Divide $x^3+\dfrac{1}{x^3}-3\left(\dfrac{1}{x^2}-x^2\right)+4\left(x+\dfrac{1}{x}\right)$ by $x+\dfrac{1}{x}.$

42. If $s=a+b+c+\ldots$ to n terms, shew that
$$\frac{s-a}{s}+\frac{s-b}{s}+\frac{s-c}{s}+\ldots=n-1.$$

43. Divide $\left(\dfrac{x}{x-y}-\dfrac{y}{x+y}\right)$ by $\left(\dfrac{x^2}{x^2+y^2}+\dfrac{y^2}{x^2-y^2}\right).$

44. Simplify $\dfrac{1-\dfrac{2xy}{(x+y)^2}}{1+\dfrac{2xy}{(x-y)^2}}\div\left(\dfrac{1-\dfrac{y}{x}}{1+\dfrac{y}{x}}\right)^2.$

45. If $\dfrac{a+b}{1-ab}=\dfrac{c+d}{cd-1}$, prove that $\dfrac{a+b+c+d}{\dfrac{1}{a}+\dfrac{1}{b}+\dfrac{1}{c}+\dfrac{1}{d}}=abcd.$

46. Simplify
$$\frac{p^4+4p^3q+6p^2q^2+4pq^3+q^4}{p^4-4p^3q+6p^2q^2-4pq^3+q^4}\div\frac{p^3+3p^2q+3pq^2+q^3}{p^3-3p^2q+3pq^2-q^3}.$$

47. Reduce $\dfrac{1-2x}{3(x^2-x+1)}+\dfrac{x+1}{2(x^2+1)}+\dfrac{1}{6(x+1)}.$

48. Simplify $\dfrac{1}{x+\dfrac{1}{y+\dfrac{1}{z}}} \div \dfrac{1}{x+\dfrac{1}{y}} - \dfrac{1}{y(xyz+x+z)}$.

49. Simplify $\dfrac{\dfrac{1}{a-x}-\dfrac{1}{a-y}+\dfrac{x}{(a-x)^2}-\dfrac{y}{(a-y)^2}}{\dfrac{1}{(a-y)(a-x)^2}-\dfrac{1}{(a-y)^2(a-x)}}$.

50. Simplify $\dfrac{\dfrac{3}{abc}}{\dfrac{1}{bc}+\dfrac{1}{ca}-\dfrac{1}{ab}} - \dfrac{3-a-b-c}{a+b-c}$.

51. Simplify $\dfrac{a+\dfrac{b}{1+\dfrac{a}{b}}}{a-\dfrac{b}{1-\dfrac{a}{b}}}(a^6-b^6)$.

XV. SIMULTANEOUS EQUATIONS OF THE FIRST DEGREE.

207. To determine several unknown quantities we must have as many independent equations as there are unknown quantities.

Thus if we had this equation given,
$$x+y=6,$$
we could determine no definite values of x and y, for
$$\left.\begin{array}{l}x=2\\y=4\end{array}\right\}, \text{ or } \left.\begin{array}{l}x=4\\y=2\end{array}\right\}, \text{ or } \left.\begin{array}{l}x=3\\y=3\end{array}\right\},$$
or other values might be given to x and y, consistently with the equation. In fact we can find as many pairs of values of x and y as we please, which will satisfy the equation.

We must have *a second equation* independent of the first, and then we may find a pair of values of x and y which will *satisfy both equations*.

Thus, if besides the equation $x+y=6$, we had another equation $x-y=2$, it is evident that the values of x and y which will *satisfy both equations* are

$$\left.\begin{array}{l} x=4 \\ y=2 \end{array}\right\},$$

since $4+2=6$, and $4-2=2$.

Also, of all the pairs of values of x and y which will satisfy one of the equations, there is but one pair which will satisfy the other equation.

We proceed to shew how this pair of values may be found.

208. Let the proposed equations be

$$2x + 7y = 34$$
$$5x + 9y = 51.$$

Multiply the first equation by 5 and the second equation by 2, we then get

$$10x + 35y = 170$$
$$10x + 18y = 102.$$

The coefficients of x are thus made *alike* in both equations.

If we now subtract each member of the second equation from the corresponding member of the first equation, we shall get (Ax. II. page 58)

$$35y - 18y = 170 - 102,$$

or
$$17y = 68;$$
$$\therefore y = 4.$$

We have thus obtained the value of *one* of the **unknown** symbols. The value of the *other* may be found thus:

Take *one* of the original equations, thus

$$2x + 7y = 34.$$

Now, since $y = 4$, $7y = 28$;
$$\therefore 2x + 28 = 34;$$
$$\therefore x = 3.$$

Hence the pair of values of x and y which satisfy the equations is 3 and 4.

NOTE. The process of thus obtaining from two or more equations an equation, from which one of the unknown quantities has disappeared, is called *Elimination*.

209. We worked out the steps fully in the example given in the last article. We shall now work an example in the form in which the process is usually given.

Ex. *To solve the equations*
$$3x + 7y = 67$$
$$5x + 4y = 58.$$

Multiplying the first equation by 5 and the second by 3,
$$15x + 35y = 335$$
$$15x + 12y = 174.$$

Subtracting, $\qquad 23y = 161,$
and therefore $\qquad y = 7.$
Now, since $\qquad 3x + 7y = 67,$
$$3x + 49 = 67,$$
$$\therefore 3x = 18,$$
$$\therefore x = 6.$$

Hence $x = 6$ and $y = 7$ are the values required.

210. In the examples given in the two preceding articles we made the coefficients of x alike. Sometimes it is more convenient to make the coefficients of y alike. Thus if we have to solve the equations
$$29x + 2y = 64$$
$$13x + y = 29,$$
we leave the first equation as it stands, and multiply the second equation by 2, thus
$$29x + 2y = 64$$
$$26x + 2y = 58.$$

Subtracting, $\qquad 3x = 6,$
and therefore $\qquad x = 2.$
Now, since $\qquad 13x + y = 29,$
$$26 + y = 29,$$
$$\therefore y = 3.$$

Hence $x = 2$ and $y = 3$ are the values required.

EXAMPLES —lxx.

1. $2x + 7y = 41$
   $3x + 4y = 42.$

2. $5x + 8y = 101$
   $9x + 2y = 95.$

3. $13x + 17y = 189$
   $2x + y = 21.$

4. $14x + 9y = 156$
   $7x + 2y = 58.$

5. $x + 15y = 49$
   $3x + 7y = 71.$

6. $15x + 19y = 132$
   $35x + 17y = 226.$

7. $6x + 4y = 236$
   $3x + 15y = 573.$

8. $39x + 27y = 105$
   $52x + 29y = 133.$

9. $72x + 14y = 330$
   $63x + 7y = 273.$

211. We shall now give some examples in which *negative* signs occur attached to the coefficient of y in one or both of the equations.

Ex. *To solve the equations:*
$$6x + 35y = 177$$
$$8x - 21y = 33.$$

Multiply the first equation by 4 and the second by 3.
$$24x + 140y = 708$$
$$24x - 63y = 99.$$

Subtracting, $203y = 609,$

and therefore $y = 3.$

The value of x may then be found.

EXAMPLES.—lxxi.

1. $2x + 7y = 52$
   $3x - 5y = 16.$

2. $7x - 4y = 55$
   $15x - 13y = 109.$

3. $x + y = 96$
   $x - y = 2.$

4. $4x + 9y = 79$
   $7x - 17y = 40.$

5. $x + 19y = 97$
   $7x - 53y = 121.$

6. $29x - 14y = 175$
   $87x - 56y = 497.$

7. $171x - 213y = 642$
   $114x - 326y = 244.$
   [s.a.]

8. $43x + 2y = 266$
   $12x - 17y = 4.$

9. $5x + 9y = 188$
   $13x - 2y = 57.$

212. We have hitherto taken examples in which the coefficients of x are both *positive*. Let us now take the following equations :

$$5x - 7y = 6$$
$$9y - 2x = 10.$$

Change all the signs of the second equation, so that we get

$$5x - 7y = 6$$
$$2x - 9y = -10.$$

Multiplying by 2 and 5,

$$10x - 14y = 12$$
$$10x - 45y = -50.$$

Subtracting,

$$-14y + 45y = 12 + 50,$$
$$\text{or, } 31y = 62,$$
$$\text{or, } y = 2.$$

The value of x may then be found.

EXAMPLES.—lxxii.

1. $4x - 7y = 22$
   $7y - 3x = 1.$

2. $9x - 5y = 52$
   $8y - 3x = 8.$

3. $17x + 3y = 57$
   $16y - 3x = 23.$

4. $7y + 3x = 78$
   $19y - 7x = 136.$

5. $5x - 3y = 4$
   $12y - 7x = 10.$

6. $3x + 2y = 39$
   $3y - 2x = 13.$

7. $5y - 2x = 21$
   $13x - 4y = 120.$

8. $9y - 7x = 13$
   $15x - 7y = 9.$

9. $12x + 7y = 176$
   $3y - 19x = 3.$

213. In the preceding examples the values of x and y have been *positive*. We shall now give some equations in which x or y or both have *negative* values.

Ex. *To solve the equations:*

$$2x - 9y = 11$$
$$3x - 4y = 7.$$

Multiplying the equations by 3 and 2 respectively, we get

$$6x - 27y = 33$$
$$6x - 8y = 14.$$

Subtracting,
$$-19y = 19,$$
$$\text{or,} \quad 19y = -19,$$
$$\text{or,} \quad y = -1.$$

Now since $9y = -9$,

$2x - 9y$ will be equivalent to $2x - (-9)$ or, $2x + 9$.

Hence, from the first equation,
$$2x + 9 = 11,$$
$$\therefore x = 1.$$

EXAMPLES.—lxxiii

1. $2x + 3y = 8$
   $3x + 7y = 7.$

2. $5x - 2y = 51$
   $19x - 3y = 180.$

3. $3x - 5y = 51$
   $2x + 7y = 3.$

4. $7y - 3x = 139$
   $2x + 5y = 91.$

5. $4x + 9y = 106$
   $8x + 17y = 198.$

6. $2x - 7y = 8$
   $4y - 9x = 19.$

7. $17x + 12y = 59$
   $19x - 4y = 153.$

8. $8x + 3y = 3$
   $12x + 9y = 3.$

9. $69y - 17x = 103$
   $14x - 13y = -41.$

214. We shall now take the case of *Fractional* Equations involving two unknown quantities.

Ex. To solve the equations,
$$2x - \frac{y-3}{5} = 4$$
$$3y = 9 - \frac{x-2}{3}.$$

First, clearing the equations of fractions, we get
$$10x - y + 3 = 20$$
$$9y = 27 - x + 2,$$

from which we obtain,
$$10x - y = 17$$
$$x + 9y = 29,$$

and hence we may find $x = 2$, $y = 3$.

EXAMPLES.—lxxiv.

1. $\dfrac{x}{2}+\dfrac{y}{3}=7$
   $\dfrac{x}{3}+\dfrac{y}{2}=8.$

2. $10x+\dfrac{y}{3}=210$
   $10y-\dfrac{x}{2}=290.$

3. $\dfrac{x}{7}+7y=251$
   $\dfrac{y}{7}+7x=299.$

4. $\dfrac{x+y}{3}+5=10$
   $\dfrac{x-y}{2}+7=9\tfrac{1}{2}.$

5. $7x+\dfrac{5y}{2}=413$
   $39x=14y-1609$

6. $\dfrac{2x+3y}{5}=10-\dfrac{y}{3}$
   $\dfrac{4y-3x}{6}=\dfrac{3x}{4}+1.$

7. $x-\dfrac{y-2}{7}=5$
   $4y-\dfrac{x+10}{3}=3.$

10. $\dfrac{x+2}{3}+8y=31$
    $\dfrac{y+5}{4}+10x=192.$

8. $\dfrac{x}{4}+8=\dfrac{y}{2}-12$
   $\dfrac{x+y}{5}+\dfrac{y}{3}=\dfrac{2x-y}{4}+35.$

11. $\dfrac{2x-y}{7}+3x=2y-6$
    $\dfrac{y+3}{5}+\dfrac{y-x}{6}=2x-8.$

9. $\dfrac{3x-5y}{2}+3=\dfrac{2x+y}{5}$
   $8-\dfrac{x-2y}{4}=\dfrac{x}{2}+\dfrac{y}{3}.$

12. $\dfrac{x-2}{5}-\dfrac{10-x}{3}=\dfrac{y-10}{4}$
    $\dfrac{2y+4}{3}=\dfrac{4x+y+13}{8}.$

13. $\dfrac{5x-6y}{13}+3x=4y-2$
    $\dfrac{5x+6y}{6}-\dfrac{3x-2y}{4}=2y-2.$

14. $\dfrac{5x-3}{2}-\dfrac{3x-19}{2}=4-\dfrac{3y-x}{3}$
    $\dfrac{2x+y}{2}-\dfrac{9x-7}{8}=\dfrac{3y+9}{4}-\dfrac{4x+5y}{16}.$

15. $\dfrac{4x+5y}{40}=x-y$
    $\dfrac{2x-y}{3}+2y=\dfrac{1}{2}.$

215. We have now to explain the method of solving *Literal Equations* involving two unknown quantities.

Ex. *To solve the equations,*

$$ax + by = c$$
$$px + qy = r.$$

Multiplying the first equation by p and the second by a, we get

$$apx + bpy = cp$$
$$apx + aqy = ar.$$

Subtracting, $\quad bpy - aqy = cp - ar,$

or, $(bp - aq)y = cp - ar$;

$$\therefore y = \frac{cp - ar}{bp - aq}.$$

We might then find x by substituting this value of y in one of the original equations, but usually the safest course is to begin afresh and make the coefficients of y alike in the original equations, multiplying the first by q and the second by b, which gives

$$aqx + bqy = cq$$
$$bpx + bqy = br.$$

Subtracting, $\quad aqx - bpx = cq - br,$

or, $(aq - bp)x = cq - br$;

$$\therefore x = \frac{cq - br}{aq - bp}.$$

EXAMPLES.—lxxv.

1. $mx + ny = e,$
   $px + qy = f.$

2. $ax + by = c$
   $dx - ey = f.$

3. $ax - by = m$
   $cx + ey = n.$

4. $cx \;\;\;= dy$
   $x + y = e.$

5. $mx - ny = r$
   $m'x + n'y = r$

6. $x + y = a$
   $x - y = b.$

7. $ax + by = c$
   $dx + fy = c^2.$

8. $abx + cdy = 2$
   $ax - cy = \dfrac{d-b}{bd}.$

9. $\dfrac{a}{b+y} = \dfrac{b}{3a+x}$
   $ax + 2by = d.$

10. $bcx + 2b - cy = 0$

$b^2 y + \dfrac{a(c^3 - b^3)}{bc} = \dfrac{2b^3}{c} + c^3 x.$

11. $(b+c)(x+c-b) + a(y+a) = 2a^2$

$\dfrac{ay}{(b-c)x} = \dfrac{(b+c)^2}{a^2}.$

12. $3x + 5y = \dfrac{(8b - 2m)bm}{b^2 - m^2}$

$b^2 x - \dfrac{bcm^2}{b+m} + (b+c+m)my = m^2 x + (b+2m)bm.$

216. We now proceed to the solution of a particular class of Simultaneous Equations in which the unknown symbols appear as the denominators of fractions, of which the following are examples.

Ex. 1. *To solve the equations,*

$$\dfrac{a}{x} + \dfrac{b}{y} = c$$

$$\dfrac{m}{x} - \dfrac{n}{y} = d.$$

Multiplying the first by m and the second by a, we get

$$\dfrac{am}{x} + \dfrac{bm}{y} = cm$$

$$\dfrac{am}{x} - \dfrac{an}{y} = ad.$$

Subtracting, $\qquad \dfrac{bm}{y} + \dfrac{an}{y} = cm - ad.$

or, $\qquad \dfrac{bm + an}{y} = cm - ad,$

or, $\qquad bm + an = (cm - ad)y,$

$$\therefore y = \dfrac{bm + an}{cm - ad}.$$

Then the value of x may be found by substituting this value of y in one of the original equations, or by making the terms containing y alike, as in the example given in Art. 215.

Ex. 2. *To solve the equations:*

$$\frac{2}{x} - \frac{5}{3y} = \frac{4}{27}$$

$$\frac{1}{4x} + \frac{1}{y} = \frac{11}{72}.$$

Multiplying the second equation by 8, we get

$$\frac{2}{x} - \frac{5}{3y} = \frac{4}{27}$$

$$\frac{2}{x} + \frac{8}{y} = \frac{11}{9}.$$

Subtracting, $\quad -\frac{5}{3y} - \frac{8}{y} = \frac{4}{27} - \frac{11}{9}.$

Changing signs, $\frac{5}{3y} + \frac{8}{y} = \frac{11}{9} - \frac{4}{27},$

or, $\qquad \frac{5+24}{3y} = \frac{33-4}{27},$

whence we find $\qquad y = 9,$

and then the value of x may be found by substituting 9 for y in one of the original equations.

EXAMPLES.—lxxvi.

1. $\dfrac{1}{x} + \dfrac{2}{y} = 10$
   $\dfrac{4}{x} + \dfrac{3}{y} = 20.$

2. $\dfrac{1}{x} + \dfrac{2}{y} = a$
   $\dfrac{3}{x} + \dfrac{4}{y} = b.$

3. $\dfrac{a}{x} + \dfrac{b}{y} = c$
   $\dfrac{b}{x} + \dfrac{a}{y} = d.$

4. $\dfrac{a}{x} + \dfrac{b}{y} = m$
   $\dfrac{a}{x} - \dfrac{b}{y} = n.$

5. $\dfrac{7}{x} + \dfrac{5}{y} = 19$
   $\dfrac{8}{x} - \dfrac{3}{y} = 7.$

6. $\dfrac{5}{3x} + \dfrac{2}{5y} = 7$
   $\dfrac{7}{6x} - \dfrac{1}{10y} = 3.$

7. $\dfrac{2}{ax} + \dfrac{3}{by} = 5$
   $\dfrac{5}{ax} - \dfrac{2}{by} = 3.$

8. $\dfrac{m}{nx} + \dfrac{n}{my} = m + n$
   $\dfrac{n}{x} + \dfrac{m}{y} = m^2 + n^2.$

217. There are two other methods of solving Simultaneous Equations of which we have hitherto made no mention, because they are not generally so convenient and simple as the method which we have explained. They are

I. The method of Substitution.

If we have to solve the equations

$$x + 3y = 7$$
$$2x + 4y = 12$$

we may find the value of x in terms of y from the first equation, thus

$$x = 7 - 3y,$$

and *substitute* this value for x in the second equation, thus

$$2(7 - 3y) + 4y = 12,$$

from which we find

$$y = 1.$$

We may then find the value of x from one of the original equations.

II. The method of Comparison.

If we have to solve the equations

$$5x + 2y = 16$$
$$7x - 3y = 5$$

we may find the values of x in terms of y from each equation, thus

$$x = \frac{16 - 2y}{5}, \text{ from the first equation.}$$

$$x = \frac{5 + 3y}{7}, \text{ from the second equation.}$$

Hence, equating these values of x, we get

$$\frac{16 - 2y}{5} = \frac{5 + 3y}{7},$$

an equation involving only one unknown symbol, from which we obtain

$$y = 3,$$

and then the value of x may be found from one of the original equations.

218. If there be *three* unknown symbols, their values may be found from three independent equations.

For from two of the equations *a third*, which involves only *two* of the unknown symbols, may be found.

And from the remaining equation and one of the others *a fourth*, containing only *the same two* unknown symbols, may be found.

So from these two equations, which involve only two unknown symbols, the value of these symbols may be found, and by substituting these values in one of the original equations the value of the third unknown symbol may be found.

Ex.
$$5x - 6y + 4z = 15$$
$$7x + 4y - 3z = 19$$
$$2x + y + 6z = 46.$$

Multiplying the first by 7 and the second by 5, we get
$$35x - 42y + 28z = 105$$
$$35x + 20y - 15z = 95.$$

Subtracting,
$$-62y + 43z = 10 \ldots\ldots\ldots\ldots\ldots(1).$$

Again, multiplying the first of the original equations by 2 and the third by 5, we get
$$10x - 12y + 8z = 30,$$
$$10x + 5y + 30z = 230.$$

Subtracting, $\quad -17y - 22z = -200 \ldots\ldots\ldots\ldots(2).$

Then, from (1) and (2) we have
$$62y - 43z = -10$$
$$17y + 22z = 200,$$
from which we can find $y = 4$ and $z = 6$.

Then substituting these values for y and z in the first equation we find the value of x to be 3.

EXAMPLES.—lxxvii.

1. $5x + 7y - 2z = 13$
    $8x + 3y + z = 17$
    $x - 4y + 10z = 23.$

2. $5x + 3y - 6z = 4$
    $3x - y + 2z = 8$
    $x - 2y + 2z = 2.$

3. $5x - 3y + 2z = 21$
    $8x - y - 3z = 3$
    $2x + 3y + 2z = 39.$

4. $4x - 5y + 2z = 6$
    $2x + 3y - z = 20$
    $7x - 4y + 3z = 35.$

5. $\quad x+\ y+\ z=\ 6$
$\quad 5x+\ 4y+3z=22$
$\quad 15x+10y+6z=53.$

6. $\quad 8x+4y-3z=6$
$\quad x+3y-\ z=7$
$\quad 4x-5y+4z=8.$

7. $\quad x+\ y+\ z=30$
$\quad 8x+4y+2z=50$
$\quad 27x+9y+3z=64.$

8. $\quad 4x-3y+\ z=\ 9$
$\quad 9x+\ y-5z=16$
$\quad x-4y+3z=\ 2.$

9. $\quad 12x+5y-4z=29$
$\quad 13x-2y+5z=58$
$\quad 17x-\ y-\ z=15.$

10. $\quad y-x+z=-\ 5$
$\quad z-y-x=-25$
$\quad x+y+z=35.$

XVI. PROBLEMS RESULTING IN SIMULTANEOUS EQUATIONS.

219. In the Solution of Problems in which we represent *two* of the numbers sought by unknown symbols, usually x and y, we must obtain two *independent* equations from the conditions of the question, and then we may obtain the values of the two unknown symbols by one of the processes described in Chapter XV.

Ex. If one of two numbers be multiplied by 3 and the other by 4, the sum of the products is 43; and if the former be multiplied by 7 and the latter by 3, the difference between the results is 14. Find the numbers.

Let x and y represent the numbers.

Then $\qquad 3x+4y=43,$
and $\qquad 7x-3y=14.$

From these equations we have
$\qquad 21x+28y=301,$
$\qquad 21x-\ 9y=42.$

Subtracting, $\qquad 37y=259.$

Therefore $\qquad y=7,$
and then the value of x may be found to be 5.

Hence the numbers are 5 and 7.

EXAMPLES.—lxxviii.

1. The sum of two numbers is 28, and their difference is 4, find the numbers.

2. The sum of two numbers is 256, and their difference is 10, find the numbers.

3. The sum of two numbers is 13·5, and their difference is 1, find the numbers.

4. Find two numbers such that the sum of 7 times the greater and 5 times the less may be 332, and the product of their difference into 51 may be 408.

5. Seven years ago the age of a father was four times that of his son, and seven years hence the age of the father will be double that of the son. Find their ages.

6. Find three numbers such that the sum of the first and second shall be 70, of the first and third 80, and of the second and third 90.

7. Three persons A, B, and C make a joint contribution which in the whole amounts to £400. Of this sum B contributes twice as much as A and £20 more; and C as much as A and B together. What sum did each contribute?

8. If A gives B ten shillings, B will have three times as much money as A. If B gives A ten shillings, A will have twice as much money as B. What has each?

9. The sum of £760 is divided between A, B, C. The shares of A and B together exceed the share of C by £240, and the shares of B and C together exceed the share of A by £360. What is the share of each?

10. The sum of two numbers divided by 2, gives as a quotient 24, and the difference between them divided by 2, gives as a quotient 17. What are the numbers?

11. Find two numbers such that when the greater is divided by the less the quotient is 4 and the remainder 3, and when the sum of the two numbers is increased by 38 and the result divided by the greater of the two numbers, the quotient is 2 and the remainder 2.

12. Divide the number 144 into three such parts, that when the first is divided by the second the quotient is 3 and the remainder 2, and when the third is divided by the sum of the other two parts, the quotient is 2 and the remainder 6.

13. *A* and *B* buy a horse for £120. *A* can pay for it if *B* will advance half the money he has in his pocket. *B* can pay for it if *A* will advance two-thirds of the money he has in his pocket. How much has each?

14. "How old are you?" said a son to his father. The father replied, "Twelve years hence you will be as old as I was twelve years ago, and I shall be three times as old as you were twelve years ago." Find the age of each.

15. Required two numbers such that three times the greater exceeds twice the less by 10, and twice the greater together with three times the less is 24.

16. The sum of the ages of a father and son is half what it will be in 25 years. The difference is one-third what the sum will be in 20 years. Find their ages.

17. If I divide the smaller of two numbers by the greater, the quotient is ·21 and the remainder ·0157. If I divide the greater number by the smaller, the quotient is 4 and the remainder ·742. Find the numbers.

18. The cost of 6 barrels of beer and 10 of porter is £51; the cost of 3 barrels of beer and 7 of porter is £32, 2s. How much beer can be bought for £30?

19. The cost of 7 lbs. of tea and 5 lbs. of coffee is £1, 9s. 4d.: the cost of 4 lbs. of tea and 9 lbs. of coffee is £1, 7s.: what is the cost of 1 lb. of each?

20. The cost of 12 horses and 14 cows is £380: the cost of 5 horses and 3 cows is £130: what is the cost of a horse and a cow respectively?

21. The cost of 8 yards of silk and 19 yards of cloth is £18, 4s. 2d.: the cost of 20 yards of silk and 16 yards of cloth, each of the same quality as the former, is £25, 16s. 8d. How much does a yard of each cost?

22. Ten men and six women earn £18, 18s. in 6 days, and four men and eight women earn £6, 6s. in 3 days. What are the earnings of a man and a woman daily?

23. A farmer bought 100 acres of land for £4220, part at £37 an acre and part at £45 an acre. How many acres had he of each kind?

SIMULTANEOUS EQUATIONS.

Note I. A number consisting of two digits may be represented algebraically by $10x+y$, where x and y represent the significant digits.

For consider such a number as 76. Here the significant digits are 7 and 6, of which the former has in consequence of its position a *local* value ten times as great as its *natural* value, and the number represented by 76 is equivalent to *ten times* 7, increased by 6.

So also a number of which x and y are the significant digits will be represented by *ten times* x, increased by y.

If the digits composing a number $10x+y$ be *inverted*, the resulting number will be $10y+x$. Thus if we invert the digits composing the number 76, we get 67, that is, *ten times* 6, increased by 7.

If a number be represented by $10x+y$, the sum of the digits will be represented by $x+y$.

A number consisting of *three* digits may be represented algebraically by
$$100x + 10y + z.$$

Ex. The sum of the digits composing a certain number is 5, and if 9 be added to the number the digits will be inverted. Find the number.

Let $10x+y$ represent the number.
Then $x+y$ will represent the sum of the digits,
and $10y+x$ will represent the number with the digits inverted.
Then our equations will be
$$x+y=5,$$
$$10x+y+9=10y+x,$$
from which we may find $x=2$ and $y=3$;
$\therefore$ 23 is the number required.

24. The sum of two digits composing a number is 8, and if 36 be added to the number the digits will be inverted. Find the number.

25. The sum of the two digits composing a number is 10, and if 54 be added to the number the digits will be inverted. What is the number?

26. The sum of the digits of a number less than 100 is 9, and if 9 be added to the number the digits will be inverted. What is the number?

27. The sum of the two digits composing a number is 6, and if the number be divided by the sum of the digits the quotient is 4. What is the number?

28. The sum of the two digits composing a number is 9, and if the number be divided by the sum of the digits the quotient is 5. What is the number?

29. If I divide a certain number by the sum of the two digits of which it is composed the quotient is 7. If I invert the order of the digits and then divide the resulting number diminished by 12 by the difference of the digits of the original number the quotient is 9. What is the number?

30. If I divide a certain number by the sum of its two digits the quotient is 6 and the remainder 3. If I invert the digits and divide the resulting number by the sum of the digits the quotient is 4 and the remainder 9. Find the number.

31. If I divide a certain number by the sum of its two digits diminished by 2 the quotient is 5 and the remainder 1. If I invert the digits and divide the resulting number by the sum of the digits increased by 2 the quotient is 5 and the remainder 8. Find the number.

32. Two digits which form a number change places on the addition of 9, and the sum of these two numbers is 33. Find the numbers.

33. A number consisting of three digits, the absolute value of each digit being the same, is 37 times the square of any digit. Find the number.

34. Of the three digits composing a number the second is double of the third: the sum of the first and third is 9: the sum of all the digits is 17. Find the number.

35. A number is composed of three digits. The sum of the digits is 21: the sum of the first and second is greater than the third by 3; and if 198 be added to the number the digits will be inverted. Find the number.

NOTE II. A fraction of which the terms are unknown may be represented by $\dfrac{x}{y}$.

Ex. A certain fraction becomes $\dfrac{1}{2}$ when 7 is added to its denominator, and 2 when 13 is added to its numerator. Find the fraction.

Let $\dfrac{x}{y}$ represent the fraction

Then
$$\dfrac{x}{y+7} = \dfrac{1}{2},$$
$$\dfrac{x+13}{y} = 2,$$

are the equations; from which we may find $x=9$ and $y=11$.

That is, the fraction is $\dfrac{9}{11}$.

36. A certain fraction becomes 2 when 7 is added to its numerator, and 1 when 1 is subtracted from its denominator. What is the fraction?

37. Find such a fraction that when 1 is added to its numerator its value becomes $\dfrac{1}{3}$, and when 1 is added to the denominator the value is $\dfrac{1}{4}$.

38. What fraction is that to the numerator of which if 1 be added the value will be $\dfrac{1}{2}$: but if 1 be added to the denominator, the value will be $\dfrac{1}{3}$?

39. The numerator of a fraction is made equal to its denominator by the addition of 1, and is half of the denominator increased by 1. Find the fraction.

40. A certain fraction becomes $\dfrac{1}{4}$ when 3 is taken from the numerator and the denominator, and it becomes $\dfrac{1}{2}$ when 5

is added to the numerator and the denominator. What is the fraction?

41. A certain **fraction** becomes $\frac{7}{9}$ when the denominator is increased by 4, and $\frac{20}{41}$ when the numerator is diminished by 15: determine the fraction.

42. What fraction is that to the numerator of which if 1 be added it becomes $\frac{1}{2}$, and to the denominator of which if 17 be added it becomes $\frac{1}{3}$?

NOTE III. In questions relating to money put out at *simple* interest we are to observe that
$$\text{Interest} = \frac{\text{Principal} \times \text{Rate} \times \text{Time}}{100},$$
where Rate means the number of pounds paid for the use of £100 for one year, and Time means the number of years for which the money is lent.

43. A man puts out £2000 in two investments. For the first he gets 5 per cent., for the second 4 per cent. on the sum invested, and by the first investment he has an income of £10 more than on the second. Find how much he invests in each case.

44. A sum of money, put out at simple interest, amounted in 10 months to £5250, and in 18 months to £5450. What was the sum and the rate of interest?

45. A sum of money, put out at simple interest, amounted in 6 years to £5200, and in 10 years to £6000. Find the sum and the rate of interest.

NOTE IV. When tea, spirits, wine, beer, and such commodities are *mixed*, it must be observed that
quantity of ingredients = quantity of mixture,
cost of ingredients = cost of mixture.

EX. I mix wine which cost 10 shillings a gallon with another sort which cost 6 shillings a gallon, to make 100 gallons, which I may sell at 7 shillings a gallon without profit or loss. How much of each do I take?

Let x represent the number of gallons at 10 shillings a gallon, and y .... 6 ....

Then $\qquad x + y = 100,$
and $\qquad 10x + 6y = 700,$

are the two equations from which we may find the values of x and y to be 25 and 75 respectively.

46. A wine-merchant has two kinds of wine, the one costs 36 pence a quart, the other 20 pence. How much of each must he put in a mixture of 50 quarts, so that the cost price of it may be 30 pence a quart?

47. A grocer mixes tea which cost him 1s. 2d. per lb. with tea that cost him 1s. 6d. per lb. He has 30 lbs. of the mixture, and by selling it at the rate of 1s. 8d. per lb. he gained as much as 10 lbs. of the cheaper tea cost him. How many lbs. of each did he put in the mixture?

NOTE V. If a man can row at the rate of x miles an hour in still water, and if he be rowing on a stream that runs at the rate of y miles an hour, then

$x + y$ will represent his rate *down* the stream,
$x - y$ .... *up* ....

48. A crew which can pull at the rate of twelve miles an hour down the stream, finds that it takes twice as long to come up a river as to go down. At what rate does the stream flow?

49. A man sculls down a stream, which runs at the rate of 4 miles an hour, for a certain distance in 1 hour and 40 minutes. In returning it takes him 4 hours and 15 minutes to arrive at a point 3 miles short of his starting-place. Find the distance he pulled down the stream, and the rate of his pulling.

50. A dog pursues a hare. The hare gets a start of 50 of her own leaps. The hare makes six leaps while the dog makes 5, and 7 of the dog's leaps are equal to 9 of the hare's. How many leaps will the hare take before she is caught?

51. A greyhound starts in pursuit of a hare, at the distance of 50 of his own leaps from her. He makes 3 leaps while the hare makes 4, and he covers as much ground in two leaps as the hare does in three. How many leaps does each make before the hare is caught?

52. I lay out half-a-crown in apples and pears, buying the apples at 4 a penny and the pears at 5 a penny. I then sell half the apples and a third of the pears for thirteen pence, which was the price at which I bought them. How many of each did I buy?

53. A company at a tavern found, when they came to pay their reckoning, that if there had been 3 more persons, each would have paid a shilling less, but had there been 2 less, each would have paid a shilling more. Find the number of the company, and each man's share of the reckoning.

54. At a contested election there are two members to be returned and three candidates, A, B, and C. A obtains 1056 votes, B, 987, C, 933. Now 85 voted for B and C, 744 for B only, 98 for C only. How many voted for A and C, for A and B, and for A only?

55. A man walks a certain distance: had his rate been half a mile an hour faster, he would have been $1\frac{1}{2}$ hours less on the road; and had it been half a mile an hour slower, he would have been $2\frac{1}{2}$ hours more on the road. Find the distance and rate.

56. A certain crew pull 9 strokes to 8 of a certain other crew, but 79 of the latter are equal to 90 of the former. Which is the faster crew?

Also, if the faster crew start at a distance equivalent to four of their own strokes behind the other, how many strokes will they take before they bump them?

57. A person, sculling in a thick fog, meets one barge and overtakes another which is going at the same rate as the former; shew that if a be the greatest distance to which he can see, and b, b' the distances that he sculls between the times of his first seeing and passing the barges,

$$\frac{2}{a} = \frac{1}{b} + \frac{1}{b'}.$$

58. Two trains, 92 feet long and 84 feet long respectively, are moving with uniform velocities on parallel rails in opposite directions, and are observed to pass each other in one second and a half; but when they are moving in the same direction, their velocities being the same as before, the faster train is observed to pass the other in six seconds; find the rate in miles per hour at which each train moves.

59. The fore-wheel of a carriage makes six revolutions more than the hind-wheel in 120 yards; but only four revolutions more when the circumference of the fore-wheel is increased one-fourth, and that of the hind-wheel one-fifth. Find the circumference of each wheel.

60. A person rows from Cambridge to Ely (a distance of 20 miles) and back again in 10 hours, and finds he can row 2 miles against the stream in the same time that he rows 3 miles with it. Find the rate of the stream, and the time of his going and returning.

61. A number consists of 6 digits, of which the last to the left hand is 1. If this number is altered by removing the 1 and putting it in the unit's place, the new number is three times as great as the original one. Find the number.

XVII. ON SQUARE ROOT.

220. In Art. 97 we defined the *Square Root*, and explained the method of taking the square root of expressions consisting of a single term.

The square root of a positive quantity may be, as we explained in Art. 97, either *positive* or *negative*.

Thus the square root of $4a^2$ is $2a$ or $-2a$, and this ambiguity is expressed thus,
$$\sqrt{4a^2} = \pm 2a.$$

In our examples in this chapter we shall in all cases regard the square root of a single term as a *positive* quantity.

221. The square root of a *product* may be found by taking the square root of each factor, and multiplying the roots, so taken, together.

Thus
$$\sqrt{a^2b^2} = ab,$$
$$\sqrt{81x^4y^2z^6} = 9x^2yz^3.$$

222. The square root of a fraction may be found by taking the square root of the numerator and the square root of the denominator, and making them the numerator and denominator of a new fraction, thus

$$\sqrt{\frac{4a^2}{81b^2}} = \frac{2a}{9b},$$

$$\sqrt{\frac{25x^2y^4}{49z^6}} = \frac{5xy^2}{7z^3}.$$

EXAMPLES.—lxxix.

Find the Square Root of each of the following expressions:

1. $4x^2y^2$.
2. $81a^6b^8$.
3. $121m^{10}n^{12}r^{14}$.
4. $64a^4b^{10}c^2$.
5. $71289a^4b^2x^6$.
6. $169a^{16}b^8c^{12}$.
7. $\dfrac{9a^2}{16b^2}$.
8. $\dfrac{1}{4a^2c^4}$.
9. $\dfrac{25a^4b^6}{121x^8y^{10}}$.
10. $\dfrac{256x^{12}}{289y^4}$.
11. $\dfrac{625a^2}{324b^2}$.

223. We may now proceed to investigate a Rule for the extraction of the square root of a compound algebraical expression.

We know that the square of $a + b$ is $a^2 + 2ab + b^2$, and therefore $a + b$ is the square root of $a^2 + 2ab + b^2$.

If we can devise an operation by which we can derive $a + b$ from $a^2 + 2ab + b^2$, we shall be able to give a rule for the extraction of the square root.

Now the first term of the root is the square root of the first term of the square, i.e. a is the square root of a^2.

Hence our rule begins:

"*Arrange the terms in the order of magnitude of the indices of one of the quantities involved, then take the square root of the first term and set down the result as the first term of the root: subtract its square from the given expression, and bring down the remainder:*" thus

$$a^2 + 2ab + b^2 (a$$
$$a^2$$
$$\overline{2ab + b^2}$$

Now this remainder may be represented thus $b(2a+b)$: hence if we divide $2ab + b^2$ by $2a + b$ we shall obtain $+b$, the second term of the root.

Hence our rule proceeds:

"*Double the first term of the root and set down the result as the first term of a divisor:*" thus our process up to this point will stand thus:

$$a^2 + 2ab + b^2 (\boldsymbol{a}$$
$$a^2$$
$$2a \quad | \quad 2ab + b^2$$

Now if we divide $2ab$ by $2a$ the result is b, and hence we obtain the second term of the root, and if we add this to $2a$ we obtain the full divisor $2a + b$.

Hence our rule proceeds thus:

"*Divide the first term of the remainder by this first term of the divisor, and add the result to the first term of the root and also to the first term of the divisor:*" thus our process up to this point will stand thus:

$$a^2 + 2ab + b^2 (a + b$$
$$a^2$$
$$2a + b \quad | \quad 2ab + b^2$$

If now we multiply $2a + b$ by b we obtain $2ab + b^2$, which we subtract from the first remainder.

Hence our rule proceeds thus:

"*Multiply the divisor by the second term of the root and subtract the result from the first remainder:*" thus our process will stand thus:

$$a^2 + 2ab + b^2 \,(\, a + b$$
$$a^2$$

$2a+b$ | $\quad 2ab + b^2$
$ \quad\;\; 2ab + b^2$

If there is now no remainder, the root has been found.

If there be a remainder, consider the two terms of the root already found as one, and proceed as before.

224. The following examples worked out will make the process more clear.

(1) $\qquad a^2 - 2ab + b^2 \,(\, a - b$
$\qquad\qquad a^2$

$2a-b$ | $\quad -2ab + b^2$
$\quad\;\; -2ab + b^2$

Here the second term of the root, and consequently the second term of the divisor, will have a *negative* sign prefixed, because $\dfrac{-2ab}{2a} = -b$.

(2) $\qquad 9p^2 + 24pq + 16q^2 \,(\, 3p + 4q$
$\qquad\qquad 9p^2$

$6p+4q$ | $\quad 24pq + 16q^2$
$\quad\;\; 24pq + 16q^2$

(3) $\qquad 25x^2 - 60x + 36 \,(\, 5x - 6$
$\qquad\qquad 25x^2$

$10x-6$ | $\quad -60x + 36$
$\quad\;\; -60x + 36$

Next take a case in which the root contains *three* terms.

$$a^2 + 2ab + b^2 - 2ac - 2bc + c^2 \,(\, a + b - c$$
$$a^2$$

$2a+b$ | $\quad 2ab + b^2 - 2ac - 2bc + c^2$
$\quad\;\; 2ab + b^2$

$2a+2b-c$ | $\quad\qquad\qquad -2ac - 2bc + c^2$
$\quad\qquad\qquad\;\; -2ac - 2bc + c^2$

When we obtained the *second remainder*, we took the double of $a+b$, considered as a single term, and set down the result as the first part of the *second divisor*. We then divided the first term of the remainder, $-2ac$, by the first term of the new divisor, $2a$, and set down the result, $-c$, attached to the part of the root already found and also to the new divisor, and then multiplied the completed divisor by $-c$.

Similarly we may proceed when the root contains 4, 5 or more terms.

EXAMPLES.—lxxx.

Extract the Square Root of the following expressions:

1. $4a^2 + 12ab + 9b^2$.
2. $16k^{10} - 2k^5l^3 + 9l^6$.
3. $a^2b^2 + 162ab + 6561$.
4. $y^6 - 38y^3 + 361$.
5. $9a^2b^2c^2 - 102abc + 289$.
6. $x^4 - 6x^3 + 19x^2 - 30x + 25$.
7. $9x^4 + 12x^3 + 10x^2 + 4x + 1$.
8. $4r^4 - 12r^3 + 13r^2 - 6r + 1$.
9. $4n^4 + 4n^3 - 7n^2 - 4n + 4$.
10. $1 - 6x + 13x^2 - 12x^3 + 4x^4$.
11. $x^6 - 4x^5 + 10x^4 - 12x^3 + 9x^2$.
12. $4y^4 - 12y^3z + 25y^2z^2 - 24yz^3 + 16z^4$.
13. $a^2 + 4ab + 4b^2 + 9c^2 + 6ac + 12bc$.
14. $a^6 + 2a^5b + 3a^4b^2 + 4a^3b^3 + 3a^2b^4 + 2ab^5 + b^6$.
15. $x^6 - 4x^5 + 6x^3 + 8x^2 + 4x + 1$.
16. $4x^4 + 8ax^3 + 4a^2x^2 + 16b^2x^2 + 16ab^2x + 16b^4$.
17. $9 - 24x + 58x^2 - 116x^3 + 129x^4 - 140x^5 + 100x^6$.
18. $16a^4 - 40a^3b + 25a^2b^2 - 80ab^2x + 64b^2x^2 + 64a^2bx$.
19. $9a^4 - 24a^3p^3 - 30a^2t + 16a^2p^6 + 40ap^3t + 25t^2$.
20. $4y^4x^2 - 12y^3x^3 + 17y^2x^4 - 12yx^5 + 4x^6$.
21. $25x^4y^2 - 30x^3y^3 + 29x^2y^4 - 12xy^5 + 4y^6$.
22. $16x^4 - 24x^3y + 25x^2y^2 - 12xy^3 + 4y^4$.
23. $9a^2 - 12ab + 24ac - 16bc + 4b^2 + 16c^2$.
24. $x^4 + 9x^2 + 25 - 6x^3 + 10x^2 - 30x$.
25. $25x^2 - 20xy + 4y^2 + 9z^2 - 12yz + 30xz$.
26. $4x^2(x^2-y) + y^3(y-2) + y^2(4x^2+1)$.

225. When any *fractional* terms are in the expression of which we have to find the Square Root, we may proceed as in the Examples just given, taking care to treat the fractional terms in accordance with the rules relating to fractions.

Thus to find the square root of $x^2 - \frac{8}{9}x + \frac{16}{81}$.

$$x^2 - \frac{8}{9}x + \frac{16}{81} \left(\, x - \frac{4}{9}\right.$$

$$\underline{x^2}$$

$$2x - \frac{4}{9} \quad \left|\; \begin{array}{l} -\frac{8}{9}x + \frac{16}{81} \\ -\frac{8}{9}x + \frac{16}{81} \end{array} \right.$$

Since $\quad \dfrac{8}{9} \div 2 = \dfrac{8}{9} \div \dfrac{2}{1} = \dfrac{8}{9} \times \dfrac{1}{2} = \dfrac{4}{9}.$

Or we might reduce $x^2 - \frac{8}{9}x + \frac{16}{81}$ to a single fraction, which would be $\dfrac{81x^2 - 72x + 16}{81}$,

and then take the square root of each of the terms of the fraction, with the followin result:

$\dfrac{9x-4}{9}$, which is the same as $x - \dfrac{4}{9}.$

EXAMPLES.—lxxxi.

1. $4a^6 + \dfrac{a^2 b^4}{16} - a^4 b^2.$

2. $\dfrac{9}{a^2} - 2 + \dfrac{a^2}{9}.$

3. $a^4 - 2 + \dfrac{1}{a^4}.$

4. $\dfrac{a^2}{b^2} + 2 + \dfrac{b^2}{a^2}.$

5. $x^4 - 2x^3 + 2x^2 - x + \dfrac{1}{4}.$

6. $x^4 + 2x^3 - x + \dfrac{1}{4}.$

7. $4a^2 - 12ab + ab^2 + 9b^2 - \dfrac{3b^3}{2} + \dfrac{b^4}{16}.$

8. $x^4 + 8x^2 + 24 + \dfrac{16}{x^4} + \dfrac{32}{x^2}$.

9. $\dfrac{9}{16} + 4a^4 + \dfrac{16}{9}a^6x^2 - 3a^2 - 2a^3x + \dfrac{16}{3}a^5x$.

10. $\dfrac{1}{x^2} + \dfrac{4}{y^2} + \dfrac{9}{z^2} - \dfrac{4}{xy} + \dfrac{6}{xz} - \dfrac{12}{yz}$.

11. $36m^2 - \dfrac{48m}{n} + \dfrac{12mp}{5} + \dfrac{16}{n^2} + \dfrac{p^2}{25} - \dfrac{8p}{5n}$.

12. $a^2b^2 - 6abcd + \dfrac{2abef}{7} + 9c^2d^2 + \dfrac{e^2f^2}{49} - \dfrac{6cdef}{7}$.

13. $\dfrac{4x^2}{z^2} + \dfrac{z^2}{x^2} + \dfrac{9y^2}{z^2} + 4 - \dfrac{6y}{x} - \dfrac{12xy}{z^2}$.

14. $\dfrac{4m^2}{n^2} + \dfrac{9n^2}{m^2} + 4 - \dfrac{16m}{n} + \dfrac{24n}{m}$.

15. $\dfrac{a^2}{9} + \dfrac{b^2}{16} + \dfrac{c^2}{25} + \dfrac{d^2}{4} - \dfrac{ab}{6} + \dfrac{2ac}{15} - \dfrac{ad}{3} - \dfrac{bc}{10} + \dfrac{bd}{4} - \dfrac{cd}{5}$.

16. $49x^4 - 28x^3 - 17x^2 + 6x + \dfrac{9}{4}$.

17. $9x^4 - 3ax^3 + 6bx^3 + \dfrac{a^2x^2}{4} - abx^2 + b^2x^2$.

18. $9x^4 - 2x^3 - \dfrac{161}{9}x^2 + 2x + 9$.

XVIII. ON CUBE ROOT.

226. THE CUBE ROOT of any expression is that expression whose *cube* or third power gives the proposed expression.

Thus a is the cube root of a^3,
$3b$ is the cube root of $27b^3$.

The cube root of a negative expression will be negative, for since
$$(-a)^3 = -a \times -a \times -a = -a^3,$$
the cube root of $-a^3$ is $-a$.

So also

$$-3x \text{ is the cube root of } -27x^3,$$
and $\quad -4a^2b \text{ is the cube root of } -64a^6b^3.$

The symbol $\sqrt[3]{}$ is used to denote the operation of extracting the cube root.

EXAMPLES.—lxxxii.

Find the Cube Roots of the following expressions:

1. $8a^3$. 2. $27x^6y^6$. 3. $-125m^3n^3$.
4. $-216a^{12}b^3$. 5. $343b^{15}c^{18}$. 6. $-1000a^3b^6c^{12}$.
7. $-1728m^{21}n^{24}$. 8. $1331a^9b^{18}$.

227. We now proceed to investigate a Rule for finding the cube root of a compound algebraical expression.

We know that the cube of $a+b$ is $a^3 + 3a^2b + 3ab^2 + b^3$, and therefore $a+b$ is the cube root of $a^3 + 3a^2b + 3ab^2 + b^3$.

We observe that the first term of the root is the cube root of the first term of the cube.

Hence our rule begins:

"*Arrange the terms in the order of magnitude of the indices of one of the quantities involved, then take the cube root of the first term and set down the result as the first term of the root: subtract its cube from the given expression, and bring down the remainder;*" thus

$$a^3 + 3a^2b + 3ab^2 + b^3 \; (a$$
$$\underline{a^3 }$$
$$ 3a^2b + 3ab^2 + b^3$$

Now this remainder may be represented thus,
$$b(3a^2 + 3ab + b^2);$$
hence if we divide $3a^2b + 3ab^2 + b^3$ by $3a^2 + 3ab + b^2$, we shall obtain $+b$, the second term of the root.

Hence our rule proceeds:

"*Multiply the square of the first term of the root by 3, and set down the result as the first term of a divisor;*" thus our process up to this point will stand thus :

$$a^3 + 3a^2b + 3ab^2 + b^3 \,(a$$
$$a^3$$

$3a^2 \quad | \quad 3a^2b + 3ab^2 + b^3$

Now if we divide $3a^2b$ by $3a^2$ the result is b, and so we obtain the second term of the root, and if we add to $3a^2$ the expression $3ab + b^2$ we obtain the full divisor $3a^2 + 3ab + b^2$.

Hence our rule proceeds thus:

"*Divide the first term of the remainder by the first term of the divisor, and add the result to the first term of the root. Then take three times the product of the first and second terms of the root, and also the square of the second term, and add these results to the first term of the divisor.*" Thus our process up to this point will stand thus:

$$a^3 + 3a^2b + 3ab^2 + b^3 \,(a + b$$
$$a^3$$

$3a^2 + 3ab + b^2 \quad | \quad 3a^2b + 3ab^2 + b^3$

If we now multiply the divisor by b, we obtain
$$3a^2b + 3ab^2 + b^3,$$
which we subtract from the first remainder.

Hence our rule proceeds thus:

"*Multiply the divisor by the second term of the root, and subtract the result from the first remainder:*" thus our process will stand thus:

$$a^3 + 3a^2b + 3ab^2 + b^3 \,(a + b$$
$$a^3$$

$3a^2 + 3ab + b^2 \quad | \quad 3a^2b + 3ab^2 + b^3$
$ 3a^2b + 3ab^2 + b^3$

If there is now no remainder, the root has been found.

If there be a remainder, consider the two terms of the root already found as one, and proceed as before.

228. The following Examples may render the process more clear:

Ex. 1.

$$a^3 - 12a^2 + 48a - 64\,(\,a - 4$$
$$a^3$$

$3a^2 - 12a + 16$ | $-12a^2 + 48a - 64$
| $-12a^2 + 48a - 64$

Here observe that the second term of the divisor is formed thus:

3 times the product of a and $-4 = 3 \times a \times -4 = -12a$.

Ex. 2. $\quad x^6 - 6x^5 + 15x^4 - 20x^3 + 15x^2 - 6x + 1\,(\,x^2 - 2x + 1$
$\qquad\qquad x^6$

$3x^4 - 6x^3 + 4x^2$ | $-6x^5 + 15x^4 - 20x^3 + 15x^2 - 6x + 1$
| $-6x^5 + 12x^4 - 8x^3$

$3x^4 - 12x^3$ | $3x^4 - 12x^3 + 15x^2 - 6x + 1$
$+ 15x^2 - 6x + 1$ | $3x^4 - 12x^3 + 15x^2 - 6x + 1$

Here the formation of the first divisor is similar to that in the preceding Examples.

The formation of the second divisor may be explained thus:

Regarding $x^2 - 2x$ as one term

$3\,(x^2 - 2x)^2 = 3\,(x^4 - 4x^3 + 4x^2) = 3x^4 - 12x^3 + 12x^2$
$3 \times (x^2 - 2x) \times 1 \qquad\qquad\qquad = \qquad\qquad\qquad 3x^2 - 6x$
$1^2 \qquad\qquad\qquad\qquad\qquad = \qquad\qquad\qquad\qquad\qquad 1$

and adding these results we obtain as the second divisor

$$3x^4 - 12x^3 + 15x^2 - 6x + 1.$$

EXAMPLES.—lxxxiii.

Find the Cube Root of each of the following expressions:

1. $a^3 - 3a^2b + 3ab^2 - b^3$.   2. $8a^3 + 12a^2 + 6a + 1$.
3. $a^3 + 24a^2b + 192ab^2 + 512b^3$.
4. $a^3 + 3a^2b + 3ab^2 + b^3 + 3a^2c + 6abc + 3b^2c + 3ac^2 + 3bc^2 + c^3$.
5. $x^3 - 3x^2y + 3xy^2 - y^3 + 3x^2z - 6xyz + 3y^2z + 3xz^2 - 3yz^2 + z^3$.
6. $27x^6 - 54x^5 + 63x^4 - 44x^3 + 21x^2 - 6x + 1$.

7. $1 - 3a + 6a^2 - 7a^3 + 6a^4 - 3a^5 + a^6$.

8. $x^3 - 3x^2y + 3xy^2 - y^3 + 8z^3 + 6x^2z - 12xyz + 6y^2z + 12xz^2 - 12yz^2$.

9. $a^6 - 12a^5 + 54a^4 - 112a^3 + 108a^2 - 48a + 8$.

10. $8m^6 - 36m^5 + 66m^4 - 63m^3 + 33m^2 - 9m + 1$.

11. $x^3 + 6x^2y + 12xy^2 + 8y^3 - 3x^2z - 12xyz - 12y^2z + 3xz^2 + 6yz^2 - z^3$.

12. $8m^3 - 36m^2n + 54mn^2 - 27n^3 - 12m^2r + 36mnr - 27n^2r + 6mr^2 - 9nr^2 - r^3$.

13. $m^3 + 3m^2 - 5 + \dfrac{3}{m^2} - \dfrac{1}{m^3}$.

229. The *fourth* root of an expression is found by taking the square root of the square root of the expression.

Thus $\sqrt[4]{16a^8b^4} = \sqrt{4a^4b^2} = 2a^2b$.

The *sixth* root of an expression is found by taking the cube root of the square root of the expression.

Thus $\sqrt[6]{64a^{12}b^6} = \sqrt[3]{8a^6b^3} = 2a^2b$.

EXAMPLES.—lxxxiv.

Find the fourth roots of

1. $16a^4 - 96a^3x + 216a^2x^2 - 216ax^3 + 81x^4$.

2. $1 + 24a^2 + 16a^4 - 8a - 32a^3$.

3. $625 + 2000x + 2400x^2 + 1280x^3 + 256x^4$.

Find the sixth roots of

4. $a^6 - 6a^5b + 15a^4b^2 - 20a^3b^3 + 15a^2b^4 - 6ab^5 + b^6$.

5. $x^6 + 6x^5 + 15x^4 + 20x^3 + 15x^2 + 6x + 1$.

6. $m^6 - 12m^5 + 60m^4 - 160m^3 + 240m^2 - 192m + 64$.

XIX. QUADRATIC EQUATIONS.

230. A QUADRATIC EQUATION, or an equation of *two* dimensions, is one into which the *square* of an unknown symbol enters, *without* or *with* the first power of the symbol.

Thus $\qquad x^2 = 16$
and $\qquad x^2 + 6x = 27$
are Quadratic Equations.

231. A PURE Quadratic Equation is one into which the square of an unknown symbol enters, the first power of the symbol not appearing.

Thus, $x^2 = 16$ is a *pure* Quadratic Equation.

232. An ADFECTED Quadratic Equation is one into which the square of an unknown symbol enters, and also the first power of the symbol.

Thus, $x^2 + 6x = 27$ is an *adfected* Quadratic Equation.

Pure Quadratic Equations.

233. When the terms of an equation involve the square of the unknown symbol *only*, the value of this square is either given or can be found by the processes described in Chapter XVII. If we then extract the square root of each side of the equation, the value of the unknown symbol will be determined.

234. The following are examples of the solution of Pure Quadratic Equations.

Ex. 1. $x^2 = 16$.

Taking the square root of each side

$$x = \pm 4.$$

We prefix the sign $\pm$ to the number on the right-hand side of the equation, for the reason given in Art. 220.

Every pure quadratic equation will therefore have *two roots*, equal in magnitude, but with different signs

Ex. 2. $4x^2 + 6 = 22$.

Here
$$4x^2 = 22 - 6,$$
$$\text{or} \quad 4x^2 = 16,$$
$$\text{or} \quad x^2 = 4;$$
$$\therefore x = \pm 2.$$

That is, the values of x which satisfy the equation are 2 and -2.

Ex. 3. $\dfrac{128}{3x^2 - 4} = \dfrac{216}{5x^2 - 6}$.

Here
$$128(5x^2 - 6) = 216(3x^2 - 4),$$
$$\text{or} \quad 640x^2 - 768 = 648x^2 - 864,$$
$$\text{or} \quad x^2 = 12;$$
$$\therefore x = \pm \sqrt{12}.$$

EXAMPLES.—lxxxv.

1. $x^2 = 64$.
2. $x^2 = a^2 b^2$.
3. $x^2 - 10000 = 0$.
4. $x^2 - 3 = 46$.
5. $5x^2 - 9 = 2x^2 + 24$.
6. $3ax^2 = 192a^3c^6$.
7. $\dfrac{x^2 - 12}{3} = \dfrac{x^2 - 4}{4}$.
8. $(500 + x)(500 - x) = 233359$.
9. $\dfrac{8112}{x} = 3x$.
10. $5\dfrac{1}{2}x^2 - 18x + 65 = (3x - 3)^2$.
11. $mx^2 + n = q$.
12. $x^2 - ax + b = ax(x - 1)$.
13. $\dfrac{45}{2x^2 + 3} = \dfrac{57}{4x^2 - 5}$.
14. $\dfrac{42}{x^2 - 2} = \dfrac{35}{x^2 - 3}$.

Adfected Quadratic Equations.

235. Adfected Quadratic Equations are solved by adding a certain term to both sides of the equation so as to make the left-hand side a perfect square.

Having arranged the equation so that the first term on the left-hand side is the square of the unknown symbol, and the second term the one containing the first power of the unknown quantity (the known symbols being on the right of the equation), we *add to both sides of the equation the square of half the coefficient of the second term*. The left-hand side of the equation then becomes *a perfect square*. If we then take the square root of both sides of the equation, we shall obtain *two simple equations*, from which the values of the **unknown symbol** may be determined.

236. The process in the solution of Adfected Quadratic Equations will be learnt by the examples which we shall give in this chapter, but before we proceed to them, it is desirable that the student should be satisfied as to the way in which an expression of the form

$$x^2 + ax$$

is made a perfect square.

Our rule, as given in the preceding Article, is this : add the square of half the coefficient of the second term, that is, the square of $\dfrac{a}{2}$, that is, $\dfrac{a^2}{4}$. We have to shew then that

$$x^2 + ax + \dfrac{a^2}{4}$$

is a perfect square, whatever a may be.

This we may do by actually performing the operation of extracting the square root of $x^2 + ax + \dfrac{a^2}{4}$, and obtaining the result $x + \dfrac{a}{2}$ with no remainder.

QUADRATIC EQUATIONS.

237. Let us examine this process by the aid of *numerical* coefficients.

Take one or two examples from the perfect squares given in page 48.

We there have

$x^2 + 18x + 81$ which is the square of $x + 9$,
$x^2 + 34x + 289$ .... $x + 17$.
$x^2 - 8x + 16$ .... $x - 4$,
$x^2 - 36x + 324$ .... $x - 18$.

In all these cases the third term is *the square of half the coefficient of x.*

For
$$81 = (9)^2 = \left(\frac{18}{2}\right)^2,$$

$$289 = (17)^2 = \left(\frac{34}{2}\right)^2,$$

$$16 = (4)^2 = \left(\frac{8}{2}\right)^2,$$

$$324 = (18)^2 = \left(\frac{36}{2}\right)^2.$$

238. Now put the question in this shape. What must we add to $x^2 + ax$ to make it a perfect square?

Suppose b to represent the quantity to be added.

Then $x^2 + ax + b$ is a perfect square.

Now if we perform the operation of extracting the square root of $x^2 + ax + b$, our process is

$$
\begin{array}{r|l}
& x^2 + ax + b \;\big(\; x + \dfrac{a}{2} \\
& x^2 \\ \hline
2x + \dfrac{a}{2} & ax + b \\
& ax + \dfrac{a^2}{4} \\ \hline
& b - \dfrac{a^2}{4}
\end{array}
$$

Hence in order that $x^2 + ax + b$ may be a perfect square we must have

$$b - \frac{a^2}{4} = 0,$$

or
$$b = \frac{a^2}{4},$$

or
$$b = \left(\frac{a}{2}\right)^2.$$

That is, b is equivalent to *the square of half the coefficient of x.*

239. Before completing the square we must be careful

(1) That the square of the unknown symbol *has no coefficient but unity,*

(2) That the square of the unknown symbol *has a positive* sign.

These points will be more fully considered in Arts. 245 and 246.

240. We shall first take the case in which the coefficient of the second term is an *even* number and its sign *positive*.

Ex. $\qquad x^2 + 6x = 40.$

Here we make the left-hand side of the equation a perfect square by the following process.

Take the coefficient of the second term, that is, 6.

Take the half of this coefficient, that is, 3.

Square the result, which gives 9.

Add 9 to both sides of the equation, and we get

$$x^2 + 6x + 9 = 49.$$

Now taking the square root of both sides, we get

$$x + 3 = \pm 7.$$

QUADRATIC EQUATIONS.

Hence we have two simple equations,

$$x+3 = +7 \quad \ldots\ldots\ldots\ldots (1),$$
and
$$x+3 = -7 \quad \ldots\ldots\ldots\ldots (2).$$

From these we find the values of x, thus:

from (1) $x = 7-3$, that is, $x = 4$,
from (2) $x = -7-3$, that is, $x = -10$.

Thus the roots of the equation are 4 and -10.

EXAMPLES.—lxxxvi.

1. $x^2 + 6x = 72$.   2. $x^2 + 12x = 64$.   3. $x^2 + 14x = 15$.
4. $x^2 + 46x = 96$.   5. $x^2 + 128x = 393$.   6. $x^2 + 8x - 65 = 0$
7. $x^2 + 18x - 243 = 0$.   8. $x^2 + 16x - 420 = 0$.

241. We next take the case in which the coefficient of the second term is an even number and its sign *negative*.

Ex. $x^2 - 8x = 9$.

The term to be added to both sides is $(8 \div 2)^2$, that is, $(4)^2$, that is, 16.

Completing the square
$$x^2 - 8x + 16 = 25.$$

Taking the square root of both sides
$$x - 4 = \pm 5.$$

This gives two simple equations,

$$x - 4 = +5 \quad \ldots\ldots\ldots\ldots (1),$$
$$x - 4 = -5 \quad \ldots\ldots\ldots\ldots (2),$$

From (1) $x = 5 + 4, \quad \therefore x = 9;$
from (2) $x = -5 + 4, \quad \therefore x = -1.$

Thus the roots of the equation are 9 and -1.

EXAMPLES.—lxxxvii.

1. $x^2 - 6x = 7$. 2. $x^2 - 4x = 5$. 3. $x^2 - 20x = 21$.
4. $x^2 - 2x = 63$. 5. $x^2 - 12x + 32 = 0$. 6. $x^2 - 14x + 45 = 0$.
7. $x^2 - 234x + 13688 = 0$. 8. $(x-3)(x-2) = 3(5x+14)$.
9. $x(3x-17) - x(2x+5) + 120 = 0$.
10. $(x-5)^2 + (x-7)^2 = x(x-8) + 46$.

242. We now take the case in which the coefficient of the second term is an *odd* number.

Ex. 1. $\qquad x^2 - 7x = 8$.

The term to be added to both sides is
$$(7 \div 2)^2 = \left(\frac{7}{2}\right)^2 = \frac{49}{4}.$$

Completing the square
$$x^2 - 7x + \frac{49}{4} = 8 + \frac{49}{4},$$
$$\text{or,} \quad x^2 - 7x + \frac{49}{4} = \frac{81}{4}.$$

Taking the square root of both sides
$$x - \frac{7}{2} = \pm \frac{9}{2}.$$

This gives two simple equations,
$$x - \frac{7}{2} = + \frac{9}{2} \quad \ldots\ldots\ldots\ldots (1),$$
$$x - \frac{7}{2} = - \frac{9}{2} \quad \ldots\ldots\ldots\ldots (2).$$

From (1) $x = \frac{9}{2} + \frac{7}{2}$, or, $x = \frac{16}{2}$, $\therefore x = 8$;

from (2) $x = -\frac{9}{2} + \frac{7}{2}$, or, $x = \frac{-2}{2}$, $\therefore x = -1$.

Thus the roots of the equation are 8 and -1.

Ex. 2. $\qquad x^2 - x = 42.$

The coefficient of the second term is 1.
The term to be added to both sides is

$$(1 \div 2)^2 = \left(\frac{1}{2}\right)^2 = \frac{1}{4};$$

$$\therefore x^2 - x + \frac{1}{4} = 42 + \frac{1}{4}$$

$$\text{or, } x^2 - x + \frac{1}{4} = \frac{169}{4};$$

$$\therefore x - \frac{1}{2} = \pm \frac{13}{2}.$$

Hence the roots of the equation are 7 and -6.

EXAMPLES.—lxxxviii.

1. $x^2 + 7x = 30.$
2. $x^2 - 11x = 12.$
3. $x^2 + 9x = 43\frac{3}{4}.$
4. $x^2 - 13x = 140.$
5. $x^2 + x = \frac{5}{16}.$
6. $x^2 - x = 72.$
7. $x^2 + 37x = 3690.$
8. $x^2 = 56 + x.$
9. $x(5-x) \div 2x(x-7) - 10(x-6) = 0.$
10. $(5x-21)(7x-33) - (17x+15)(2x-3) = 448.$

243. Our next case is that in which the coefficient of the second term is a fraction *of which the numerator is an even number.*

Ex. $\qquad x^2 - \frac{4}{5}x = 21.$

The term to be added to both sides is

$$\left(\frac{4}{5} \div 2\right)^2 = \left(\frac{4}{5} \times \frac{1}{2}\right)^2 = \left(\frac{2}{5}\right)^2 = \left(\frac{4}{25}\right);$$

$$\therefore x^2 - \frac{4}{5}x + \frac{4}{25} = 21 + \frac{4}{25},$$

$$\text{or, } x^2 - \frac{4}{5}x + \frac{4}{25} = \frac{529}{25};$$

$$\therefore x - \frac{2}{5} = \pm \frac{23}{5}.$$

Hence the values of x are 5 and $-\frac{21}{5}$.

EXAMPLES.—lxxxix.

1. $x^2 - \frac{2}{3}x = \frac{35}{9}.$ 2. $x^2 + \frac{4}{5}x = -\frac{3}{25}.$ 3. $x^2 - \frac{28x}{9} + \frac{1}{3} = 0.$

4. $x^2 - \frac{8}{11}x - \frac{3}{11} = 0.$ 5. $x^2 + \frac{4}{35}x = \frac{3}{7}.$ 6. $x^2 - \frac{16}{5}x = \frac{16}{5}.$

7. $x^2 - \frac{26}{2}x + \frac{16}{3} = 0.$ 8. $x^2 - \frac{4}{7}x = 45.$

244. We now take the case in which the coefficient of the second term is a fraction *whose numerator is an odd number*.

Ex. $\qquad x^2 - \frac{7}{3}x = -\frac{136}{3}.$

The term to be added to both sides is

$$\left(\frac{7}{3} \div 2\right)^2 = \left(\frac{7}{3} \times \frac{1}{2}\right)^2 = \left(\frac{7}{6}\right)^2 = \frac{49}{36};$$

$$\therefore x^2 - \frac{7}{3}x + \frac{49}{36} = -\frac{136}{3} + \frac{49}{36},$$

or $\qquad x^2 - \frac{7}{3}x + \frac{49}{36} = -\frac{1681}{36};$

$$\therefore x - \frac{7}{6} = \pm \frac{41}{6}.$$

Hence the values of x are 8 and $-\frac{17}{3}.$

EXAMPLES.—xc.

1. $x^2 - \frac{1}{3}x = 8.$ 2. $x^2 - \frac{1}{5}x = 98.$ 3. $x^2 + \frac{1}{2}x = 39.$

4. $x^2 + \frac{3}{2}x = 76.$ 5. $x^2 - \frac{9}{5}x = 16.$ 6. $x^2 - \frac{11}{2}x + 6 = 0.$

7. $x^2 - \frac{15}{4}x - 34 = 0.$ 8. $x^2 - \frac{23}{7}x = \frac{3}{4}.$

245. The square of the unknown symbol *must not be preceded by a negative sign.*

Hence, if we have to solve the equation
$$6x - x^2 = 9,$$
we change the sign of every term, and we get
$$x^2 - 6x = -9.$$

Completing the square
$$x^2 - 6x + 9 = 9 - 9,$$
$$\text{or} \quad x^2 - 6x + 9 = 0.$$
Hence
$$x - 3 = 0,$$
$$\text{or} \quad x = 3.$$

NOTE. We are not to be surprised at finding only *one* value for x. The interpretation to be placed on such a result is, that the two roots of the equation are equal in value and alike in sign.

246. The square of the unknown symbol must have *no coefficient but unity.*

Hence, if we have to solve the equation
$$5x^2 - 3x = 2,$$
we must divide all the terms by 5, and we get
$$x^2 - \frac{3x}{5} = \frac{2}{5}.$$
From which we get $x = 1$ and $x = -\frac{2}{5}$.

247. In solving Quadratic Equations involving *literal* coefficients of the unknown symbol, the same rules will apply as in the cases of numerical coefficients.

Thus, to solve the equation
$$\frac{2a}{x} - \frac{x}{a} - 2 = 0.$$

Clearing the equation of fractions, we get
$$2a^2 - x^2 - 2ax = 0;$$
therefore
$$-x^2 - 2ax = -2a^2,$$
$$\text{or} \quad x^2 + 2ax = 2a^2.$$

Completing the square
$$x^2 + 2ax + a^2 = 3a^2,$$
whence $\quad x + a = \pm \sqrt{3}.a\,;$
therefore $\quad x = -a + \sqrt{3}.a,$ or $x = -a - \sqrt{3}.a.$

The following are Examples of *Literal Quadratic Equations.*

EXAMPLES.—xci.

1. $x^2 + 2ax = a^2.$
2. $x^2 - 4ax = 7a^2.$
3. $x^2 + 3mx = \dfrac{7m^2}{4}.$
4. $x^2 - \dfrac{5n}{2}x = \dfrac{3n^2}{2}.$
5. $x^2 + (a-1)x = a.$
6. $x^2 + (a-b)x = ab.$
7. $\dfrac{a^2}{(x+a)^2} - \dfrac{b^2}{(x-a)^2} = 0.$
8. $adx - acx^2 = bcx - bd.$
9. $cx + \dfrac{ac}{a+b} = (a+b)x^2.$
10. $\dfrac{a^2x^2}{b^2} - \dfrac{2ax}{c} + \dfrac{b^2}{c^2} = 0.$
11. $abx^2 + \dfrac{3a^2x}{c} = \dfrac{6a^2 + ab - 2b^2}{c^2} - \dfrac{b^2x}{c}.$
12. $(4a^2 - 9cd^2)x^2 + (4a^2c^2 + 4abd^2)x + (ac^2 + bd^2)^2 = 0.$

248. If both sides of an equation can be divided by the unknown symbol, divide by it, and observe that 0 is in that case one root of the equation.

Thus in solving the equation
$$x^3 - 2x^2 = 3x,$$
we may divide by x, and reduce the equation to the form
$$x^2 - 2x = 3,$$
from which we get
$$x = 3 \quad \text{or} \quad x = -1.$$

Then the *three* roots of the original equation are 0, 3 and -1.

We shall now give some Miscellaneous Examples of Quadratic Equations.

EXAMPLES.—xcii.

1. $x^2 - 7x + 2 = 10.$
2. $x^2 - 5x + 3 = 9.$
3. $x^2 - 11x - 7 = 5.$
4. $x^2 - 13x - 6 = 8.$
5. $x^2 + 7x - 18 = 0.$
6. $4x - \dfrac{12-x}{x-3} = 22.$
7. $x^2 - 9x + 20 = 0.$
8. $5x - 3\dfrac{x-1}{x-3} = \dfrac{7x-6}{2}.$
9. $x^2 - 6x - 14 = 2.$
10. $\dfrac{4x}{x+3} - \dfrac{x-3}{2x+5} = 2.$
11. $\dfrac{4x}{x+7} - \dfrac{x-7}{2x+3} = 2.$
12. $x^2 - 12 = 11x.$
13. $x^2 - 14 = 13x.$
14. $\dfrac{1}{2}x^2 - \dfrac{1}{3}x + 7\dfrac{3}{8} = 8.$
15. $3x - \dfrac{169}{x} = 26.$
16. $2x^2 = 18x - 40.$
17. $\dfrac{4+3x}{10} - \dfrac{15-x}{x-6} = \dfrac{7x-14}{20}.$
18. $3x^2 = 24x - 36.$
19. $\dfrac{3x-5}{9x} - \dfrac{6x}{3x-25} = \dfrac{1}{3}.$
20. $\dfrac{7}{4} - \dfrac{2x-5}{x+5} = \dfrac{3x-7}{2x}.$
21. $\dfrac{4x-10}{x+5} - \dfrac{7-3x}{x} = \dfrac{7}{2}.$
22. $(x-3)^2 + 4x = 44.$
23. $\dfrac{x+11}{x} = 7 - \dfrac{9+4x}{x^2}.$
24. $6x^2 + x = 2.$
25. $x^2 - \dfrac{1}{2}x = \dfrac{1}{9}.$
26. $x^2 - x = 210.$
27. $\dfrac{6}{x+1} + \dfrac{2}{x} = 3.$
28. $\dfrac{4x^2}{3} - 11 = \dfrac{x}{3}.$
29. $\dfrac{x}{x-1} = \dfrac{3}{2} + \dfrac{x-1}{x}.$
30. $15x^2 - 7x = 46.$
31. $\dfrac{1}{x-2} - \dfrac{2}{x+2} = \dfrac{3}{5}.$
32. $\dfrac{4x}{5-x} - \dfrac{20-4x}{x} = 15.$
33. $\dfrac{10}{x} - \dfrac{14-2x}{x^2} = \dfrac{22}{9}.$
34. $\dfrac{x}{x+60} = \dfrac{7}{3x-5}.$
35. $\dfrac{12}{5-x} + \dfrac{8}{4-x} = \dfrac{32}{x+2}.$
36. $\dfrac{x}{7-x} + \dfrac{7-x}{x} = 2\dfrac{9}{10}.$
37. $x^2 + (a+b)x + ab = 0.$
38. $x^2 - (b-a)x - ab = 0.$
39. $x^2 - 2ax + a^2 - b^2 = 0.$
40. $x^2 - (a^2 - a^3)x - a^5 = 0.$
41. $x^2 + \dfrac{a}{b}x - \dfrac{2a^2}{b^2} = 0.$
42. $x^2 - \dfrac{a^2+b^2}{ab}x + 1 = 0.$

XX. ON SIMULTANEOUS EQUATIONS INVOLVING QUADRATICS.

249. For the solution of Simultaneous Equations of a degree higher than the first no fixed rules can be laid down. We shall point out the methods of solution which may be adopted with advantage in particular cases.

250. If the *simple power* of one of the unknown symbols can be expressed in terms of the other symbol by means of one of the given equations, the Method of Substitution, explained in Art. 217, may be employed, thus:

Ex. To solve the equations

$$x + y = 50$$
$$xy = 600.$$

From the first equation

$$x = 50 - y.$$

Substitute this value for x in the second equation, and we get
$$(50 - y) \cdot y = 600.$$

This gives $50y - y^2 = 600.$

From which we find the values of y to be 30 and 20.

And we may then find the corresponding values of x to be 20 and 30.

251. But it is better that the student should accustom himself to work such equations *symmetrically*, thus:

To solve the equations

$$x + y = 50 \quad \ldots\ldots\ldots\ldots\ldots\ldots(1),$$
$$xy = 600 \quad \ldots\ldots\ldots\ldots\ldots\ldots(2).$$

From (1) $\quad x^2 + 2xy + y^2 = 2500.$
From (2) $\quad 4xy = 2400.$

Subtracting, $x^2 - 2xy + y^2 = 100$,

$$\therefore x - y = \pm 10.$$

Then from this equation and (1) we find

$$x = 30 \text{ or } 20 \text{ and } y = 20 \text{ or } 30.$$

EXAMPLES.—xciii.

1. $x + y = 40$
   $xy = 300$.

2. $x + y = 13$
   $xy = 36$.

3. $x + y = 29$
   $xy = 100$.

4. $x - y = 19$
   $xy = 66$.

5. $x - y = 45$
   $xy = 250$.

6. $x - y = 99$
   $xy = 100$.

252. To solve the equations

$$x - y = 12 \dots\dots\dots\dots(1),$$
$$x^2 + y^2 = 74 \dots\dots\dots\dots(2).$$

From (1) $\quad x^2 - 2xy + y^2 = 144 \dots\dots\dots\dots(3).$

Subtract this from (2), then

$$2xy = -70,$$
$$\therefore 4xy = -140.$$

Add this to (3), then

$$x^2 + 2xy + y^2 = 4,$$
$$\therefore x + y = \pm 2.$$

Then from this equation and (1) we get

$$x = 7 \text{ or } 5 \text{ and } y = -5 \text{ or } -7.$$

EXAMPLES.—xciv.

1. $x - y = 4$
   $x^2 + y^2 = 40$.

2. $x - y = 10$
   $x^2 + y^2 = 178$.

3. $x - y = 14$
   $x^2 + y^2 = 436$.

4. $x + y = 8$
   $x^2 + y^2 = 32$.

5. $x + y = 12$
   $x^2 + y^2 = 104$.

6. $x + y = 49$
   $x^2 + y^2 = 1681$.

253. To solve the equations
$$x^3 + y^3 = 35 \quad \ldots (1),$$
$$x + y = 5 \quad \ldots (2).$$

Divide (1) by (2), then we get
$$x^2 - xy + y^2 = 7 \quad \ldots (3),$$
From (2) $\quad x^2 + 2xy + y^2 = 25 \ldots (4),$

Subtracting (3) from (4),
$$3xy = 18,$$
$$\therefore 4xy = 24.$$

Then from this equation and (4) we get
$$x^2 - 2xy + y^2 = 1,$$
$$\therefore x - y = \pm 1;$$
and from this equation and (2) we find
$$x = 3 \text{ or } 2 \text{ and } y = 2 \text{ or } 3.$$

EXAMPLES.—XCV

1. $x^3 + y^3 = 91$
   $x + y = 7.$

2. $x^3 + y^3 = 341$
   $x + y = 11.$

3. $x^3 + y^3 = 1008$
   $x + y = 12.$

4. $x^3 - y^3 = 56$
   $x - y = 2.$

5. $x^3 - y^3 = 98$
   $x - y = 2.$

6. $x^3 - y^3 = 279$
   $x - y = 3.$

254. To solve the equations
$$\frac{1}{x} + \frac{1}{y} = \frac{5}{6} \quad \ldots (1),$$
$$\frac{1}{x^2} + \frac{1}{y^2} = \frac{13}{36} \quad \ldots (2).$$

From (1), by squaring it, we get
$$\frac{1}{x^2} + \frac{2}{xy} + \frac{1}{y^2} = \frac{25}{36} \quad \ldots (3).$$

From this subtract (2), and we have
$$\frac{2}{xy} = \frac{12}{36};$$
$$\therefore \frac{4}{xy} = \frac{24}{36}.$$

INVOLVING QUADRATICS.

Now subtract this from (3), and we get

$$\frac{1}{x^2} - \frac{2}{xy} + \frac{1}{y^2} = \frac{1}{36};$$

$$\therefore \frac{1}{x} - \frac{1}{y} = \pm \frac{1}{6};$$

and from this equation and (1) we find

$$x = 2 \text{ or } 3 \text{ and } y = 3 \text{ or } 2.$$

EXAMPLES.—XCVI.

1. $\dfrac{1}{x} + \dfrac{1}{y} = \dfrac{9}{20}.$
   $\dfrac{1}{x^2} + \dfrac{1}{y^2} = \dfrac{41}{400}.$

2. $\dfrac{1}{x} + \dfrac{1}{y} = \dfrac{3}{4}.$
   $\dfrac{1}{x^2} + \dfrac{1}{y^2} = \dfrac{5}{16}.$

3. $\dfrac{1}{x} + \dfrac{1}{y} = 5.$
   $\dfrac{1}{x^2} + \dfrac{1}{y^2} = 13.$

4. $\dfrac{1}{x} - \dfrac{1}{y} = \dfrac{1}{12}.$
   $\dfrac{1}{x^2} - \dfrac{1}{y^2} = \dfrac{7}{144}.$

5. $\dfrac{1}{x} - \dfrac{1}{y} = 2\dfrac{1}{2}.$
   $\dfrac{1}{x^2} - \dfrac{1}{y^2} = 8\dfrac{3}{4}.$

6. $\dfrac{1}{x} - \dfrac{1}{y} = 3.$
   $\dfrac{1}{x^2} - \dfrac{1}{y^2} = 21.$

255. To solve the equations

$$x^2 + 3xy = 7 \ \dotfill (1),$$
$$xy + 4y^2 = 18 \dotfill (2).$$

If we *add* the equations we get

$$x^2 + 4xy + 4y^2 = 25.$$

Taking the square root of each side, and taking only the positive root of the right-hand side into account,

$$x + 2y = 5;$$
$$\therefore x = 5 - 2y.$$

Substituting this value for x in (2) we get

$$(5 - 2y) y + 4y^2 = 18,$$

an equation by which y may be determined.

NOTE. In some examples we must *subtract* the second equation from the first in order to get a perfect square.

256. To solve the equations
$$x^3 - y^3 = 26 \dots \dots \dots (1),$$
$$x^2 + xy + y^2 = 13 \dots \dots \dots (2).$$

Dividing (1) by (2) we get $x - y = 2 \dots \dots \dots (3)$,
squaring, $\qquad x^2 - 2xy + y^2 = 4 \dots \dots \dots (4).$

Subtract this from (2), and we have
$$3xy = 9;$$
$$\therefore 4xy = 12.$$

Adding this to (4), we get $x^2 + 2xy + y^2 = 16$;
$$\therefore x + y = \pm 4.$$

Then from this equation and (3) we find
$$x = 3 \text{ or } -1, \text{ and } y = 1 \text{ or } -3.$$

257. To solve the equations
$$x^2 + y^2 = 65 \dots \dots \dots (1),$$
$$xy = 28 \dots \dots \dots (2).$$

Multiplying (2) by 2, we have
$$\left. \begin{array}{l} x^2 + y^2 = 65 \\ 2xy = 56 \end{array} \right\};$$
$$\therefore \left. \begin{array}{l} x^2 + 2xy + y^2 = 121 \\ x^2 - 2xy + y^2 = 9 \end{array} \right\};$$
$$\therefore x + y = \pm 11 \dots \dots \dots (A),$$
$$x - y = \pm 3 \dots \dots \dots (B).$$

The equations A and B furnish *four* pairs of simple equations,

$x + y = 11,$	$x + y = 11,$	$x + y = -11,$	$x + y = -11,$
$x - y = 3,$	$x - y = -3,$	$x - y = 3,$	$x - y = -3.$

from which we find the values of x to be $7, 4, -7$ and -4, and the corresponding values of y to be $4, 7, -4$ and -7.

258. The artifice, by which the solution of the equations given in this article is effected, is applicable to cases in which the equations *are homogeneous and of the same order.*

To solve the equations
$$x^2 + xy = 15,$$
$$xy - y^2 = 2.$$

Suppose $y = mx$.

Then $x^2 + mx^2 = 15$, from the first equation,
and $mx^2 - m^2x^2 = 2$, from the second equation.

Dividing one of these equations by the other,
$$\frac{x^2 + mx^2}{mx^2 - m^2x^2} = \frac{15}{2},$$
or
$$\frac{x^2(1+m)}{x^2(m - m^2)} = \frac{15}{2},$$
or
$$\frac{1+m}{m - m^2} = \frac{15}{2}.$$

From this equation we can determine the values of m.

One of these values is $\frac{2}{3}$, and putting this for m in the equation $x^2 + mx^2 = 15$, we get $x^2 + \frac{2}{3}x^2 = 15$.

From which we find $x = \pm 3$,
and then we can find y from one of the original equations.

259. The examples which we shall now give are intended as an exercise on the methods of solution explained in the four preceding articles.

EXAMPLES.—XCVII.

1. $x^3 - y^3 = 37$
   $x^2 + xy + y^2 = 37.$

2. $x^2 + 6xy = 144$
   $6xy + 36y^2 = 432.$

3. $x^2 + xy = 210$
   $y^2 + xy = 231.$

4. $x^2 + y^2 = 68$
   $xy = 16.$

5. $x^3 + y^3 = 152$
   $x^2 - xy + y^2 = 19.$

6. $4x^2 + 9xy = 190.$
   $4x - 5y = 10.$

7. $x^2 + xy + y^2 = 39$
   $3y^2 - 5xy = 25.$

8. $x^2 + xy = 66$
   $xy - y^2 = 5.$

9. $3x^2 + 4xy = 20.$
   $5xy + 2y^2 = 12.$

10. $x^2 - xy + y^2 = 7$
    $3x^2 + 13xy + 8y^2 = 162.$

11. $x^2 - xy = 35$
    $xy + y^2 = 18.$

12. $3x^2 + 4xy + 5y^2 = 71.$
    $5x + 7y = 29.$

13. $x^3 + y^3 = 2728$
    $x^2 - xy + y^2 = 124.$

14. $x^2 + 9xy = 340$
    $7xy - y^2 = 171.$

15. $x^2 + y^2 = 225$
    $xy = 108.$

XXI. ON PROBLEMS RESULTING IN QUADRATIC EQUATIONS.

260. The method of stating problems resulting in Quadratic Equations does not require any general explanation.

Some of the Examples which we shall give involve *one* unknown symbol, others involve *two*.

Ex. 1. What number is that whose square exceeds the number by 42?

Let x represent the number.

Then $$x^2 = x + 42,$$
or, $$x^2 - x = 42;$$
therefore $$x^2 - x + \frac{1}{4} = \frac{169}{4};$$
whence $$x - \frac{1}{2} = \pm \frac{13}{2}.$$

And we find the values of x to be 7 or -6.

Ex. 2. The sum of two numbers is 14 and the sum of their squares is 100. Find the numbers.

Let x and y represent the numbers.

Then $$x + y = 14,$$
and $$x^2 + y^2 = 100.$$

Proceeding as in Art. 252, we find

$$x = 8 \text{ or } 6, \quad y = 6 \text{ or } 8.$$

Hence the numbers are 8 and 6.

EXAMPLES.—XCVIII.

1. What number is that whose half multiplied by its third part gives 864?

2. What is the number of which the seventh and eighth parts being multiplied together and the product divided by 3 the quotient is $298\frac{2}{3}$?

3. I take a certain number from 94. I then add the number to 94. I multiply the two results together, and the result is 8512. What is the number?

4. What are the numbers whose product is 750 and the quotient of one by the other $3\frac{1}{3}$?

5. The sum of the squares of two numbers is 13001, and the difference of the same squares is 1449. Find the numbers.

6. The product of two numbers, one of which is as much above 21 as the other is below 21, is 377. Find the numbers.

7. The half, the third, the fourth and the fifth parts of a certain number being multiplied together the product is 6750. Find the number.

8. By what number must 11500 be divided, so that the quotient may be the same as the divisor, and the remainder 51?

9. Find a number to which 20 being added, and from which 10 being subtracted, the square of the first result added to twice the square of the second result gives 17475.

10. The sum of two numbers is 26, and the sum of their squares is 436. Find the numbers.

11. The difference between two numbers is 17, and the sum of their squares is 325. What are the numbers?

12. What two numbers are they whose product is 255 and the sum of whose squares is 514?

13. Divide 16 into two parts such that their product added to the sum of their squares may be 208.

[S.A.]

14. What number added to its square root gives as a result 1332?

15. What number exceeds its square root by $48\frac{3}{4}$?

16. What number exceeds its square root by 2550?

17. The product of two numbers is 24, and their sum multiplied by their difference is 20. Find the numbers.

18. What two numbers are those whose sum multiplied by the greater is 204, and whose difference multiplied by the less is 35?

19. What two numbers are those whose difference is 5 and their sum multiplied by the greater 228?

20. Find three consecutive numbers whose product is equal to 3 times the middle number.

21. The difference between the squares of two consecutive numbers is 15. Find the numbers.

22. The sum of the squares of two consecutive numbers is 481. Find the numbers.

23. The sum of the squares of three consecutive numbers is 365. Find the numbers.

NOTE. If I buy x apples for y pence,

$\frac{y}{x}$ will represent the cost of an apple in pence.

If I buy x sheep for z pounds,

$\frac{z}{x}$ will represent the cost of a sheep in pounds.

EX. A boy bought a number of oranges for 16d. Had he bought 4 more for the same money, he would have paid one-third of a penny less for each orange. How many did he buy?

Let x represent the number of oranges.

Then $\frac{16}{x}$ will represent the cost of an orange in pence.

Hence
$$\frac{16}{x} = \frac{16}{x+4} + \frac{1}{3},$$
or $16(3x+12) = 48x + x^2 + 4x,$
or $x^2 + 4x = 192,$

from which we find the values of x to be 12 or -16.

Therefore he bought 12 oranges.

24. I buy a number of handkerchiefs for £3. Had I bought 3 more for the same money, they would have cost one shilling each less. How many did I buy?

25. A dealer bought a number of calves for £80. Had he bought 4 more for the same money, each calf would have cost £1 less. How many did he buy?

26. A man bought some pieces of cloth for £33. 15s., which he sold again for £2. 8s. the piece, and gained as much as one piece cost him. What did he give for each piece?

27. A merchant bought some pieces of silk for £180. Had he bought 3 pieces more, he would have paid £3 less for each piece. How many did he buy?

---

28. For a journey of 108 miles 6 hours less would have sufficed had one gone 3 miles an hour faster. How many miles an hour did one go?

29. A grazier bought as many sheep as cost him £60. Out of these he kept 15, and selling the remainder for £54, gained 2 shillings a head by them. How many sheep did he buy?

30. A cistern can be filled by two pipes running together in 2 hours, 55 minutes. The larger pipe by itself will fill it sooner than the smaller by 2 hours. What time will each pipe take separately to fill it?

31. The length of a rectangular field exceeds its breadth by one yard, and the area contains ten thousand and one hundred square yards. Find the length of the sides.

32. A certain number consists of two digits. The left-hand digit is double of the right-hand digit, and if the digits be inverted the product of the number thus formed and the original number is 2268. Find the number.

33. A ladder, whose foot rests in a given position, just reaches a window on one side of a street, and when turned about its foot, just reaches a window on the other side. If the two positions of the ladder be at right angles to each other, and the heights of the windows be 36 and 27 feet respectively, find the width of the street and the length of the ladder.

34. Cloth, being wetted, shrinks up $\frac{1}{8}$ in its length and $\frac{1}{16}$ in its width. If the surface of a piece of cloth is diminished by $5\frac{3}{4}$ square yards, and the length of the 4 sides by $4\frac{1}{4}$ yards, what was the length and width of the cloth?

35. A certain number, less than 50, consists of two digits whose difference is 4. If the digits be inverted, the difference between the squares of the number thus formed and of the original number is 3960. Find the number.

36. A plantation in rows consists of 10000 trees. If there had been 20 less rows, there would have been 25 more trees in a row. How many rows are there?

37. A colonel wished to form a solid square of his men. The first time he had 39 men over; the second time he increased the side of the square by one man, and then he found that he wanted 50 men to complete it. How many men were there in the regiment?

XXII. INDETERMINATE EQUATIONS.

261. WHEN the number of unknown symbols exceeds that of the independent equations, the number of simultaneous values of the symbols will be indefinite. We propose to explain in this Chapter how a certain number of these values may be found in the case of Simultaneous Equations involving two unknown quantities.

Ex. To find the *integral* values of x and y which will satisfy the equation
$$3x + 7y = 10.$$
Here
$$3x = 10 - 7y;$$
$$\therefore x = 3 - 2y + \frac{1-y}{3}.$$

Now if x and y are integers, $\frac{1-y}{3}$ must also be an integer.

$$\therefore y = 1 - 3m,$$
and $\quad x = 3 - 2y + m = 3 - 2 + 6m + m = 1 + 7m;$
or the *general* solution of the equation in whole numbers is
$$x = 1 + 7m \text{ and } y = 1 - 3m,$$
where m may be 0, 1, 2...... or any integer, positive or negative.

If $\quad\quad\quad m = 0, x = 1, y = 1;$
if $\quad\quad\quad m = 1, x = 8, y = -2;$
if $\quad\quad\quad m = 2, x = 15, y = -5;$

and so on, from which it appears that the only *positive* integral values of x and y which satisfy the equation are 1 and 1.

262. It is next to be observed that it is desirable to divide both sides of the equation by the *smaller* of the two coefficients of the unknown symbols.

Ex. To find integral solutions of the equation
$$7x + 5y = 31.$$
Here $\quad\quad\quad 5y = 31 - 7x;$
$$\therefore y = 6 - x + \frac{1 - 2x}{5}.$$

Let $\dfrac{1 - 2x}{5} = m$, an integer.

Then $1 - 2x = 5m$, whence $2x = 1 - 5m;$
$$\therefore x = \frac{1 - m}{2} - 2m.$$

Let $\dfrac{1 - m}{2} = n$, an integer.

Then $1 - m = 2n$, whence $m = 1 - 2n.$

Hence $\quad x = n - 2m = n - 2 + 4n = 5n - 2;$
$\quad\quad\quad y = 6 - x + m = 6 - 5n + 2 + 1 - 2n = \mathbf{9 - 7n}.$

Now if $\quad\quad n = 0, x = -2, y = 9;$
if $\quad\quad\quad n = 1, x = 3, y = 2;$
if $\quad\quad\quad n = 2, x = 8, y = -5;$

and so on.

263. In how many ways can a person pay a bill of £13 with crowns and guineas?

Let x and y denote the number of crowns and guineas.

Then
$$5x + 21y = 260;$$
$$\therefore 5x = 260 - 21y;$$
$$x = 52 - 4y - \frac{y}{5}.$$

Let $\frac{y}{5} = m$, an integer.

Then $\qquad y = 5m,$

and $\qquad x = 52 - 4y - m = 52 - 21m.$

If
$$m = 0, \ x = 52, \ y = \ 0;$$
$$m = 1, \ x = 31, \ y = \ 5;$$
$$m = 2, \ x = 10, \ y = 10;$$

and higher values of m will give *negative* values of x.

Thus the number of ways is *three*.

264. To find a number which when divided by 7 and 5 will give remainders 2 and 3 respectively.

Let x be the number.

Then $\frac{x-2}{7} =$ an integer, suppose m;

and $\qquad \frac{x-3}{5} =$ an integer, suppose n.

Then $x = 7m + 2$ and $x = 5n + 3$;
$$\therefore 7m + 2 = 5n + 3;$$
$$\therefore 5n = 7m - 1, \text{ whence } n = m + \frac{2m-1}{5}.$$

Let $\frac{2m-1}{5} = p$, an integer.

Then $2m = 5p + 1$, whence $m = 2p + \frac{p+1}{2}$.

Let $\frac{p+1}{2} = q$, an integer.

Then
$$p = 2q - 1,$$
$$m = 2p + q = 4q - 2 + q = 5q - 2,$$
$$x = 7m + 2 = 35q - 12.$$

Hence if $y=0, x=-12$;
if $y=1, x=23$;
if $y=2, x=58$;
and so on.

EXAMPLES.—XCIX.

Find positive integral solutions of

1. $5x+7y=29$.
2. $7x+19y=92$.
3. $13x+19y=1170$.
4. $3x+5y=26$.
5. $14x-5y=7$.
6. $11x+15y=1031$.
7. $11x+7y=308$.
8. $4x-19y=23$.
9. $20x-9y=683$.
10. $3x+7y=383$.
11. $27x+4y=54$.
12. $7x+9y=653$.

13. Find two fractions with denominators 7 and 9 and their sum $\dfrac{57}{63}$.

14. Find two proper fractions with denominators 11 and 13 and their difference $\dfrac{82}{143}$.

15. In how many ways can a debt of £1. 9s. be paid in florins and half-crowns?

16. In how many ways can £20 be paid in half-guineas and half-crowns?

17. What number divided by 5 gives a remainder 2 and by 9 a remainder 3?

18. In how many different ways may £11. 15s. be paid in guineas and crowns?

19. In how many different ways may £4. 11s. 6d. be paid with half-guineas and half-crowns?

20. Shew that $323x-527y=1000$ cannot be satisfied by integral values of x and y.

21. A farmer buys oxen, sheep, and hens. The whole number bought was 100, and the whole price £100. If the oxen cost £5, the sheep £1, and the hens 1s. each, how many of each had he? Of how many solutions does this Problem admit?

22. A owes B 4s. 10d.; if A has only sixpences in his pocket and B only fourpenny pieces, how can they best settle the matter?

23. A person has £12. 4s. in half-crowns, florins, and shillings; the number of half-crowns and florins together is four times the number of shillings, and the number of coins is the greatest possible. Find the number of coins of each kind.

24. In how many ways can the sum of £5 be paid in exactly 50 coins, consisting of half-crowns, florins, and fourpenny pieces?

25. A owes B a shilling. A has only sovereigns, and B has only dollars worth 4s. 3d. each. How can A most easily pay B?

26. Divide 25 into two parts such that one of them is divisible by 2 and the other by 3.

27. In how many ways can I pay a debt of £2. 9s. with crowns and florins?

28. Divide 100 into two parts such that one is a multiple of 7 and the other of 11.

29. The sum of two numbers is 100. The first divided by 5 gives 2 as a remainder, and if we divide the second by 7 the remainder is 4. Find the numbers.

30. Find a number less than 400 which is a multiple of 7, and which when divided by 2, 3, 4, 5, 6, gives as a remainder in each case 1.

XXIII. THE THEORY OF INDICES.

265. The number placed over a symbol to express the power of the symbol is called the **Index**.

Up to this point our *indices* have in all cases been Positive Whole Numbers.

We have now to treat of Fractional and Negative indices; and to put this part of the subject in a clearer light, we shall commence from the elementary principles laid down in Arts. 45, 46.

266. First, we must carefully observe the following results:
$$a^3 \times a^2 = a^5,$$
$$(a^3)^2 = a^6.$$

For $\quad a^3 \times a^2 = a \cdot a \cdot a \cdot a \cdot a = a^5,$
and $\quad (a^3)^2 = a^3 \cdot a^3 = a \cdot a \cdot a \cdot a \cdot a \cdot a = a^6.$

These are examples of the Two Rules which govern all combinations of Indices. The general proof of these Rules we shall now proceed to give.

267. DEF. When m is a positive integer, a^m means $a \cdot a \cdot a \ldots\ldots$ with a written m times as a factor.

268. There are two rules for the combination of indices.

Rule I. $a^m \times a^n = a^{m+n}.$
Rule II. $(a^m)^n = a^{mn}.$

269. *To prove* RULE I.
$$a^m = a \cdot a \cdot a \ldots\ldots \text{ to } m \text{ factors,}$$
$$a^n = a \cdot a \cdot a \ldots\ldots \text{ to } n \text{ factors.}$$

Therefore

$a^m \times a^n = (a.a.a \ldots\ldots \text{ to } m \text{ factors}) \times (a.a.a \ldots\ldots \text{ to } n \text{ factors})$
$= a.a.a \ldots\ldots \text{ to } (m+n) \text{ factors}.$
$= a^{m+n}$, by the Definition.

To prove RULE II.

$(a^m)^n = a^m . a^m . a^m \ldots\ldots \text{ to } n \text{ factors},$
$= (a.a.a \ldots\ldots \text{ to } m \text{ factors})(a.a.a \ldots \text{ to } m \text{ factors}) \ldots$
repeated n times,
$= a.a.a \ldots\ldots \text{ to } mn \text{ factors},$
$= a^{mn}$, by the Definition.

270. We have deduced *immediately* from the Definition that when m and n are positive integers $a^m \times a^n = a^{m+n}$. When m and n are not positive integers, the Definition has no meaning. We therefore *extend* the Definition by saying that a^m and a^n, whatever m and n may be, shall be such that $a^m \times a^n = a^{m+n}$, and we shall now proceed to shew what meanings we assign to a^m, in consequence of this definition, in the following cases.

271. **Case I.** *To find the meaning of* $a^{\frac{p}{q}}$, p *and* q *being positive integers*.

$$a^{\frac{p}{q}} \times a^{\frac{p}{q}} = a^{\frac{p}{q}+\frac{p}{q}},$$
$$a^{\frac{p}{q}} \times a^{\frac{p}{q}} \times a^{\frac{p}{q}} = a^{\frac{p}{q}+\frac{p}{q}} \times a^{\frac{p}{q}} = a^{\frac{p}{q}+\frac{p}{q}+\frac{p}{q}};$$

and by continuing this process,

$$a^{\frac{p}{q}} \times a^{\frac{p}{q}} \times \ldots\ldots \text{ to } q \text{ factors} = a^{\frac{p}{q}+\frac{p}{q}+\frac{p}{q}+\ldots \text{ to } q \text{ terms}}$$
$$= a^p.$$

But by the nature of the symbol $\sqrt[q]{\ }$

$$\sqrt[q]{a^p} \times \sqrt[q]{a^p} \times \ldots\ldots \text{ to } q \text{ factors} = a^p;$$
$$\therefore a^{\frac{p}{q}} \times a^{\frac{p}{q}} \times \ldots\ldots \text{ to } q \text{ factors} = \sqrt[q]{a^p} \times \sqrt[q]{a^p} \times \ldots \text{ to } q \text{ factors}:$$
$$\therefore a^{\frac{p}{q}} = \sqrt[q]{a^p}.$$

THE THEORY OF INDICES.

272. Case II. *To find the meaning of a^{-s}, s being a positive number, whole or fractional.*

We must first find the meaning of a^0.

We have
$$a^m \times a^0 = a^{m+0}$$
$$= a^m;$$
$$\therefore a^0 = 1.$$

Now
$$a^s \times a^{-s} = a^{s-s}$$
$$= a^0$$
$$= 1;$$
$$\therefore a^{-s} = \frac{1}{a^s}.$$

273. Thus the interpretation of a^m has been deduced from Rule I. It remains to be proved that this interpretation agrees with Rule II. This we shall do by shewing that Rule II. follows from Rule I., whatever m and n may be.

274. *To shew that $(a^m)^n = a^{mn}$ for all values of m and n.*

(1) Let n be a positive integer: then, whatever m may be,
$$(a^m)^n = a^m . a^m . a^m \ldots \text{ to } n \text{ factors}$$
$$= a^{m+m+m+\ldots \text{to } n \text{ terms}}$$
$$= a^{mn}.$$

(2) Let n be a positive fraction, and equal to $\frac{p}{q}$, p and q being positive integers; then, whatever be the value of m,
$$(a^m)^{\frac{p}{q}} \times (a^m)^{\frac{p}{q}} \times \ldots \text{ to } q \text{ factors} = (a^m)^{\frac{p}{q}+\frac{p}{q}+\ldots \text{ to } q \text{ terms}}$$
$$= (a^m)^p$$
$$= a^{mp}, \text{ by (1).}$$

But $a^{\frac{mp}{q}} \times a^{\frac{mp}{q}} \times \ldots \text{ to } q \text{ factors} = a^{\frac{mp}{q}+\frac{mp}{q}+\ldots \text{to } q \text{ terms}}$
$$= a^{mp};$$
$$\therefore (a^m)^{\frac{p}{q}} = a^{\frac{mp}{q}};$$

that is,
$$(a^m)^n = a^{mn}.$$

(3) Let $n = -s$, s being a positive number, whole or fractional: then, whatever m may be,

$$(a^m)^{-s} = \frac{1}{(a^m)^s}, \text{ by Art. 272,}$$

$$= \frac{1}{a^{ms}}, \text{ by (1) and (2) of this Article;}$$

that is, $(a^m)^n = \frac{1}{a^{-mn}}$

$$= a^{mn}.$$

275. We shall now give some examples of the mode in which the Theorems established in the preceding articles are applied to particular cases. We shall commence with examples of the combination of the indices of two single terms.

276. Since $x^m \times x^n = x^{m+n}$,

(1) $x^c \times x^{a-c} = x^{c+a-c} = x^a$.

(2) $x^c \times x = x^{c+1}$.

(3) $x^{a+b-c} \cdot x^{a-b+c} \cdot x^{b-a+c} = x^{a+b-c+a-b+c+b-a+c}$ &c. = greater.

(4) $a^{m-n} \cdot b^{n-p} \times a^{n-m} \cdot b^{p-n} \cdot c$

$= a^{m-n+n-m} \cdot b^{n-p+p-n} \cdot c$

$= a^0 \cdot b^0 \cdot c$

$= 1 \cdot 1 \cdot c$

$= c$.

277. Since $(x^m)^n = x^{mn}$,

(1) $(x^6)^3 = x^{6 \times 3} = x^{18}$.

(2) $(x^4)^{\frac{1}{2}} = x^{4 \times \frac{1}{2}} = x^2$.

(3) $(a^{6x})^{\frac{1}{3}} = a^{6x \times \frac{1}{3}} = a^{2x}$.

278. Since $x^{\frac{r}{q}} = \sqrt[q]{x^r}$,

(1) $x^{\frac{3}{2}} = \sqrt{x^3}$.

(2) $x^{\frac{2}{3}} = \sqrt[3]{x^2}$.

NOTE. When Examples are given of actual numbers raised to fractional powers, they may often be put in a form more fit for easy solution, thus:

(1) $144^{\frac{3}{2}} = (144^{\frac{1}{2}})^3 = (\sqrt{144})^3 = 12^3 = 1728.$

(2) $125^{\frac{2}{3}} = (125^{\frac{1}{3}})^2 = (\sqrt[3]{125})^2 = 5^2 = 25.$

279. Since $(x^m)^n = x^{mn}$,

(1) $\{(x^m)^n\}^p = (x^{mn})^p = x^{mnp}.$

(2) $\{(a^{-m})^{-n}\}^p = (a^{mn})^p = a^{mnp}.$

(3) $\{(x^{-m})^n\}^p = (x^{-mn})^p = x^{-mnp}.$

280. Since $x^{-n} = \dfrac{1}{x^n}$,

we may replace an expression raised to a negative power by the reciprocal (Art. 199) of the expression raised to the same positive power: thus

(1) $a^{-1} = \dfrac{1}{a}.$ (2) $a^{-2} = \dfrac{1}{a^2}.$ (3) $a^{-\frac{2}{3}} = \dfrac{1}{a^{\frac{2}{3}}}.$

EXAMPLES.—C.

(1) Express with fractional indices:

1. $\sqrt{x^5} + \sqrt[3]{x^2} + (\sqrt{x})^7.$ 3. $\sqrt[3]{a^4} + (\sqrt[3]{a})^5 + a\sqrt{a^3}.$

2. $\sqrt{x^2 y^3} + \sqrt[4]{x^9 y^2} + \sqrt[7]{x^2 y^5}.$ 4. $\sqrt[3]{xy^3z^2} + \sqrt[4]{a^2y^3z^4} + \sqrt[5]{ay^5z^2}.$

(2) Express with negative indices so as to remove all powers from the denominators:

1. $\dfrac{1}{x} + \dfrac{a}{x^2} + \dfrac{b^2}{x^3} + \dfrac{3}{x^4}.$ 3. $\dfrac{x^3}{4y^2z^2} + \dfrac{5x^2}{7yz^3} + \dfrac{x}{yz}.$

2. $\dfrac{x^2}{y^2} + \dfrac{3x}{y^3} + \dfrac{4}{y^4}.$ 4. $\dfrac{xy}{3z^2} + \dfrac{1}{5x^2y^2} + \dfrac{z}{x^5y^4}.$

(3) Express with negative indices so as to remove all powers from the numerators:

THE THEORY OF INDICES.

1. $\dfrac{1}{a} + \dfrac{x}{a^2} + \dfrac{x^3}{b^2} + \dfrac{x^4}{3}.$

2. $\dfrac{y^2}{x^2} + \dfrac{y^3}{3x} + \dfrac{y^5}{5}.$

3. $\dfrac{4a^2b^2}{c^3} + \dfrac{3a}{\sqrt{bc}} + \dfrac{\sqrt[3]{x}}{y}.$

4. $-\dfrac{\sqrt{xy}}{3z} + \dfrac{\sqrt[3]{a^3b^2}}{c^2} + \dfrac{\sqrt[5]{a^{10}b^5}}{c}.$

(4) Express with root-symbols and positive indices:

1. $2x^{\frac{2}{3}} + 3x^{\frac{1}{3}}y^{\frac{2}{3}} + x^{-4}y^{-2}.$

2. $x^{-\frac{1}{3}} + y^{-\frac{2}{3}} + z^{-3}.$

3. $\dfrac{x^{-\frac{1}{3}}}{y^{-\frac{2}{3}}} + \dfrac{3x^{-2}}{y^{-\frac{3}{4}}} + \dfrac{x^{-\frac{2}{3}}}{3y^{-\frac{1}{3}}}.$

4. $\dfrac{x^{-2}}{y^{\frac{1}{3}}} + \dfrac{x^{-\frac{1}{3}}}{y^{-1}} + \dfrac{x^{-\frac{2}{3}}}{-\frac{1}{3}}.$

281. Since $x^m \div x^n = \dfrac{x^m}{x^n} = x^m \cdot x^{-n} = x^{m-n},$

(1) $x^8 \div x^3 = x^{8-3} = x^5.$

(2) $x^3 \div x^8 = x^{3-8} = x^{-5} = \dfrac{1}{x^5}.$

(3) $x^m \div x^{m-n} = x^{m-(m-n)} = x^{m-m+n} = x^n.$

(4) $a^b \div a^{b+c} = a^{b-(b+c)} = a^{b-b-c} = a^{-c} = \dfrac{1}{x^c}.$

(5) $x^{\frac{2}{3}} \div x^{\frac{1}{3}} = x^{\frac{2}{3}-\frac{1}{3}} = x^{\frac{1}{3}}.$

(6) $x^{\frac{1}{2}} \div x^{\frac{5}{6}} = x^{\frac{1}{2}-\frac{5}{6}} = x^{\frac{3}{6}-\frac{5}{6}} = x^{-\frac{2}{6}} = x^{-\frac{1}{3}} = \dfrac{1}{x^{\frac{1}{3}}}.$

282. Ex. *Multiply* $a^{3r} - a^{2r} + a^r - 1$ *by* $a^r + 1.$

$a^{3r} - a^{2r} + a^r - 1$
$a^r + 1$
$\overline{a^{4r} - a^{3r} + a^{2r} - a^r}$
$\phantom{a^{4r}} + a^{3r} - a^{2r} + a^r - 1$
$\overline{a^{4r} - 1}$

EXAMPLES.—ci.

Multiply

1. $x^{2p} + x^p y^p + y^{2p}$ by $x^{2p} - x^p y^p + y^{2p}.$

2. $a^{3m} + 3a^{2m}y^n + 9a^m y^{2n} + 27y^{3n}$ by $a^m - 3y^n.$

3. $x^{4d} - 2ax^{2d} + 4a^2$ by $x^{4d} + 2ax^{2d} + 4a^2.$

4. $a^m + b^n + c^r$ by $a^m - b^n + c^r$.
5. $a^m + b^n - 2c^r$ by $2a^m - b + c^2$.
6. $x^{mn-n} - y^m$ by $x^n + y^{mn-n}$.
7. $x^{2n} - x^n y^n + y^{2n}$ by $x^{2n} + x^n y^n + y^{2n}$.
8. $a^{p^2+p} - b^{p^2} + c^p$ by $a^{p^2-p} + b^{1-p^2} + c^{1-p}$.
9. Form the square of $x^{2p} + x^p + 1$.
10. Form the square of $x^{2p} - x^p + 1$.

283. **Ex.** Divide $x^{4p} - 1$ by $x^p - 1$.

$$x^p - 1 \,)\, x^{4p} - 1 \,(\, x^{3p} + x^{2p} + x^p + 1$$
$$\underline{x^{4p} - x^{3p}}$$
$$x^{3p} - 1$$
$$\underline{x^{3p} - x^{2p}}$$
$$x^{2p} - 1$$
$$\underline{x^{2p} - x^p}$$
$$x^p - 1$$
$$\underline{x^p - 1}$$

EXAMPLES.—cii.

Divide

1. $x^{4m} - y^{4m}$ by $x^m - y^m$.
2. $x^{5n} + y^{5n}$ by $x^n + y^n$.
3. $x^{6r} - y^{6r}$ by $x^r - y^r$.
4. $a^{15p} + b^{10q}$ by $a^{3p} + b^{2q}$.
5. $x^{5d} - 243$ by $x^d - 3$.
6. $a^{4m} + 4a^{2m}x^{2n} + 16x^{4n}$ by $a^{2m} + 2a^m x^n + 4x^{2n}$.
7. $9x^p + 3x^{4p} + 14x^{3p} + 2$ by $1 + 5x^p + x^{2p}$.
8. $14b^{4m}c^m - 13b^{3m}c^{2m} - 5b^{5m} + 4b^{2m}c^{3m}$ by $b^{3m} + b^m c^{2m} - 2b^{2m}c^m$.
9. Find the square root of
    $a^{6m} + 6a^{5m} + 15a^{4m} + 20a^{3m} + 15a^{2m} + 6a^m + 1$.
10. Find the square root of
    $a^{2m} + b^{2n} + c^{2r} + 2a^m b^n + 2a^m c^r + 2b^n c^r$.

Fractional Indices.

284. Ex. *Multiply* $a^{\frac{2}{3}} - a^{\frac{1}{3}}b^{\frac{1}{3}} + b^{\frac{2}{3}}$ *by* $a^{\frac{1}{3}} + b^{\frac{1}{3}}$.

$$a^{\frac{2}{3}} - a^{\frac{1}{3}}b^{\frac{1}{3}} + b^{\frac{2}{3}}$$
$$a^{\frac{1}{3}} + b^{\frac{1}{3}}$$
$$\overline{}$$
$$a \;\; - a^{\frac{2}{3}}b^{\frac{1}{3}} + a^{\frac{1}{3}}b^{\frac{2}{3}}$$
$$+ a^{\frac{2}{3}}b^{\frac{1}{3}} - a^{\frac{1}{3}}b^{\frac{2}{3}} + b$$
$$\overline{}$$
$$a \;\; + b$$

EXAMPLES.—ciii.

Multiply

1. $x^{\frac{2}{3}} - 2x^{\frac{1}{3}} + 1$ by $x^{\frac{1}{3}} - 1$.
2. $y^{\frac{3}{4}} + y^{\frac{1}{2}} + y^{\frac{1}{4}} + 1$ by $y^{\frac{1}{4}} - 1$.
3. $a^{\frac{2}{3}} - x^{\frac{2}{3}}$ by $a^{\frac{4}{3}} + a^{\frac{2}{3}}x^{\frac{2}{3}} + x^{\frac{4}{3}}$.
4. $a^{\frac{2}{3}} + b^{\frac{2}{3}} + c^{\frac{2}{3}} - a^{\frac{1}{3}}b^{\frac{1}{3}} - a^{\frac{1}{3}}c^{\frac{1}{3}} - b^{\frac{1}{3}}c^{\frac{1}{3}}$ by $a^{\frac{1}{3}} + b^{\frac{1}{3}} + c^{\frac{1}{3}}$.
5. $5x^{\frac{3}{4}} + 2x^{\frac{1}{2}}y^{\frac{1}{4}} + 3x^{\frac{1}{4}}y^{\frac{1}{2}} + 7y^{\frac{3}{4}}$ by $2x^{\frac{1}{4}} - 3y^{\frac{1}{4}}$.
6. $m^{\frac{4}{5}} + m^{\frac{3}{5}}n^{\frac{1}{5}} + m^{\frac{2}{5}}n^{\frac{2}{5}} + m^{\frac{1}{5}}n^{\frac{3}{5}} + n^{\frac{4}{5}}$ by $m^{\frac{1}{5}} - n^{\frac{1}{5}}$.
7. $m^{\frac{2}{3}} - 2d^{\frac{1}{4}}m^{\frac{1}{3}} + 4d^{\frac{1}{2}}$ by $m^{\frac{2}{3}} + 2d^{\frac{1}{4}}m^{\frac{1}{3}} + 4d^{\frac{1}{2}}$.
8. $8a^{\frac{3}{7}} + 4a^{\frac{2}{7}}b^{\frac{1}{7}} + 5a^{\frac{1}{7}}b^{\frac{2}{7}} + 9b^{\frac{3}{7}}$ by $2a^{\frac{1}{7}} - 3b^{\frac{1}{7}}$.

Form the square of each of the following expressions:

9. $x^{\frac{1}{3}} + a^{\frac{1}{3}}$.
10. $x^{\frac{1}{3}} - a^{\frac{1}{3}}$.
11. $x^{\frac{2}{5}} + y^{\frac{2}{5}}$.
12. $a + b^{\frac{1}{4}}$.
13. $x^{\frac{1}{2}} - 2x^{\frac{1}{4}} + 3$.
14. $2x^{\frac{2}{7}} + 3x^{\frac{1}{7}} + 4$.
15. $x^{\frac{1}{3}} - y^{\frac{1}{3}} + z^{\frac{1}{3}}$.
16. $x^{\frac{1}{4}} + 2y^{\frac{1}{4}} - z^{\frac{1}{4}}$.

285. **Ex.** *Divide $a-b$ by $\sqrt[4]{a} - \sqrt[4]{b}$.*

Putting $a^{\frac{1}{4}}$ for $\sqrt[4]{a}$, and $b^{\frac{1}{4}}$ for $\sqrt[4]{b}$, we proceed thus:

$$a^{\frac{1}{4}} - b^{\frac{1}{4}} \big) a - b \, (a^{\frac{3}{4}} + a^{\frac{1}{2}}b^{\frac{1}{4}} + a^{\frac{1}{4}}b^{\frac{1}{2}} + b^{\frac{3}{4}}$$

$$\underline{a \div a^{\frac{3}{4}}b^{\frac{1}{4}}}$$

$$a^{\frac{3}{4}}b^{\frac{1}{4}} - b$$

$$\underline{a^{\frac{3}{4}}b^{\frac{1}{4}} - a^{\frac{1}{2}}b^{\frac{1}{2}}}$$

$$a^{\frac{1}{2}}b^{\frac{1}{2}} - b$$

$$\underline{a^{\frac{1}{2}}b^{\frac{1}{2}} - a^{\frac{1}{4}}b^{\frac{3}{4}}}$$

$$a^{\frac{1}{4}}b^{\frac{3}{4}} - b$$

$$\underline{a^{\frac{1}{4}}b^{\frac{3}{4}} - b}$$

EXAMPLES.—civ.

Divide

1. $x-y$ by $x^{\frac{1}{2}} - y^{\frac{1}{2}}$.
2. $a-b$ by $a^{\frac{1}{2}} + b^{\frac{1}{2}}$.
3. $x-y$ by $x^{\frac{1}{3}} - y^{\frac{1}{3}}$.
4. $a+b$ by $a^{\frac{1}{3}} + b^{\frac{1}{3}}$.
5. $x+y$ by $x^{\frac{1}{5}} + y^{\frac{1}{5}}$.
6. $m-n$ by $m^{\frac{1}{6}} - n^{\frac{1}{6}}$.
7. $x-81y$ by $x^{\frac{1}{4}} - 3y^{\frac{1}{4}}$.
8. $81a - 16b$ by $3a^{\frac{1}{4}} - 2b^{\frac{1}{4}}$.
9. $a-x$ by $x^{\frac{1}{2}} + a^{\frac{1}{2}}$.
10. $m-243$ by $m^{\frac{1}{5}} - 3$.
11. $x + 17x^{\frac{1}{2}} + 70$ by $x^{\frac{1}{2}} + 7$.
12. $x^{\frac{2}{3}} + x^{\frac{1}{3}} - 12$ by $x^{\frac{1}{3}} - 3$.
13. $b^{\frac{1}{3}} - 3b^{\frac{2}{3}} + 3b - b^{\frac{4}{3}}$ by $b^{\frac{1}{3}} - 1$.
14. $x + y + z - 3x^{\frac{1}{3}}y^{\frac{1}{3}}z^{\frac{1}{3}}$ by $x^{\frac{1}{3}} + y^{\frac{1}{3}} + z^{\frac{1}{3}}$.
15. $x - 5x^{\frac{2}{3}} - 46x^{\frac{1}{3}} - 40$ by $x^{\frac{1}{3}} + 4$.
16. $m + m^{\frac{1}{2}}n^{\frac{1}{2}} + n$ by $m^{\frac{1}{2}} - m^{\frac{1}{4}}n^{\frac{1}{4}} + n^{\frac{1}{2}}$.
17. $p - 4p^{\frac{3}{4}} + 6p^{\frac{1}{2}} - 4p^{\frac{1}{4}} + 1$ by $p^{\frac{1}{2}} - 2p^{\frac{1}{4}} + 1$.
18. $2x + x^{\frac{1}{2}}y^{\frac{1}{2}} - 3y - 4y^{\frac{1}{2}}z^{\frac{1}{2}} - x^{\frac{1}{2}}z^{\frac{1}{2}} - z$ by $2x^{\frac{1}{2}} + 3y^{\frac{1}{2}} + z^{\frac{1}{2}}$.
19. $x+y$ by $x^{\frac{4}{5}} - x^{\frac{3}{5}}y^{\frac{1}{5}} + x^{\frac{2}{5}}y^{\frac{2}{5}} - x^{\frac{1}{5}}y^{\frac{3}{5}} + y^{\frac{4}{5}}$.

[S.A.]

Negative Indices.

286. Ex. Multiply $x^{-3}+x^{-2}y^{-1}+x^{-1}y^{-2}+y^{-3}$ by $x^{-1}-y^{-1}$.

$$x^{-3}+x^{-2}y^{-1}+x^{-1}y^{-2}+y^{-3}$$
$$x^{-1}-y^{-1}$$
$$\overline{x^{-4}+x^{-3}y^{-1}+x^{-2}y^{-2}+x^{-1}y^{-3}}$$
$$\phantom{x^{-4}+}-x^{-3}y^{-1}-x^{-2}y^{-2}-x^{-1}y^{-3}-y^{-4}$$
$$\overline{x^{-4}-y^{-4}}$$

EXAMPLES.—CV.

Multiply

1. $a^{-1}+b^{-1}$ by $a^{-1}-b^{-1}$.   2. $x^{-3}+b^{-2}$ by $x^{-3}-b^{-2}$.
3. $x^3+x+x^{-1}+x^{-3}$ by $x-x^{-1}$.   4. x^2-1+x^{-2} by x^2+1+x^{-2}.
5. $a^{-2}+b^{-2}$ by $a^{-2}-b^{-2}$.   6. $a^{-1}-b^{-1}+c^{-1}$ by $a^{-1}+b^{-1}+c^{-1}$.
7. $1+ab^{-1}+a^2b^{-2}$ by $1-ab^{-1}+a^2b^{-2}$.
8. $a^2b^{-2}+2+a^{-2}b^2$ by $a^2b^{-2}-2-a^{-2}b^2$.
9. $4x^{-3}+3x^{-2}+2x^{-1}+1$ by $x^{-2}-x^{-1}+1$.
10. $\frac{5}{2}x^{-2}+3x^{-1}-\frac{7}{3}$ by $2x^{-2}-x^{-1}-\frac{1}{2}$.

287. Ex. Divide x^2+1+x^{-2} by $x-1+x^{-1}$.

$$x-1+x^{-1})\ x^2+1+x^{-2}\ (x+1+x^{-1}$$
$$\phantom{x-1+x^{-1})\ }x^2-x+1$$
$$\overline{\phantom{x-1+x^{-1})\ }x+x^{-2}}$$
$$\phantom{x-1+x^{-1})\ }x-1+x^{-1}$$
$$\overline{\phantom{x-1+x^{-1})\ xx}1-x^{-1}+x^{-2}}$$
$$\phantom{x-1+x^{-1})\ xx}1-x^{-1}+x^{-2}$$

NOTE. The order of the powers of a is

$$\ldots\ldots a^3,\ a^2,\ a^1,\ a^0,\ a^{-1},\ a^{-2},\ a^{-3}\ldots\ldots$$

a series which may be written thus

$$\ldots\ldots a^3,\ a^2,\ a,\ 1,\ \frac{1}{a},\ \frac{1}{a^2},\ \frac{1}{a^3}\ldots\ldots$$

EXAMPLES.—cvi.

Divide

1. $x^2 - x^{-2}$ by $x + 1$
2. $x^2 - b^{-2}$ by $a - b^{-1}$.
3. $m^3 + n^{-3}$ by $m + n^{-1}$.
4. $c^5 - d^{-5}$ by $c - d^{-1}$.
5. $x^2y^{-2} + 2 + x^{-2}y^2$ by $xy^{-1} + x^{-1}y$.
6. $a^{-4} + a^{-2}b^{-2} + b^{-4}$ by $a^{-2} - a^{-1}b^{-1} + b^{-2}$.
7. $x^3y^{-3} - x^{-3}y^3 - 3xy^{-1} + 3x^{-1}y$ by $xy^{-1} - x^{-1}y$.
8. $\dfrac{3x^{-5}}{4} - 4x^{-4} + \dfrac{77x^{-3}}{8} - \dfrac{43x^{-2}}{4} - \dfrac{33x^{-1}}{4} + 27$

  by $\dfrac{x^{-2}}{2} - x^{-1} + 3$.

9. $a^3b^{-3} + a^{-3}b^3$ by $ab^{-1} + a^{-1}b$.
10. $a^{-3} + b^{-3} + c^{-3} - 3a^{-1}b^{-1}c^{-1}$ by $a^{-1} + b^{-1} + c^{-1}$.

288. To shew that $(ab)^n = a^n \cdot b^n$.

$(ab)^n = ab \cdot ab \cdot ab \ldots$ to n factors
$= (a \cdot a \cdot a \ldots$ to n factors$) \times (b \cdot b \cdot b \ldots$ to n factors$)$
$= a^n \cdot b^n$.

We shall now give a series of Examples to introduce the various forms of combination of indices explained in this Chapter.

EXAMPLES.—cvii.

1. Divide $x^{\frac{4}{3}} - 4xy + 4x^{\frac{2}{3}}y + 4y^2$ by $x^{\frac{2}{3}} + 2x^{\frac{1}{3}}y^{\frac{1}{2}} + 2y$.
2. Simplify $\{(x^{3ab})^3 \cdot (x^6)^2\}^{\frac{1}{3a-2}}$.
3. Simplify $(x^{10b} \cdot x^{18a})^{\frac{1}{3a-2}}$.
4. Simplify $\left\{ \dfrac{1}{x^2 - a^2} - \dfrac{1}{x^2 + a^2} - \dfrac{\dfrac{1}{x+a} - \dfrac{1}{x-a}}{\dfrac{x^2 + a^2}{a}} \right\}^{\frac{1}{2}}$.

5. Multiply $\frac{7}{3}x^{-2} + 4c^{-1} - \frac{2}{7}$ by $3x^{-2} - 2x^{-1} - \frac{1}{2}$.

6. Simplify $\dfrac{x^{a+b} \cdot x^{a-b} \cdot x^{c-2a}}{x^{c-a}}$. 7. Divide $x^{2n} - y^{2n}$ by $x^n + y^n$.

8. Multiply $(a^{\frac{1}{2}} + b^{\frac{3}{5}})^3$ by $a^{\frac{1}{2}} - b^{\frac{2}{5}}$.

9. Divide $a - b$ by $\sqrt[3]{a} - \sqrt[3]{b}$.   10. Prove that $(a^2)^m = (a^m)^2$.

11. If $a^{m^n} = (a^m)^n$, find m in terms of n.

12. Simplify $x^{a+b+c} \cdot x^{a+b-c} \cdot x^{a-b+c} \cdot x^{b+c-a}$.

13. Simplify $\left(\dfrac{x^{p+q}}{x^q}\right)^p \div \left(\dfrac{x^q}{x^{q-p}}\right)^{p-q}$.   14. Divide $4a^x$ by $\dfrac{a^{-x}}{4}$.

15. Simplify $[\{(a^{-m})^{-n}\}^r] \div [\{(a^m)^n\}^{-r}]$.

16. Multiply $a^m + b^p - 2c^n$ by $2a^m - 3b$.

17. Multiply $a^{m-n} b^{n-p}$ by $a^{n-m} b^{p-n} c$.

18. Shew that $\dfrac{a + (b^2 a)^{\frac{1}{3}} - (a^2 b)^{\frac{1}{3}}}{a+b} = \dfrac{a^{\frac{1}{3}}}{a^{\frac{1}{3}} + b^{\frac{1}{3}}}$.

19. Multiply $x^{\frac{1}{3}} + x^{\frac{1}{6}} + 1$ by $x^{\frac{1}{3}} - x^{\frac{1}{6}} + 1$
 and their product by $x^{\frac{2}{3}} - x^{\frac{1}{3}} + 1$.

20. Multiply $a^m - ba^{m-1}x + ca^{m-2}x^2$ by $a^n + ba^{n-1}x - ca^{n-2}x^2$.

21. Divide $x^{2p(q-1)} - y^{2q(p-1)}$ by $x^{p(q-1)} + y^{q(p-1)}$.

22. Simplify $\{(a^m)^{n-\frac{1}{m}}\}^{\frac{1}{m+1}}$.

23. Multiply $x^{3r} + x^{2r}y^p + x^r y^{2p} + y^{3p}$ by $x^r - y^p$.

24. Write down the values of $625^{\frac{1}{4}}$ and 12^{-2}.

25. Multiply $x^{(m-1)n} - y^{(n-1)m}$ by $x^n - y^m$.

26. Multiply $x^{\frac{3}{4}} + 2x^{\frac{1}{2}} - 1$ by $x^{\frac{1}{4}} - 2x^{-\frac{1}{4}}$.

XXIV. ON SURDS.

289. All numbers which we cannot exactly determine, because they are not multiples of a Primary or Subordinate Unit, are called **Surds**.

290. We shall confine our attention to those Surds which originate *in the Extraction of roots* where the results cannot be exhibited in whole or fractional numbers.

For example, if we perform the operation of extracting the square root of 2, we obtain 1·4142..., and though we may carry on the process to any required extent, we shall never be able to stop at any particular point and to say that we have found the exact number which is equivalent to the Square Root of 2.

291. We can approximate to the real value of a surd by finding two numbers *between which it lies*, differing from each other by a fraction as small as we please.

Thus, since $\sqrt{2} = 1\cdot4142\ldots$

$\sqrt{2}$ lies between $\frac{14}{10}$ and $\frac{15}{10}$, which differ by $\frac{1}{10}$;

also between $\frac{141}{100}$ and $\frac{142}{100}$, which differ by $\frac{1}{100}$;

also between $\frac{1414}{1000}$ and $\frac{1415}{1000}$, which differ by $\frac{1}{1000}$.

And, generally, if we find the square root of 2 to n places of decimals, we shall find two numbers between which $\sqrt{2}$ lies, differing from each other by the fraction $\frac{1}{10^n}$.

292. Next, we can always find a fraction differing from the real value of a surd by less than any *assigned* quantity.

For example, suppose it required to find a fraction differing from $\sqrt{2}$ by less than $\frac{1}{12}$.

Now $2(12)^2$, that is 288, lies between $(16)^2$ and $(17)^2$;

$\therefore$ 2 lies between $\left(\frac{16}{12}\right)^2$ and $\left(\frac{17}{12}\right)^2$;

$\therefore \sqrt{2}$ lies between $\frac{16}{12}$ and $\frac{17}{12}$;

$\therefore \sqrt{2}$ differs from $\frac{16}{12}$ by less than $\frac{1}{12}$.

293. Surds, though they cannot be expressed by whole or fractional numbers, are nevertheless numbers of which we may form an approximate idea, and we may make three assertions respecting them.

(1) Surds may be compared so far as asserting that one is greater or less than another. Thus $\sqrt{3}$ is clearly greater than $\sqrt{2}$, and $\sqrt[3]{9}$ is greater than $\sqrt[3]{8}$.

(2) Surds may be multiples of other surds: thus $2\sqrt{2}$ is the double of $\sqrt{2}$.

(3) Surds, when multiplied together, may produce as a result a whole or fractional number: thus

$$\sqrt{2} \times \sqrt{2} = 2,$$

and $\sqrt[3]{\frac{3}{4}} \times \sqrt[3]{\frac{3}{4}} \times \sqrt[3]{\frac{3}{4}} = \frac{3}{4}.$

294. The symbols $\sqrt{a}, \sqrt[3]{a}, \sqrt[4]{a}, \sqrt[n]{a}$, in cases where the second, third, fourth, and n^{th} roots respectively of a cannot be exhibited as whole or fractional numbers, will represent surds of the second, third, fourth, and n^{th} order.

These symbols we may, in accordance with the principles laid down in Chapter XXIII., replace by $a^{\frac{1}{2}}, a^{\frac{1}{3}}, a^{\frac{1}{4}}, a^{\frac{1}{n}}$.

ON SURDS.   215

295. Surds *of the same order* are those for which the root-symbol or surd-index is the same.

Thus $\sqrt{a}$, $3\sqrt{(3b)}$, $4\sqrt{(mn)}$, $r^{\frac{1}{2}}$ are surds of the same order.

Like surds are those in which the same root-symbol or surd-index appears *over the same quantity*.

Thus $2\sqrt{a}$, $3\sqrt{a}$, $4a^{\frac{1}{2}}$ are like surds.

296. A whole or fractional number may be expressed in the form of a surd, by raising the number to the power denoted by the order of the surd, and placing the result under the symbol of evolution that corresponds to the surd-index.

Thus
$$a = \sqrt{a^2},$$
$$\frac{b}{c} = \sqrt[3]{\frac{b^3}{c^3}}.$$

297. Surds of *different orders* may be transformed into surds of the same order by reducing the surd-indices to fractions with the same denominator.

Thus we may transform $\sqrt[3]{x}$ and $\sqrt[4]{y}$ into surds of the same order, for
$$\sqrt[3]{x} = x^{\frac{1}{3}} = x^{\frac{4}{12}} = \sqrt[12]{x^4},$$
and
$$\sqrt[4]{y} = y^{\frac{1}{4}} = y^{\frac{3}{12}} = \sqrt[12]{y^3},$$
and thus both surds are transformed into surds of the twelfth order.

EXAMPLES.—cviii.

Transform into Surds of the same order:

1. $\sqrt{x}$ and $\sqrt[3]{y}$. 2. $\sqrt[3]{4}$ and $\sqrt[4]{2}$. 3. $\sqrt{(18)}$ and $\sqrt[3]{(50)}$.
4. $\sqrt[m]{2}$ and $\sqrt[n]{2}$. 5. $\sqrt[m]{a}$ and $\sqrt[n]{b}$. 6. $\sqrt[3]{(a+b)}$ and $\sqrt{(a-b)}$.

298. If a whole or fractional number be multiplied into a surd, the product will be represented by placing the multiplier and the multiplicand side by side with no sign, or with a dot (.) between them.

Thus the product of 3 and $\sqrt{2}$ is represented by $3\sqrt{2}$,
.... of 4 and $5\sqrt{2}$ .... by $20\sqrt{2}$,
.... of a and $\sqrt{c}$ .... by $a\sqrt{c}$.

299. Like surds may be combined by the ordinary processes of addition and subtraction, that is, by adding the coefficients of the surd and placing the result as a coefficient of the surd.

Thus
$$\sqrt{a} + \sqrt{a} = 2\sqrt{a},$$
$$5\sqrt{b} - 3\sqrt{b} = 2\sqrt{b},$$
$$x\sqrt{c} - \sqrt{c} = (x-1)\sqrt{c}.$$

300. We now proceed to prove a Theorem of great importance, which may be thus stated.

The root of any expression is the same as the product of the roots of the separate factors of the expression, that is

$$\sqrt{(ab)} = \sqrt{a} \cdot \sqrt{b},$$
$$\sqrt[3]{(xyz)} = \sqrt[3]{x} \cdot \sqrt[3]{y} \cdot \sqrt[3]{z},$$
$$\sqrt[n]{(pqr)} = \sqrt[n]{p} \cdot \sqrt[n]{q} \cdot \sqrt[n]{r}.$$

We have in fact to shew from the Theory of Indices that

$$(ab)^{\frac{1}{n}} = a^{\frac{1}{n}} \cdot b^{\frac{1}{n}}.$$

Now
$$\{(ab)^{\frac{1}{n}}\}^n = (ab)^{\frac{n}{n}} = ab,$$

and
$$\{a^{\frac{1}{n}} \cdot b^{\frac{1}{n}}\}^n = (a^{\frac{1}{n}})^n \cdot (b^{\frac{1}{n}})^n = a^{\frac{n}{n}} \cdot b^{\frac{n}{n}} = a \cdot b;$$

$$\therefore \{(ab)^{\frac{1}{n}}\}^n = \{a^{\frac{1}{n}} \cdot b^{\frac{1}{n}}\}^n;$$

$$\therefore (ab)^{\frac{1}{n}} = a^{\frac{1}{n}} \cdot b^{\frac{1}{n}}.$$

301. We can sometimes reduce an expression in the form of a surd to an equivalent expression with a whole or fractional number as one factor.

Thus
$$\sqrt{(72)} = \sqrt{(36 \times 2)} = \sqrt{(36)} \cdot \sqrt{2} = 6\sqrt{2},$$
$$\sqrt[3]{(128)} = \sqrt[3]{(64 \times 2)} = \sqrt[3]{(64)} \cdot \sqrt[3]{2} = 4\sqrt[3]{2},$$
$$\sqrt[n]{(a^n x)} = \sqrt[n]{a^n} \cdot \sqrt[n]{x} = a \cdot \sqrt[n]{x}.$$

EXAMPLES.—cix.

Reduce to equivalent expressions with a whole or fractional number as one factor:

1. $\sqrt{(24)}$.
2. $\sqrt{(50)}$.
3. $\sqrt{(4a^3)}$.
4. $\sqrt{(125a^4b^3)}$.
5. $\sqrt{(32yz^3)}$.
6. $\sqrt{(1000a)}$.
7. $\sqrt{(720c^2)}$.
8. $7 \cdot \sqrt{(396x)}$.
9. $18 \cdot \sqrt{\left(\frac{5}{27}x^3\right)}$.
10. $a \cdot \sqrt{\frac{a^3}{b}}$.
11. $\sqrt{(a^3 + 2a^2x + ax^2)}$.
12. $\sqrt{(x^3 - 2x^2y + xy^2)}$.
13. $\sqrt{(50a^2 - 100ab + 50b^2)}$.
14. $\sqrt{(63c^4y - 42c^2y^2 + 7y^3)}$.
15. $\sqrt[3]{(54a^6b^2)}$.
16. $\sqrt[3]{(160x^4y^7)}$.
17. $\sqrt[3]{(108m^9n^{10})}$.
18. $\sqrt[3]{(1372a^{15}b^{16})}$.
19. $\sqrt[3]{(x^4 + 3x^3y + 3x^2y^2 + xy^3)}$.
20. $\sqrt[3]{(a^4 - 3a^3b + 3a^2b^2 - ab^3)}$.

302. An expression containing two factors, one a surd, the other a whole or fractional number, as $3\sqrt{2}$, $a\sqrt[3]{x}$, may be transformed into a complete surd.

Thus
$$3\sqrt{2} = (3^2)^{\frac{1}{2}} \cdot \sqrt{2} = \sqrt{9} \cdot \sqrt{2} = \sqrt{(18)},$$
$$a\sqrt[3]{x} = (a^3)^{\frac{1}{3}} \cdot \sqrt[3]{x} = \sqrt[3]{a^3} \cdot \sqrt[3]{x} = \sqrt[3]{(a^3x)}.$$

EXAMPLES.—cx.

Reduce to complete Surds:

1. $4\sqrt{3}$.
2. $3\sqrt{7}$.
3. $5\sqrt[3]{9}$.
4. $2\sqrt[4]{6}$.
5. $3\sqrt[3]{\frac{3}{7}}$.
6. $3\sqrt{a}$.
7. $4a\sqrt{(3x)}$.
8. $2ax\sqrt{\left(\frac{3a}{bc}\right)}$.
9. $(m+n) \cdot \sqrt{\left(\frac{m-n}{m+n}\right)}$.
10. $(a+b)\left(\frac{1}{a^2-b^2}\right)^{\frac{1}{2}}$.
11. $\left(\frac{x-y}{x+y}\right) \cdot \left(\frac{x^2+xy}{x^2-2xy+y^2}\right)^{\frac{1}{2}}$.

303. Surds may be *compared* by transforming them into surds of the same order. Thus if it be required to determine whether $\sqrt{2}$ be greater or less than $\sqrt[3]{3}$, we proceed thus:

$$\sqrt{2} = 2^{\frac{1}{2}} = 2^{\frac{3}{6}} = \sqrt[6]{2^3} = \sqrt[6]{8},$$

$$\sqrt[3]{3} = 3^{\frac{1}{3}} = 3^{\frac{2}{6}} = \sqrt[6]{3^2} = \sqrt[6]{9}.$$

And since $\sqrt[6]{9}$ is greater than $\sqrt[6]{8}$,

$\sqrt[3]{3}$ is greater than $\sqrt{2}$.

EXAMPLES.—cxi.

Arrange in order of magnitude the following Surds:

1. $\sqrt{3}$ and $\sqrt[3]{4}$.
2. $\sqrt{10}$ and $\sqrt[3]{15}$.
3. $2\sqrt{3}$ and $3\sqrt{2}$.
4. $\sqrt{\frac{3}{5}}$ and $\sqrt[3]{\left(\frac{14}{15}\right)}$.
5. $3\sqrt{7}$ and $4\sqrt{3}$.
6. $2\sqrt{87}$ and $3\sqrt{33}$.
7. $2\sqrt[3]{22}$, $3\sqrt[3]{7}$ and $4\sqrt{2}$.
8. $3\sqrt{19}$, $5\sqrt[3]{18}$ and $3\sqrt[3]{82}$.
9. $2\sqrt[3]{14}$, $5\sqrt[3]{2}$ and $3\sqrt[3]{3}$.
10. $\frac{1}{2}\sqrt{2}, \frac{1}{3}\sqrt{3}$ and $\frac{1}{4}\sqrt{4}$.

304. The following are examples in the application of the rules of Addition, Subtraction, Multiplication, and Division to Surds of the same order.

1. Find the sum of $\sqrt{18}$, $\sqrt{128}$, and $\sqrt{32}$.

$$\sqrt{(18)} + \sqrt{(128)} + \sqrt{(32)} = \sqrt{(9 \times 2)} + \sqrt{(64 \times 2)} + \sqrt{(16 \times 2)}$$
$$= 3\sqrt{2} + 8\sqrt{2} + 4\sqrt{2}$$
$$= 15\sqrt{2}.$$

2. From $3\sqrt{(75)}$ take $4\sqrt{(12)}$.

$$3\sqrt{(75)} - 4\sqrt{(12)} = 3\sqrt{(25 \times 3)} - 4\sqrt{(4 \times 3)}$$
$$= 3 \cdot 5 \cdot \sqrt{3} - 4 \cdot 2\sqrt{3}$$
$$= 15\sqrt{3} - 8\sqrt{3}$$
$$= 7\sqrt{3}.$$

ON SURDS.

3. Multiply $\sqrt{8}$ by $\sqrt{(12)}$.

$$\sqrt{8} \times \sqrt{(12)} = \sqrt{(8 \times 12)}$$
$$= \sqrt{(96)}$$
$$= \sqrt{(16 \times 6)}$$
$$= 4\sqrt{6}.$$

4. Divide $\sqrt{32}$ by $\sqrt{18}$.

$$\frac{\sqrt{(32)}}{\sqrt{(18)}} = \frac{\sqrt{(16 \times 2)}}{\sqrt{(9 \times 2)}} = \frac{4\sqrt{2}}{3\sqrt{2}} = \frac{4}{3}.$$

EXAMPLES.—cxii.

Simplify

1. $\sqrt{(27)} + 2\sqrt{(48)} + 3\sqrt{(108)}$.
2. $3\sqrt{(1000)} + 4\sqrt{(50)} + 12\sqrt{(288)}$.
3. $a\sqrt{(a^2x)} + b\sqrt{(b^2x)} + c\sqrt{(c^2x)}$.
4. $\sqrt[3]{(128)} + \sqrt[3]{(686)} + \sqrt[3]{(16)}$.
5. $7\sqrt[3]{(54)} + 3\sqrt[3]{(16)} + \sqrt[3]{(432)}$.
6. $\sqrt{(96)} - \sqrt{(54)}$.
7. $\sqrt{(243)} - \sqrt{(48)}$.
8. $12\sqrt{(72)} - 3\sqrt{(128)}$.
9. $5\sqrt[3]{(16)} - 2\sqrt[3]{(54)}$.
10. $7\sqrt[3]{(81)} - 3\sqrt[3]{(1029)}$.
11. $\sqrt{6} \times \sqrt{8}$.
12. $\sqrt{(14)} \times \sqrt{(20)}$.
13. $\sqrt{(50)} \times \sqrt{(200)}$.
14. $\sqrt[3]{(3a^2b)} \times \sqrt[3]{(9ab^2)}$.
15. $\sqrt[3]{(12ab)} \times \sqrt[3]{(8a^2b^3)}$.
16. $\sqrt{(12)} \div \sqrt{3}$.
17. $\sqrt{(18)} \div \sqrt{(50)}$.
18. $\sqrt[3]{(a^2b)} \div \sqrt[3]{(ab^2)}$.
19. $\sqrt[3]{(a^3b)} \div \sqrt[3]{(ab^3)}$.
20. $\sqrt{(x^2 + x^3y)} \div \sqrt{(x + 2x^2y + x^3y^2)}$.

305. We now proceed to treat of the Multiplication of Compound Surds, an operation which will be frequently required in a later part of the subject.

The Student must bear in mind the two following Rules:

Rule I. $\sqrt{a} \times \sqrt{b} = \sqrt{(ab)}$,

Rule II. $\sqrt{a} \times \sqrt{a} = a$,

which will be true for all values of a and b.

EXAMPLES.—CXiii.

Multiply

1. $\sqrt{x}$ by $\sqrt{y}$.
2. $\sqrt{(x-y)}$ by $\sqrt{y}$.
3. $\sqrt{(x+y)}$ by $\sqrt{(x+y)}$.
4. $\sqrt{(x-y)}$ by $\sqrt{(x+y)}$.
5. $6\sqrt{x}$ by $3\sqrt{x}$.
6. $7\sqrt{(x+1)}$ by $8\sqrt{(x+1)}$.
7. $10\sqrt{x}$ by $9\sqrt{(x-1)}$.
8. $\sqrt{(3x)}$ by $\sqrt{(4x)}$.
9. $\sqrt{x}$ by $-\sqrt{x}$.
10. $\sqrt{(x-1)}$ by $-\sqrt{(x-1)}$.
11. $3\sqrt{x}$ by $-4\sqrt{x}$.
12. $-2\sqrt{a}$ by $-3\sqrt{a}$.
13. $\sqrt{(x-7)}$ by $-\sqrt{x}$.
14. $-2\sqrt{(x+7)}$ by $-3\sqrt{x}$.
15. $-4\sqrt{(a^2-1)}$ by $-2\sqrt{(a^2-1)}$.
16. $2\sqrt{(a^2-2a+3)}$ by $-3\sqrt{(a^2-2a+3)}$.

306. The following Examples will illustrate the way of proceeding in forming the products of Compound Surds.

Ex. 1. To multiply $\sqrt{x}+3$ by $\sqrt{x}+2$.

$$\begin{array}{r} \sqrt{x}+3 \\ \sqrt{x}+2 \\ \hline x+3\sqrt{x} \\ +2\sqrt{x}+6 \\ \hline x+5\sqrt{x}+6 \end{array}$$

Ex. 2. To multiply $4\sqrt{x}+3\sqrt{y}$ by $4\sqrt{x}-3\sqrt{y}$.

$$\begin{array}{r} 4\sqrt{x}+3\sqrt{y} \\ 4\sqrt{x}-3\sqrt{y} \\ \hline 16x+12\sqrt{(xy)} \\ -12\sqrt{(xy)}-9y \\ \hline 16x-9y \end{array}$$

Ex. 3. To form the square of $\sqrt{(x-7)}-\sqrt{x}$.

$$\begin{array}{r} \sqrt{(x-7)}-\sqrt{x} \\ \sqrt{(x-7)}-\sqrt{x} \\ \hline x-7-\sqrt{(x^2-7x)} \\ -\sqrt{(x^2-7x)}+x \\ \hline 2x-7-2\sqrt{(x^2-7x)} \end{array}$$

EXAMPLES.—CXIV.

Multiply

1. $\sqrt{x}+7$ by $\sqrt{x}+2$.
2. $\sqrt{x}-5$ by $\sqrt{x}+3$.
3. $\sqrt{(a+9)}+3$ by $\sqrt{(a+9)}-3$.
4. $\sqrt{(a-4)}-7$ by $\sqrt{(a-4)}+7$.
5. $3\sqrt{x}-7$ by $\sqrt{x}+4$.
6. $2\sqrt{(x-5)}+4$ by $3\sqrt{(x-5)}-6$.
7. $\sqrt{(6+x)}+\sqrt{x}$ by $\sqrt{(6+x)}-\sqrt{x}$.
8. $\sqrt{(3x+1)}+\sqrt{(2x-1)}$ by $\sqrt{3x}-\sqrt{(2x-1)}$.
9. $\sqrt{a}+\sqrt{(a-x)}$ by $\sqrt{x}-\sqrt{(a-x)}$.
10. $\sqrt{(3+x)}+\sqrt{x}$ by $\sqrt{(3+x)}$.
11. $\sqrt{x}+\sqrt{y}+\sqrt{z}$ by $\sqrt{x}-\sqrt{y}+\sqrt{z}$.
12. $\sqrt{a}+\sqrt{(a-x)}+\sqrt{x}$ by $\sqrt{a}-\sqrt{(a-x)}+\sqrt{x}$.

Form the squares of the following expressions:

13. $21+\sqrt{(x^2-9)}$.
14. $\sqrt{(x+3)}+\sqrt{(x+8)}$.
15. $\sqrt{x}+\sqrt{(x-4)}$.
16. $\sqrt{(x-6)}+\sqrt{x}$.
17. $2\sqrt{x}-3$.
18. $\sqrt{(x+y)}-\sqrt{(x-y)}$.
19. $\sqrt{x}\cdot\sqrt{(x+1)}-\sqrt{(x-1)}$.
20. $\sqrt{(x+1)}+\sqrt{x}\cdot\sqrt{(x-1)}$.

307. We may now extend the Theorem explained in Art. 101. We there shewed how to resolve expressions of the form

$$a^2 - b^2$$

into factors, restricting our observations to the case of *perfect squares*.

The Theorem extends to the difference between *any two* quantities.

Thus

$$a - b = (\sqrt{a} + \sqrt{b})(\sqrt{a} - \sqrt{b}).$$
$$x^2 - y = (x + \sqrt{y})(x - \sqrt{y}).$$
$$1 - x = (1 + \sqrt{x})(1 - \sqrt{x}).$$

308. Hence we can always find a multiplier which will free from surds an expression of any of the *four* forms

1. $a + \sqrt{b}$ or 2. $\sqrt{a} + \sqrt{b}$,
3. $a - \sqrt{b}$ or 4. $\sqrt{a} - \sqrt{b}$.

For since the *first* and *third* of these expressions give as a product $a^2 - b$, which is free from surds, and since the *second* and *fourth* give as a product $a - b$, which is free from surds, it follows that the required multiplier may be in all cases found.

Ex. 1. To find the multiplier which will free from surds each of the following expressions:

1. $5 + \sqrt{3}$. 2. $\sqrt{6} + \sqrt{5}$. 3. $2 - \sqrt{5}$. 4. $\sqrt{7} - \sqrt{2}$.

The multipliers will be

1. $5 - \sqrt{3}$. 2. $\sqrt{6} - \sqrt{5}$. 3. $2 + \sqrt{5}$. 4. $\sqrt{7} + \sqrt{2}$.

The products will be

1. $25 - 3$. 2. $6 - 5$. 3. $4 - 5$. 4. $7 - 2$.

That is, 22, 1, -1, and 5.

Ex. 2. To reduce the fraction $\dfrac{a}{b - \sqrt{c}}$ to an equivalent fraction with a denominator free from surds.

Multiply both terms of the fraction by $b + \sqrt{c}$, and it becomes

$$\frac{ab + a\sqrt{c}}{b^2 - c},$$

which is in the required form.

EXAMPLES.—CXV.

Express in factors:

1. $c - d$. 2. $c^2 - d$. 3. $c - d^2$.
4. $1 - y$. 5. $1 - 3x^2$. 6. $5m^2 - 1$.
7. $4a^2 - 3c$. 8. $9 - 8n$. 9. $11n^2 - 16$.
10. $p^2 - 4r$. 11. $p - 3q^2$. 12. $a^{2m} - b^n$.

Reduce the following fractions to equivalent fractions with denominators free from surds.

13. $\dfrac{1}{a - \sqrt{b}}$

14. $\dfrac{\sqrt{a}}{\sqrt{a} - \sqrt{b}}$

15. $\dfrac{4 + 3\sqrt{2}}{2\sqrt{2}}$

16. $\dfrac{2}{2 - \sqrt{2}}$

17. $\dfrac{\sqrt{3}}{2 - \sqrt{3}}$

18. $\dfrac{2 - \sqrt{2}}{2 + \sqrt{2}}$

19. $\dfrac{\sqrt{a} + \sqrt{x}}{\sqrt{a} - \sqrt{x}}$

20. $\dfrac{1 + \sqrt{x}}{1 - \sqrt{x}}$

21. $\dfrac{\sqrt{(a+x)} + \sqrt{(a-x)}}{\sqrt{(a+x)} - \sqrt{(a-x)}}$

22. $\dfrac{\sqrt{(m^2 + 1)} - \sqrt{(m^2 - 1)}}{\sqrt{(m^2 + 1)} + \sqrt{(m^2 - 1)}}$

23. $\dfrac{a + \sqrt{(a^2 - 1)}}{a - \sqrt{(a^2 - 1)}}$

24. $\dfrac{a + \sqrt{(a^2 - x^2)}}{a - \sqrt{(a^2 - x^2)}}$

309. The *squares* of all numbers, negative as well as positive, are *positive*.

Since there is no assignable number the square of which would be a negative quantity, we conclude that an expression which appears under the form $\sqrt{(-a^2)}$ represents an impossible quantity.

310. All impossible square roots may be reduced to one common form, thus

$$\sqrt{(-a^2)} = \sqrt{\{a^2 \times (-1)\}} = \sqrt{a^2} \cdot \sqrt{(-1)} = a \cdot \sqrt{(-1)}$$
$$\sqrt{(-x)} = \sqrt{\{x \times (-1)\}} = \sqrt{x} \cdot \sqrt{(-1)}.$$

Where, since a and $\sqrt{x}$ are possible numbers, the whole impossibility of the expressions is reduced to the appearance of $\sqrt{(-1)}$ as a factor.

311. DEF. By $\sqrt{(-1)}$ we understand an expression which when multiplied by itself produces -1.

Therefore

$\{\sqrt{(-1)}\}^2 = -1,$

$\{\sqrt{(-1)}\}^3 = \{\sqrt{(-1)}\}^2 \cdot \sqrt{(-1)} = (-1) \cdot \sqrt{(-1)} = -\sqrt{(-1)},$

$\{\sqrt{(-1)}\}^4 = \{\sqrt{(-1)}\}^2 \cdot \{\sqrt{(-1)}\}^2 = (-1) \cdot (-1) = 1,$

and so on.

EXAMPLES.—CXVI.

Multiply, observing that
$$\sqrt{-a} \times \sqrt{-b} = -\sqrt{ab}.$$

1. $4 + \sqrt{(-3)}$ by $4 - \sqrt{(-3)}$.
2. $\sqrt{3} - 2\sqrt{(-2)}$ by $\sqrt{3} + 2\sqrt{(-2)}$.
3. $4\sqrt{(-2)} - 2\sqrt{2}$ by $\frac{1}{2}\sqrt{(-2)} - 3\sqrt{2}$.
4. $\sqrt{(-2)} + \sqrt{(-3)} + \sqrt{(-4)}$ by $\sqrt{(-2)} - \sqrt{(-3)} - \sqrt{(-4)}$.
5. $3\sqrt{(-a)} + \sqrt{(-b)}$ by $4\sqrt{(-a)} - 2\sqrt{(-b)}$.
6. $a + \sqrt{(-a)}$ by $a - \sqrt{(-a)}$.
7. $a\sqrt{(-a)} + b\sqrt{(-b)}$ by $a\sqrt{(-a)} - b\sqrt{(-b)}$.
8. $a + \beta\sqrt{(-1)}$ by $a - \beta\sqrt{(-1)}$.
9. $1 - \sqrt{(1-e^2)}$ by $1 + \sqrt{(1-e^2)}$.
10. $e^{p\sqrt{(-1)}} + e^{-p\sqrt{(-1)}}$ by $e^{p\sqrt{(-1)}} - e^{-p\sqrt{(-1)}}$.

312. We shall now give a few Miscellaneous Examples to illustrate the principles explained in this Chapter.

EXAMPLES.—CXVII.

1. Simplify $\dfrac{\sqrt{x} + \sqrt{y}}{3\sqrt{y}} - \dfrac{\sqrt{x} - \sqrt{y}}{3\sqrt{x}}$.
2. Prove that $\{1 + \sqrt{(-1)}\}^2 + \{1 - \sqrt{(-1)}\}^2 = 0$.
3. Simplify $\dfrac{\sqrt{x} + \sqrt{y}}{2\sqrt{x}} + \dfrac{\sqrt{x} - \sqrt{y}}{2\sqrt{y}}$.
4. Prove that $\{1 + \sqrt{(-1)}\}^2 - \{1 - \sqrt{(-1)}\}^2 = \sqrt{(-16)}$.
5. Divide $x^4 + a^4$ by $x^2 + \sqrt{2ax} + a^2$.
6. Divide $m^4 + n^4$ by $m^2 - \sqrt{2mn} + n^2$.
7. Simplify $\sqrt{(x^3 + 2x^2y + xy^2)} + \sqrt{(x^3 - 2x^2y + xy^2)}$.
8. Simplify $\dfrac{a-b}{\sqrt{a} - \sqrt{b}} - \dfrac{a+b}{\sqrt{a} + \sqrt{b}}$, and verify by putting $a = 9$ and $b = 4$.

9. Find the square of $a\sqrt{\dfrac{c}{b}} - \sqrt{(cd)}$.

10. Find the square of $a\sqrt{2} - \dfrac{1}{a\sqrt{2}}$.

11. Simplify
$$\dfrac{\sqrt{(x^2+a^2)} + \sqrt{(x^2-a^2)}}{\sqrt{(x^2+a^2)} - \sqrt{(x^2-a^2)}} + \dfrac{\sqrt{(x^2+a^2)} - \sqrt{(x^2-a^2)}}{\sqrt{(x^2+a^2)} + \sqrt{(x^2-a^2)}}.$$

12. Simplify $\dfrac{\sqrt{(1-x)} + \dfrac{1}{\sqrt{(1+x)}}}{1 + \dfrac{1}{\sqrt{(1-x^2)}}}$.

13. Simplify $\dfrac{x-1}{x+1}\left\{\dfrac{x-1}{\sqrt{x}-1} + \dfrac{1-x}{x+\sqrt{x}}\right\}$.

14. Form the square of $\sqrt{\left(\dfrac{x}{4}+3\right)} - \sqrt{\left(\dfrac{x}{4}-3\right)}$.

15. Form the square of $\sqrt{(x+a)} - \sqrt{(x-a)}$.

16. Multiply $\sqrt[m]{(a^{2m-n}b^{5m+1}c^{3p})}$ by $\sqrt[m]{(a^{n}b^{m-1}c^{m-3p})}$.

17. Raise to the 5th power $-1 - a\sqrt{(-1)}$.

18. Simplify $\sqrt[3]{(81)} - \sqrt[3]{(-512)} + \sqrt[3]{(192)}$.

19. Simplify $\dfrac{6c^2}{x-1}\sqrt{\left(\dfrac{4x^3 - 8x^2 + 4x}{3c^4}\right)}$.

20. Simplify $\dfrac{x}{x-7}\left\{\sqrt[18]{(3p^2x^3 - 63p^2x^2 + 441p^2x - 1029p^2)}\right\}$.

21. Simplify $2(n-1)\sqrt[3]{\left(-\dfrac{1}{2n^4 - 6n^3 + 6n^2 - 2n}\right)}$.

22. Simplify $2(n-1)\sqrt{(63)} + \dfrac{1}{3}\sqrt{(112)} - \dfrac{\sqrt{(28n^4)}}{n^2}$
$\qquad + \sqrt{\{175(n-1)^2 c^2\}} \times \dfrac{2}{3c} - 2\sqrt{\left(\dfrac{7n^2}{36}\right)}$.

23. What is the difference between
$$\sqrt{\{17 - \sqrt{(33)}\}} \times \sqrt{\{17 + \sqrt{(33)}\}}$$
$$\sqrt[3]{\{65 + \sqrt{(129)}\}} \times \sqrt[3]{\{65 - \sqrt{(129)}\}} ?$$

and

[S.A.]

313. We have now to treat of the method of finding the Square Root of a Binomial Surd, that is, of an expression of one of the following forms:

$$m + \sqrt{n},\ m - \sqrt{n},$$

where m stands for a whole or fractional number, and $\sqrt{n}$ for a surd of the *second* order.

314. We have first to prove two Theorems.

Theorem I. *If* $\sqrt{a} = m + \sqrt{n}$, m *must be zero.*

Squaring both sides,

$$a = m^2 + 2m\sqrt{n} + n;$$
$$\therefore 2m\sqrt{n} = a - m^2 - n;$$
$$\therefore \sqrt{n} = \frac{a - m^2 - n}{2m};$$

that is, $\sqrt{n}$, a surd, is equal to a whole or fractional number, which is impossible.

Hence the assumed equality can never hold unless $m = 0$, in which case $\sqrt{a} = \sqrt{n}$.

Theorem II. *If* $b + \sqrt{a} = m + \sqrt{n}$, *then must* $b = m$, *and* $\sqrt{a} = \sqrt{n}$.

For, if not, let $b = m + x$.

Then $m + x + \sqrt{a} = m + \sqrt{n}$,

or $x + \sqrt{a} = \sqrt{n}$;

which, by Theorem I., is impossible unless $x = 0$, in which case $b = m$ and $\sqrt{a} = \sqrt{n}$.

315. *To find the Square Root of* $a + \sqrt{b}$.

Assume $\sqrt{(a + \sqrt{b})} = \sqrt{x} + \sqrt{y}$.

Then $a + \sqrt{b} = x + 2\sqrt{(xy)} + y$;

$$\therefore x + y = a \dotfill (1),$$
$$2\sqrt{(xy)} = \sqrt{b} \dotfill (2),$$

from which we have to find x and y.

Now from (1) $\quad x^2 + 2xy + y^2 = a^2,$

and from (2) $\quad 4xy = b;$

$\therefore x^2 - 2xy + y^2 = a^2 - b;$

$\therefore x - y = \sqrt{(a^2 - b)}.$

Also, $\quad x + y = a.$

From these equations we find

$$x = \frac{a + \sqrt{(a^2 - b)}}{2} \text{ and } y = \frac{a - \sqrt{(a^2 - b)}}{2};$$

$$\therefore \sqrt{(a + \sqrt{b})} = \sqrt{\left\{\frac{a + \sqrt{(a^2 - b)}}{2}\right\}} + \sqrt{\left\{\frac{a - \sqrt{(a^2 - b)}}{2}\right\}}.$$

Similarly we may show that

$$\sqrt{(a - \sqrt{b})} = \sqrt{\left\{\frac{a + \sqrt{(a^2 - b)}}{2}\right\}} - \sqrt{\left\{\frac{a - \sqrt{(a^2 - b)}}{2}\right\}}.$$

316. The practical use of this method will be more clearly seen from the following example.

Find the Square Root of $18 + 2\sqrt{(77)}$.

Assume $\quad \sqrt{\{18 + 2\sqrt{(77)}\}} = \sqrt{x} + \sqrt{y}.$

Then $\quad 18 + 2\sqrt{(77)} = x + 2\sqrt{(xy)} + y;$

$\therefore \left. \begin{array}{r} x + y = 18 \\ 2\sqrt{(xy)} = 2\sqrt{(77)} \end{array} \right\}.$

Hence $\quad \left. \begin{array}{r} x^2 + 2xy + y^2 = 324 \\ 4xy = 308 \end{array} \right\};$

$\therefore x^2 - 2xy + y^2 = 16;$

$\therefore x - y = \pm 4;$

also, $\quad x + y = 18.$

Hence $\quad x = 11 \text{ or } 7, \text{ and } y = 7 \text{ or } 11.$

That is, the square root required is $\sqrt{(11)} + \sqrt{7}.$

EXAMPLES.—CXVIII.

Find the square roots of the following Binomial Surds:

1. $10 + 2\sqrt{(21)}$.
2. $16 + 2\sqrt{(55)}$.
3. $9 - 2\sqrt{(14)}$.
4. $94 - 42\sqrt{5}$.
5. $13 - 2\sqrt{(30)}$.
6. $38 - 12\sqrt{(10)}$.
7. $14 - 4\sqrt{6}$.
8. $103 - 12\sqrt{(11)}$.
9. $75 - 12\sqrt{(21)}$.
10. $87 - 12\sqrt{(42)}$.
11. $3\tfrac{1}{2} - \sqrt{(10)}$.
12. $57 - 12\sqrt{(15)}$.

317. It is often easy to determine the square roots of expressions such as those given in the preceding set of Examples *by inspection*.

Take for instance the expression $18 + 2\sqrt{(77)}$.

What we want is to find two numbers whose sum is 18 and whose product is 77: these are evidently 11 and 7.

Then $\quad 18 + 2\sqrt{(77)} = 11 + 7 + 2\sqrt{(11 \times 7)}$
$\qquad\qquad\qquad = \{\sqrt{(11)} + \sqrt{7}\}^2.$

That is $\sqrt{(11)} + \sqrt{7}$ is the square root of $18 + 2\sqrt{(77)}$.

To effect this resolution by inspection it is necessary *that the coefficient of the surd should be* 2, and this we can always ensure.

For example, if the proposed expression be $4 + \sqrt{(15)}$, we proceed thus:

$$4 + \sqrt{(15)} = \frac{8 + 2\sqrt{(15)}}{2} = \frac{5 + 3 + 2\sqrt{(5 \times 3)}}{2}$$

$$= \left(\frac{\sqrt{5} + \sqrt{3}}{\sqrt{2}}\right)^2;$$

$\therefore \dfrac{\sqrt{5} + \sqrt{3}}{\sqrt{2}}$ is the square root of $4 + \sqrt{(15)}$.

Again, to find the Square Root of $28 - 10\sqrt{3}$.

$$28 - 10\sqrt{3} = 28 - 2\sqrt{(75)}$$
$$= 25 + 3 - 2\sqrt{(25 \times 3)}$$
$$= (5 - \sqrt{3})^2;$$

$\therefore 5 - \sqrt{3}$ is the square root required.

XXV. ON EQUATIONS INVOLVING SURDS.

318. Any equation may be cleared of a *single* surd, by transposing all the other terms to the contrary side of the equation, and then raising each side to the power corresponding to the order of the surd.

The process will be explained by the following Examples.

Ex. 1. $\sqrt{x}=4$.

Raising both sides to the *second* power,
$$x=16.$$

Ex. 2. $\sqrt[3]{x}=3$.

Raising both sides to the *third* power,
$$x=27.$$

Ex. 3. $\sqrt{(x^2+7)}-x=1$.

Transposing the second term,
$$\sqrt{(x^2+7)}=1+x.$$

Raising both sides to the *second* power,
$$x^2+7=1+2x+x^2,$$
$$\therefore x=3.$$

EXAMPLES.—CXIX.

1. $\sqrt{x}=7$.
2. $\sqrt{x}=9$.
3. $x^{\frac{1}{2}}=5$.
4. $\sqrt[3]{x}=2$.
5. $x^{\frac{1}{3}}=3$.
6. $\sqrt[3]{x}=4$.
7. $\sqrt{(x+9)}=6$.
8. $\sqrt{(x-7)}=7$.
9. $\sqrt{(x-15)}=8$.
10. $(x-9)^{\frac{1}{2}}=12$.
11. $\sqrt[3]{(4x-16)}=2$.
12. $20-3\sqrt{x}=2$.

13. $\sqrt[3]{(2x+3)}+4=7.$
14. $b+c\sqrt{x}=a.$
15. $\sqrt{(x^2-9)}+x=9.$
16. $\sqrt{(x^2-11)}=x-1.$
17. $\sqrt{(4x^2+5x-2)}=2x+1.$
18. $\sqrt{(9x^2-12x-51)}+3=3x.$
19. $\sqrt{(x^2-ax+b)}-a=x.$
20. $\sqrt{(25x^2-3mx+n)}-5x=m.$

319. When *two* surds are involved in an equation, one at least may be made to disappear by disposing the terms in such a way, that one of the surds stands by itself on one side of the equation, and then raising each side to the power corresponding to the order of the surd. If a surd be still left, it can be made to stand by itself, and removed by raising each side to a certain power.

Ex. 1. $\sqrt{(x-16)}+\sqrt{x}=8.$

Transposing the second term, we get
$$\sqrt{(x-16)}=8-\sqrt{x}.$$

Then, squaring both sides (Art. 306),
$$x-16=64-16\sqrt{x}+x;$$
therefore $\quad 16\sqrt{x}=64+16,$
or $\quad 16\sqrt{x}=80,$
or $\quad \sqrt{x}=5;$
$\quad x=25.$

Ex. 2. $\quad\sqrt{(x-5)}+\sqrt{(x+7)}=6.$

Transposing the second term,
$$\sqrt{(x-5)}=6-\sqrt{(x+7)}.$$

Squaring both sides, $x-5=36-12\sqrt{(x+7)}+x+7;$
therefore $\quad 12\sqrt{(x+7)}=36+x+7-x+5,$
or $\quad 12\sqrt{(x+7)}=48,$
or $\quad \sqrt{(x+7)}=4.$

Squaring both sides, $\quad x+7=16;$
therefore $\quad x=9.$

EXAMPLES.—CXX.

1. $\sqrt{(16+x)} + \sqrt{x} = 8$.
2. $\sqrt{(x-16)} = 8 - \sqrt{x}$.
3. $\sqrt{(x+15)} + \sqrt{x} = 15$.
4. $\sqrt{(x-21)} = \sqrt{x} - 1$.
5. $\sqrt{(x-1)} = 3 - \sqrt{(x+4)}$.
6. $1 + \sqrt{(3x+1)} = \sqrt{(4x+4)}$.
7. $1 - \sqrt{(1-3x)} = 2\sqrt{(1-x)}$.
8. $a - \sqrt{(x-a)} = \sqrt{x}$.
9. $\sqrt{x} + \sqrt{(x-m)} = \dfrac{m}{2}$.
10. $\sqrt{(x-1)} + \sqrt{(x-4)} - 3 = 0$.

320. When surds appear in the denominators of fractions in equations, the equations may be cleared of fractional terms by the process described in Art. 186, care being taken to follow the Laws of Combination of Surd Factors given in Art. 305.

EXAMPLES.—CXXi.

1. $\sqrt{x} + \sqrt{(x-9)} = \dfrac{36}{\sqrt{(x-9)}}$.
2. $\sqrt{x} + \sqrt{(x-21)} = \dfrac{35}{\sqrt{x}}$.
3. $\sqrt{(x+7)} + \sqrt{x} = \dfrac{28}{\sqrt{(x+7)}}$.
4. $\sqrt{(x-15)} + \sqrt{x} = \dfrac{105}{\sqrt{(x-15)}}$.
5. $\sqrt{x} + \sqrt{(x-4)} = \dfrac{8}{\sqrt{(x-4)}}$.
6. $\sqrt{x} + \sqrt{(3a+x)} - \dfrac{9a}{\sqrt{(3a+x)}} = 0$.
7. $\dfrac{\sqrt{(ax)}+b}{x+b} = \dfrac{b-a}{b-\sqrt{(ax)}}$.
8. $(1+\sqrt{x})(2-\sqrt{x}) = \dfrac{4+\sqrt{x}}{2}$.
9. $\dfrac{\sqrt{x}+16}{\sqrt{x}+4} = \dfrac{\sqrt{x}+32}{\sqrt{x}+12}$.
10. $\dfrac{\sqrt{x}-8}{\sqrt{x}-6} = \dfrac{\sqrt{x}-4}{\sqrt{x}+2}$.

321. The following are examples of Surd Equations resulting in quadratics.

Ex. 1. $\quad 2\sqrt{x} + \dfrac{2}{\sqrt{x}} = 5$.

Clearing the equation of fractions, $2x + 2 = 5\sqrt{x}$.

Squaring both sides, we get $4x^2 + 8x + 4 = 25x$; whence we find $x = 4$ or $\frac{1}{4}$.

Ex. 2. $\sqrt{(x+9)} = 2\sqrt{x} - 3$.

Squaring both sides, $\quad x + 9 = 4x - 12\sqrt{x} + 9$;

therefore $\quad\quad\quad\quad\quad 12\sqrt{x} = 3x$,

or $\quad\quad\quad\quad\quad\quad\quad 4\sqrt{x} = x$.

Squaring both sides, $\quad 16x = x^2$.

Divide by x, and we get $\quad 16 = x$.

Hence the values of x which satisfy the equation are 16 and 0 (Art. 248).

Ex. 3. $\quad \sqrt{(2x+1)} + 2\sqrt{x} = \dfrac{21}{\sqrt{(2x+1)}}$.

Clearing the equation of fractions,

$$2x + 1 + 2\sqrt{(2x^2 + x)} = 21;$$

therefore $\quad\quad 2\sqrt{(2x^2 + x)} = 20 - 2x$,

or $\quad\quad\quad\quad \sqrt{(2x^2 + x)} = 10 - x$.

Squaring both sides, $\quad 2x^2 + x = 100 - 20x + x^2$,

whence $\quad\quad\quad\quad x = 4$ or -25.

322. We shall now give a set of examples of Surd Equations some of which are reducible to Simple and others to Quadratic Equations.

EXAMPLES.—cxxii.

1. $4x - 12\sqrt{x} = 16$.
2. $45 - 14\sqrt{x} = -x$.
3. $3\sqrt{(7 + 2x^2)} = 5\sqrt{(4x - 3)}$.
4. $\sqrt{(6x - 11)} = \sqrt{(249 - 2x^2)}$.
5. $\sqrt{(6 - x)} = 2 - \sqrt{(2x - 1)}$.
6. $x - 2\sqrt{(4 - 3x)} + 12 = 0$.
7. $\sqrt{(2x + 7)} + \sqrt{(3x - 18)} = \sqrt{(7x + 1)}$.
8. $2\sqrt{(204 - 5x)} = 20 - \sqrt{(3x - 68)}$.

9. $\sqrt{x-4} = \dfrac{33}{\sqrt{x+4}}$.

10. $\sqrt{x+11} = \dfrac{608}{\sqrt{x-11}}$.

11. $\sqrt{(x+5)} \cdot \sqrt{(x+12)} = 12$.

12. $\sqrt{(x+3)} + \sqrt{(x+8)} = 5\sqrt{x}$.

13. $\sqrt{(25+x)} + \sqrt{(25-x)} = 8$.

14. $\sqrt{(x+4)} + \sqrt{(2x-1)} = 6$.

15. $\sqrt{(13x-1)} - \sqrt{(2x-1)} = 5$.

16. $\sqrt{(7x+1)} - \sqrt{(3x+1)} = 2$.

17. $\sqrt{(4+x)} + \sqrt{x} = 3$.

18. $\sqrt{x} + \sqrt{(x+9975)} = \dfrac{525}{\sqrt{x}}$.

19. $\sqrt{\left(\dfrac{x}{4}+3\right)} + \sqrt{\left(\dfrac{x}{4}-3\right)} = \sqrt{\left(\dfrac{2x}{3}\right)}$.

20. $\sqrt{(x^2-1)} + 6 = \dfrac{16}{\sqrt{(x^2-1)}}$.

21. $\sqrt{\{(x-a)^2 + 2ab + b^2\}} = x - a + b$.

22. $\sqrt{\{(x+a)^2 + 2ab + b^2\}} = b - a - x$.

23. $\sqrt{(x+4)} - \sqrt{x} = \sqrt{\left(x+\dfrac{3}{2}\right)}$.

24. $\dfrac{x-1}{\sqrt{x-1}} = x + \dfrac{5}{4}$.

25. $\sqrt{(4+x)} - \sqrt{3} = \sqrt{x}$.

26. $\sqrt{(x+4)} + \sqrt{(x+5)} = 9$.

27. $\sqrt{x} + \sqrt{(x-4)} = \dfrac{8}{\sqrt{(x-4)}}$.

28. $x^2 = 21 + \sqrt{(x^2-9)}$.

29. $\sqrt{(50+x)} - \sqrt{(50-x)} = 2$.

30. $\sqrt{(2x+4)} - \sqrt{\left(\dfrac{x}{2}+6\right)} = 1$.

31. $\sqrt{(3+x)} + \sqrt{x} = \dfrac{6}{\sqrt{(3+x)}}$.

32. $\dfrac{1}{\sqrt{(x+1)}} + \dfrac{1}{\sqrt{(x-1)}} = \dfrac{1}{\sqrt{(x^2-1)}}$.

33. $\dfrac{3x + \sqrt{(4x-x^2)}}{3x - \sqrt{(4x-x^2)}} = 2$.

34. $\sqrt{x} + \sqrt{\{a - \sqrt{(ax+x^2)}\}} = \sqrt{a}$.

XXVI. ON THE ROOTS OF EQUATIONS

323. We have already proved that a Simple Equation can have only *one* root (Art. 193): we have now to prove that a Quadratic Equation can have only *two* roots.

324. We must first call attention to the following fact:

If $mn = 0$, *either* $m = 0$, *or* $n = 0$.

Thus there is an ambiguity: but if we know that m cannot be equal to 0, then we know for certain that $n = 0$, and if we know that n cannot be equal to 0, then we know for certain that $m = 0$.

Further, if $lmn = 0$, then *either* $l = 0$, *or* $m = 0$, *or* $n = 0$, and so on for any number of factors.

Ex. 1. Solve the equation $(x-3)(x+4) = 0$.

Here we must have
$$x - 3 = 0, \text{ or } x + 4 = 0,$$
that is, $\quad x = 3, \text{ or } x = -4.$

Ex. 2. $(x - 3a)(5x - 2b) = 0$.

Here we must have
$$x - 3a = 0, \text{ or } 5x - 2b = 0,$$
that is, $\quad x = 3a, \text{ or } x = \dfrac{2b}{5}.$

EXAMPLES.—CXXIII.

1. $(x-2)(x-5)=0.$ 2. $(x-3)(x+7)=0.$ 3. $(x+9)(x+2)=0.$
4. $(x-5a)(x-6b)=0.$ 6. $(19x-227)(14x+83)=0.$
5. $(2x+7)(3x-5)=0$ 7. $(5x-4m)(6x-11n)=0.$
8. $(x^2+5ax+6a^2)(x^2-7ax+12a^2)=0.$
9. $(x^2-4)(x^2-2ax+a^2)=0.$
10. $x(x^2-5x)=0.$
11. $(acx-2a+b)(bcx+3a-b)=0.$
12. $(cx-d)(cx-e)=0.$

325. The general form of a quadratic equation is
$$ax^2+bx+c=0.$$
Hence
$$a\left(x^2+\frac{b}{a}x+\frac{c}{a}\right)=0.$$
Now a cannot $=0$,
$$\therefore x^2+\frac{b}{a}x+\frac{c}{a}=0.$$

Writing p for $\frac{b}{a}$ and q for $\frac{c}{a}$, we may take the following as the type of a quadratic equation of which the coefficient of the first term is unity,
$$x^2+px+q=0.$$

326. To show that a quadratic equation has *only two* roots.

Let $x^2+px+q=0$ be the equation.
Suppose it to have *three* different roots, a, b, c.
Then
$$a^2+ap+q=0\ldots\ldots\ldots\ldots(1),$$
$$b^2+bp+q=0\ldots\ldots\ldots\ldots(2),$$
$$c^2+cp+q=0\ldots\ldots\ldots\ldots(3).$$

Subtracting (2) from (1),
$$a^2-b^2+(a-b)p=0,$$
or,
$$(a-b)(a+b+p)=0.$$

Now $a-b$ does not equal 0, since a and b are not alike,
$$\therefore a+b+p=0 \quad \ldots\ldots\ldots\ldots\ldots\ldots (4).$$

Again, subtracting (3) from (1),
$$a^2 - c^2 + (a-c)\,p = 0,$$
or,
$$(a-c)(a+c+p) = 0.$$

Now $a-c$ does not equal 0, since a and c are not alike,
$$\therefore a+c+p=0 \quad \ldots\ldots\ldots\ldots\ldots\ldots (5).$$

Then subtracting (5) from (4), we get
$$b-c=0, \text{ and therefore } b=c.$$

Hence there are not more than *two* distinct roots.

327. We now proceed to show the relations existing between the Roots of a quadratic equation and the Coefficients of the terms of the equation.

328.
$$x^2 + px + q = 0$$
is the general form of a quadratic equation, in which the coefficient of the first term is unity.

Hence
$$x^2 + px = -q$$
$$x^2 + px + \frac{p^2}{4} = \frac{p^2}{4} - q,$$
$$x + \frac{p}{2} = \pm \sqrt{\left(\frac{p^2}{4} - q\right)},$$
$$x = -\frac{p}{2} \pm \sqrt{\left(\frac{p^2}{4} - q\right)}.$$

Now if a and β be the roots of the equation,
$$a = -\frac{p}{2} + \sqrt{\left(\frac{p^2}{4} - q\right)} \quad \ldots\ldots\ldots\ldots (1),$$
$$\beta = -\frac{p}{2} - \sqrt{\left(\frac{p^2}{4} - q\right)} \quad \ldots\ldots\ldots\ldots (2).$$

Adding (1) and (2), we get
$$a + \beta = -p \quad \ldots\ldots\ldots\ldots\ldots\ldots (3).$$

Multiplying (1) and (2), we get

$$\alpha\beta = \frac{p^2}{4} - \frac{p}{2}\sqrt{\left(\frac{p^2}{4} - q\right)} + \frac{p}{2}\sqrt{\left(\frac{p^2}{4} - q\right)} - \left(\frac{p^2}{4} - q\right),$$

or $\quad \alpha\beta = \dfrac{p^2}{4} - \dfrac{p^2}{4} + q,$

or $\quad \alpha\beta = q$ ....(4).

From (3) we learn that *the sum of the roots is equal to the coefficient of the second term with its sign changed.*

From (4) we learn that *the product of the roots is equal to the last term.*

329. The equation $x^2 + px + q = 0$ has its roots real and different, real and equal, or impossible and different, according as p^2 is $>$ = or $<4q$.

For the roots are

$$-\frac{p}{2} + \sqrt{\left(\frac{p^2}{4} - q\right)}, \text{ or } \frac{-p + \sqrt{(p^2 - 4q)}}{2},$$

and $\quad -\dfrac{p}{2} - \sqrt{\left(\dfrac{p^2}{4} - q\right)}, \text{ or } \dfrac{-p - \sqrt{(p^2 - 4q)}}{2}.$

First, let p^2 be greater than $4q$, then $\sqrt{(p^2 - 4q)}$ is a possible quantity, and the roots are different in value and both real.

Next, let $p^2 = 4q$, then each of the roots is equal to the real quantity $\dfrac{-p}{2}$.

Lastly, let p^2 be less than $4q$, then $\sqrt{(p^2 - 4q)}$ is an impossible quantity and the roots are different and both impossible

EXAMPLES.—CXXIV.

1. If the equations

$$ax^2 + bx + c = 0, \text{ and } a'x^2 + b'x + c' = 0,$$

have respectively two roots, one of which is the reciprocal of the other, prove that

$$(aa' - cc')^2 = (ab' - bc')(a'b - b'c).$$

2. If a, β be the roots of the equation $ax^2 + bx + c = 0$, prove that
$$a^2 + \beta^2 = \frac{b^2 - 2ac}{a^2}.$$

3. If a, β be the roots of the equation $ax^2 + bx + c = 0$, prove that
$$acx^2 + (2ac - b^2)x + ac = ac\left(x - \frac{a}{\beta}\right)\left(x - \frac{\beta}{a}\right).$$

4. Prove that, if the roots of the equation $ax^2 + bx + c = 0$ be equal, $ax^2 + bx + c$ is a perfect square with respect to x.

5. If a, β represent the two roots of the equation
$$x^2 - (1 + a)x + \frac{1}{2}(1 + a + a^2) = 0,$$
show that $\qquad a^2 + \beta^2 = a.$

330. If a and β be the roots of the equation $x^2 + px + q = 0$, then $\qquad x^2 + px + q = (x - a)(x - \beta).$

For since $p = -(a + \beta)$ and $q = a\beta$,
$$x^2 + px + q = x^2 - (a + \beta)x + a\beta$$
$$= (x - a)(x - \beta).$$

Hence we may form a quadratic equation of which the roots are given.

Ex. 1. Form the equation whose roots are 4 and 5.

Here $x - a = x - 4$ and $x - \beta = x - 5$;

$\therefore$ the equation is $(x - 4)(x - 5) = 0$;

or, $\qquad x^2 - 9x + 20 = 0.$

Ex. 2. Form the equation whose roots are $\frac{1}{2}$ and -3.

Here $x - a = x - \frac{1}{2}$ and $x - \beta = x + 3$;

$\therefore$ the equation is $\left(x - \frac{1}{2}\right)(x + 3) = 0$;

or, $\qquad (2x - 1)(x + 3) = 0;$
or, $\qquad 2x^2 + 5x - 3 = 0.$

EXAMPLES.—CXXV.

Form the equations whose roots are

1. 5 and 6.
2. 4 and -5.
3. -2 and -7.
4. $\frac{1}{2}$ and $\frac{2}{3}$.
5. 7 and $-\frac{5}{9}$.
6. $\sqrt{3}$ and $-\sqrt{3}$.
7. $m+n$ and $m-n$.
8. $\frac{1}{a}$ and $\frac{1}{\beta}$.
9. $-\frac{a}{\beta}$ and $\frac{\beta}{a}$.

---

331. Any expression containing x is said to be A FUNCTION of x. An expression containing any symbol x is said to be a *positive integral function* of x when all the powers of x contained in it have positive integral indices.

For example, $5x^7 + 2x^5 + \frac{3}{2}x^4 + \frac{1}{10}x^2 + 3$ is a positive integral function of x, but $6x^5 + 3x^{\frac{1}{3}} + 1$ and $5x^7 - 2x^{-2} + 3x^2 + 1$ are not, because the first contains $x^{\frac{1}{3}}$, of which the index is not integral, and the second contains x^{-2}, of which the index is not positive.

332. The expression $5x^3 + 4x^2 + 2$ is said to be the expression corresponding to the equation $5x^3 + 4x^2 + 2 = 0$, and the latter is the equation corresponding to the former.

333. If a be a root of an equation, then $x - a$ is a factor of the corresponding expression, provided the equation and expression contain only positive integral powers of x. This principle is useful in resolving such an expression into factors. We have already proved it to be true in the case of a quadratic equation. The general proof of it is not suitable for the stage at which the learner is now supposed to be arrived, but we will illustrate it by some Examples.

Ex. 1. Resolve $2x^2 - 5x + 3$ into factors.

If we solve the equation $2x^2 - 5x + 3 = 0$, we shall find that its roots are 1 and $\dfrac{3}{2}$.

Now divide $2x^2 - 5x + 3$ by $x - 1$; the quotient is $2x - 3$ that is $2\left(x - \dfrac{3}{2}\right)$;

$$\therefore \text{ the given expression} = 2(x - 1)\left(x - \dfrac{3}{2}\right).$$

Ex. 2. Resolve $2x^3 + x^2 - 11x - 10$ into factors.

By trial we find that this expression vanishes if we put $x = -1$; that is, -1 is a root of the equation

$$2x^3 + x^2 - 11x - 10 = 0.$$

Divide the expression by $x + 1$: the quotient is $2x^2 - x - 10$;

$\therefore$ the expression $= (2x^2 - x - 10)(x + 1)$

$$= 2\left(x^2 - \dfrac{x}{2} - 5\right)(x + 1).$$

We must now resolve $x^2 - \dfrac{x}{2} - 5$ into factors, by solving the corresponding equation $x^2 - \dfrac{x}{2} - 5 = 0$.

The roots of this equation are -2 and $\dfrac{5}{2}$;

$$\therefore 2x^3 + x^2 - 11x - 10 = 2(x + 2)\left(x - \dfrac{5}{2}\right)(x + 1)$$
$$= (x + 2)(2x - 5)(x + 1).$$

EXAMPLES.—CXXVI.

Resolve into simple factors the following expressions:

1. $x^3 - 11x^2 + 36x - 36$.
2. $x^3 - 7x^2 + 14x - 8$.
3. $x^3 - 5x^2 - 46x - 40$.
4. $4x^3 + 6x^2 + x - 1$.
5. $6x^3 + 11x^2 - 9x - 14$.
6. $x^3 + y^3 + z^3 - 3xyz$.
7. $a^3 - b^3 - c^3 - 3abc$.
8. $3x^3 - x^2 - 23x + 21$.
9. $2x^3 - 5x^2 - 17x + 20$.
10. $15x^3 + 41x^2 + 5x - 21$.

334. If we can find one root of such an equation as
$$2x^3 + x^2 - 11x - 10 = 0,$$
we can find all the roots.

One root of the equation is -1;
$$\therefore (x+1)(2x^2 - x - 10) = 0;$$
$$\therefore x + 1 = 0, \text{ or } 2x^2 - x - 10 = 0;$$
$$\therefore x = -1, \text{ or } -2, \text{ or } \frac{5}{2}.$$

Similarly, if we can find one root of an equation involving the 4th power of x, we can derive from it an equation involving the 3rd and lower powers of x, from which we may find the other roots. And if again we can find one root of this, the other two roots can be found from a quadratic equation.

335. Any equation into which an unknown symbol or expression enters in two terms only, having its index in one of the terms *double* of its index in the other, may be solved as a quadratic equation.

Ex. Solve the equation $x^6 - 6x^3 = 7$.

Regarding x^3 as the quantity to be obtained by the solution of the equation, we get
$$x^6 - 6x^3 + 9 = 16;$$
therefore $\qquad x^3 - 3 = \pm 4;$
therefore $\qquad x^3 = 7, \text{ or } x^3 = -1.$
Hence $\qquad x = \sqrt[3]{7} \text{ or } x = \sqrt[3]{-1},$
and one value of $\sqrt[3]{-1}$ is -1.

336. In some cases by adding a certain quantity to both sides of an equation we can bring it into a form capable of solution, thus, to solve the equation
$$x^2 + 5x + 4 = 5\sqrt{(x^2 + 5x + 28)},$$
add 24 to each side.

Then $\qquad x^2 + 5x + 28 = 5\sqrt{(x^2 + 5x + 28)} + 24;$
or, $\qquad x^2 + 5x + 28 - 5\sqrt{(x^2 + 5x + 28)} = 24.$

This is now in the form of a quadratic equation, the unknown quantity being $\sqrt{(x^2 + 5x + 28)}$, and completing the square we have

$$x^2 + 5x + 28 - 5\sqrt{(x^2 + 5x + 28)} + \frac{25}{4} = \frac{121}{4};$$

$$\therefore \sqrt{(x^2 + 5x + 28)} - \frac{5}{2} = \pm\frac{11}{2};$$

whence $\sqrt{(x^2 + 5x + 28)} = 8$ or -3;

$$\therefore x^2 + 5x + 28 = 64 \text{ or } 9;$$

from which we may find four values of x, viz. $4, -9$, and $-\frac{5}{2} \pm \frac{\sqrt{(-51)}}{2}$.

EXAMPLES.—CXXVII.

Find roots of the following equations:

1. $x^4 - 12x^2 = 13$.
2. $x^6 + 14x^3 + 24 = 0$.
3. $x^8 + 22x^4 + 21 = 0$.
4. $x^{2m} + 3x^m = 4$.
5. $x^{4n} - \frac{5}{3}x^{2n} = \frac{25}{12}$.
6. $x - \frac{9}{2}x^{\frac{1}{2}} = \frac{5}{2}$.
7. $x^{-2} + 3x^{-1} = \frac{4}{9}$.
8. $x^{-2n} - x^{-n} = 20$.
9. $x^2 - 2x + 6(x^2 - 2x + 5)^{\frac{1}{2}} = 11$.
10. $x^2 - x + 5\sqrt{(2x^2 - 5x + 6)} = \frac{3x + 33}{2}$.
11. $x^2 - 2\sqrt{(3x^2 - 2ax + 4)} + 4 = \frac{2a}{3}\left(x + \frac{a}{2} + 1\right)$.
12. $ax + 2\sqrt{(x^2 - ax + a^2)} = x^2 + 2a$.

337. Every equation has as many roots as it has dimensions, and no more. This we have proved in the case of simple and quadratic equations (Arts. 193, 323). The general proof is not suited to this work, but we may illustrate it by the following Examples.

Ex. 1. To solve the equation $x^3 - 1 = 0$.

One root is clearly 1.

Dividing by $x - 1$, we obtain $x^2 + x + 1 = 0$, of which the roots are $\dfrac{-1 + \sqrt{-3}}{2}$ and $\dfrac{-1 - \sqrt{-3}}{2}$.

Hence the *three* roots are 1, $\dfrac{-1+\sqrt{-3}}{2}$ and $\dfrac{-1-\sqrt{-3}}{2}$.

Ex. 2. To solve the equation $x^4 - 1 = 0$.

Two of the roots are evidently $+1$ and -1.

Hence, dividing by $(x-1)(x+1)$, that is by x^2-1, we obtain $x^2 + 1 = 0$, of which the roots are $\sqrt{-1}$ and $-\sqrt{-1}$.

Hence the *four* roots are $1, -1, \sqrt{-1}$, and $-\sqrt{-1}$.

The equation $x^6 - 6x^3 = 7$ will in like manner have *six* roots, for it may be reduced, as in Art. 335, to two cubic equations, $x^3 - 7 = 0$ and $x^3 + 1 = 0$, each of which has *three* roots, which may be found as in Ex. 1.

XXVII. ON RATIO.

338. IF A and B stand for two unequal quantities of the same kind, we may consider their inequality in two ways. We may ask.

(1) *By what quantity* one is greater than the other?

The answer to this is made by stating the difference between the two quantities. Now since quantities are represented in Algebra by their measures (Art. 33), if a and b be the measures of A and B, the difference between A and B is represented algebraically by $a - b$.

(2) *By how many times* one is greater than the other?

The answer to this question is made by stating the number of times the one contains the other.

NOTE. The quantities must be of the *same kind*. We cannot compare inches with hours, nor lines with surfaces.

339. The second method of comparing A and B is called finding the RATIO of A to B, and we give the following definition.

DEF. Ratio is the relation which one quantity bears to another of the same kind with respect to the number of times the one contains the other.

340. The ratio of A to B is expressed thus, $A : B$.

A and B are called the TERMS of the ratio.

A is called the ANTECEDENT and B the CONSEQUENT.

341. Now since quantities are represented in Algebra by their measures, we must represent the ratio between two quantities by the ratio between their measures. Our next step then must be to show how to estimate the ratio between two *numbers*. This ratio is determined by finding how many times one contains the other, that is, by obtaining the quotient resulting from the division of one by the other. If a and b, then, be any two numbers, the fraction $\frac{a}{b}$ will express the ratio of a to b. (Art. 136.)

342. Thus if a and b be the measures of A and B respectively, the ratio of A to B is represented algebraically by the fraction $\frac{a}{b}$.

343. If a or b or both are surd numbers, the fraction $\frac{a}{b}$ may also be a surd, and its approximate value can be found by Art. 291. Suppose this value to be $\frac{m}{n}$, where m and n are whole numbers: then we should say that the ratio $A : B$ is approximately represented by $\frac{m}{n}$.

344. Ratios may be compared with each other, by comparing the fractions by which they are denoted.

Thus the ratios $3 : 4$ and $4 : 5$ may be compared by comparing the fractions $\frac{3}{4}$ and $\frac{4}{5}$.

These are equivalent to $\frac{15}{20}$ and $\frac{16}{20}$ respectively; and since $\frac{16}{20}$ is greater than $\frac{15}{20}$, the ratio $4 : 5$ is greater than the ratio $3 : 4$.

EXAMPLES.—CXXVIII.

1. Place in order of magnitude the ratios $2:3$, $6:7$, $7:9$.
2. Compare the ratios $x+3y:x+2y$ and $x+2y:x+y$.
3. Compare the ratios $x-5y:x-4y$ and $x-3y:x-2y$.
4. What number must be added to each of the terms of the ratio $a:b$, that it may become the ratio $c:d$?
5. The sum of the squares of the Antecedent and Consequent of a Ratio is 181, and the product of the Antecedent and Consequent is 90. What is the ratio?

345. A ratio of *greater inequality* is one whose antecedent is greater than its consequent.

A ratio of *less inequality* is one whose antecedent is less than its consequent.

This is the same as saying a ratio of greater inequality is represented by an Improper Fraction, and a ratio of less inequality by a Proper Fraction.

346. *A Ratio of greater inequality is diminished by adding the same number to both its terms.*

Thus if 1 be added to both terms of the ratio $5:2$ it becomes $6:3$, which is less than the former ratio, since $\frac{6}{3}$, that is, 2, is less than $\frac{5}{2}$.

And, in general, if x be added to both terms of the ratio $a:b$, where a is greater than b, we may compare the two ratios thus,

ratio $a+x:b+x$ is less than ratio $a:b$,

if $\qquad \dfrac{a+x}{b+x}$ be less than $\dfrac{a}{b}$,

if $\qquad \dfrac{ab+bx}{b^2+bx}$ be less than $\dfrac{ab+ax}{b^2+bx}$,

if $\qquad ab+bx$ be less than $ab+ax$,

if $\qquad bx$ be less than ax,

if $\qquad b$ be less than a.

Now b is less than a;

$\therefore a+x:b+x$ is less than $a:b$.

347. We may observe that Art. 346 is merely a repetition of that which we proposed as an Example at the end of the chapter on Miscellaneous Fractions. There is not indeed any necessity for us to weary the reader with examples on Ratio: for since we express a ratio by a fraction, nearly all that we might have had to say about Ratios has been anticipated in our remarks on Fractions.

348. The student may, however, work the following Theorems as Examples.

(1) If $a : b$ be a ratio of greater inequality, and x a positive quantity, the ratio $a-x : b-x$ is greater than the ratio $a : b$.

(2) If $a : b$ be a ratio of less inequality, and x a positive quantity, the ratio $a+x : b+x$ is greater than the ratio $a : b$.

(3) If $a : b$ be a ratio of less inequality, and x a positive quantity, the ratio $a-x : b-x$ is less than the ratio $a : b$.

349. In some cases we may from a single equation involving two unknown symbols determine the ratio between the two symbols. In other words we may be able to determine the *relative* values of the two symbols, though we cannot determine their *absolute* values.

Thus from the equation $4x = 3y$,

we get $$\frac{x}{y} = \frac{3}{4}.$$

Again, from the equation $3x^2 = 2y^2$,

we get $\dfrac{x^2}{y^2} = \dfrac{2}{3}$; and therefore $\dfrac{x}{y} = \dfrac{\sqrt{2}}{\sqrt{3}}.$

EXAMPLES.—CXXIX.

Find the ratio of x to y from the following equations:

1. $9x = 6y$.
2. $ax = by$.
3. $ax - by = cx + dy$.
4. $x^2 + 2xy = 5y^2$.
5. $x^2 - 12xy = 13y^2$.
6. $x^2 + mxy = n^2y^2$.

7. Find two numbers in the ratio of $3 : 4$, of which the sum is to the sum of their squares $:: 7 : 50$.

8. Two numbers are in the ratio of $6 : 7$, and when 12 is added to each the resulting numbers are in the ratio of $12 : 13$. Find the numbers.

9. The sum of two numbers is 100, and the numbers are in the ratio of 7 : 13. Find them.

10. The difference of the squares of two numbers is 48, and the sum of the numbers is to the difference of the numbers in the ratio 12 : 1. Find the numbers.

11. If 5 gold coins and 4 silver ones are worth as much as 3 gold coins and 12 silver ones, find the ratio of the value of a gold coin to that of a silver one.

12. If 8 gold coins and 9 silver ones are worth as much as 6 gold coins and 19 silver ones, find the ratio of the value of a silver coin to that of a gold one.

350. Ratios are *compounded* by multiplying together the fractions by which they are denoted.

Thus the ratio compounded of $a : b$ and $c : d$ is $ac : bd$.

EXAMPLES.—CXXX.

Write the ratios compounded of the ratios

1. $2 : 3$ and $4 : 5$.
2. $3 : 7$, $14 : 9$ and $4 : 3$.
3. $x^2 - y^2 : x^3 + y^3$ and $x^2 - xy + y^2 : x + y$.
4. $a^2 - b^2 + 2bc - c^2 : a^2 - b^2 - 2bc - c^2$ and $a + b + c : a + b - c$.
5. $m^3 + n^3 : m^3 - n^3$ and $m - n : m + n$.
6. $x^2 + 5x + 6 : y^2 - 7y + 12$, and $y^2 - 3y : x^2 + 3x$.

351. The ratio $a^2 : b^2$ is called the DUPLICATE RATIO of $a : b$.

Thus $100 : 64$ is the duplicate ratio of $10 : 8$,

and $36x^2 : 25y^2$ is the duplicate ratio of $6x : 5y$.

The ratio $a^3 : b^3$ is called the TRIPLICATE RATIO of $a : b$.

Thus $64 : 27$ is the triplicate ratio of $4 : 3$,

and $343x^3 : 1331y^3$ is the triplicate ratio of $7x : 11y$.

352. The definition of Ratio given in Euclid is the same as in Algebra, and so also is the expression for the ratio that one quantity bears to another, that is, $A : B$. But Euclid cannot employ fractions, and hence he cannot represent the value of a ratio as we do in Algebra.

XXVIII ON PROPORTION.

353. Proportion consists in the equality of two ratios.

The algebraic test of PROPORTION is *that the two fractions representing the ratios must be equal.*

Thus the ratio $a : b$ will be equal to the ratio $c : d$,
$$\text{if } \frac{a}{b} = \frac{c}{d},$$
and the *four numbers a, b, c, d* are in such a case said to be in proportion.

354. If the ratios $a : b$ and $c : d$ form a proportion, we express the fact thus:
$$a : b = c : d.$$

This is the clearest manner of expressing the equality of the ratios $a : b$ and $c : d$, but there is another way of expressing the same fact, thus
$$a : b :: c : d,$$
which is read thus,
$$a \text{ is to } b \text{ as } c \text{ is to } d.$$

The two terms a and d are called the EXTREMES.
.... b and c .... the MEANS.

355. *When four numbers are in proportion,*
product of extremes = product of means.

Let a, b, c, d be in proportion.

Then $$\frac{a}{b} = \frac{c}{d}.$$

Multiplying both sides of the equation by bd, we get

$$ad = bc.$$

Conversely, if $ad = bc$ we can show that $a : b = c : d$.

For since $ad = bc$,
dividing both sides by bd, we get

$$\frac{ad}{bd} = \frac{bc}{bd},$$

that is, $\quad\quad\quad\quad \frac{a}{b} = \frac{c}{d}$, i.e. $a : b = c : d$.

356. If $ad = bc$,

Dividing by cd, we get $\frac{a}{c} = \frac{b}{d}$, i.e. $a : c = b : d$;

Dividing by ab, we get $\frac{d}{b} = \frac{c}{a}$, i.e. $d : b = c : a$;

Dividing by ac, we get $\frac{d}{c} = \frac{b}{a}$, i.e. $d : c = b : a$.

357. From this it follows that if any 4 numbers be so related that the product of two is equal to the product of the other two, we can express the 4 numbers in the form of a proportion.

The factors of one of the products must form the extremes.

The factors of the other product must form the means.

358. *Three* quantities are said to be in CONTINUED PROPORTION when the ratio of the first to the second is equal to the ratio of the second to the third.

Thus a, b, c are in continued proportion if

$$a : b = b : c.$$

The quantity b is called a MEAN PROPORTIONAL between a and c.

Four quantities are said to be in Continued Proportion when the ratios of the first to the second, of the second to the third, and of the third to the fourth are all equal.

Thus a, b, c, d are in continued proportion when

$$a : b = b : c = c : d.$$

359. We showed in Art. 205 the process by which when two or more fractions are known to be equal, other relations between the numbers involved in them may be determined. That process is of course applicable to Examples in Ratio and Proportion, as we shall now show by particular instances.

Ex. 1. If $a : b = c : d$, prove that

$$a^2 + b^2 : a^2 - b^2 = c^2 + d^2 : c^2 - d^2.$$

Since $a : b = c : d$, $\dfrac{a}{b} = \dfrac{c}{d}$.

Let $\dfrac{a}{b} = \lambda$. Then $\dfrac{c}{d} = \lambda$;

$$\therefore a = \lambda b, \text{ and } c = \lambda d.$$

Now
$$\frac{a^2 + b^2}{a^2 - b^2} = \frac{\lambda^2 b^2 + b^2}{\lambda^2 b^2 - b^2} = \frac{b^2(\lambda^2 + 1)}{b^2(\lambda^2 - 1)} = \frac{\lambda^2 + 1}{\lambda^2 - 1},$$

and
$$\frac{c^2 + d^2}{c^2 - d^2} = \frac{\lambda^2 d^2 + d^2}{\lambda^2 d^2 - d^2} = \frac{d^2(\lambda^2 + 1)}{d^2(\lambda^2 - 1)} = \frac{\lambda^2 + 1}{\lambda^2 - 1}.$$

Hence
$$\frac{a^2 + b^2}{a^2 - b^2} = \frac{c^2 + d^2}{c^2 - d^2};$$

that is, $a^2 + b^2 : a^2 - b^2 = c^2 + d^2 : c^2 - d^2$.

Ex. 2. If $a : b :: c : d$, prove that

$$a : c :: \sqrt[4]{(a^4 + b^4)} : \sqrt[4]{(c^4 + d^4)}.$$

Let $\dfrac{a}{b} = \lambda$. Then $\dfrac{c}{d} = \lambda$;

$$\therefore a = \lambda b, \text{ and } c = \lambda d.$$

Now $\dfrac{a}{c} = \dfrac{\lambda b}{\lambda d} = \dfrac{b}{d}$,

and $\dfrac{\sqrt[4]{(a^4+b^4)}}{\sqrt[4]{(c^4+d^4)}} = \dfrac{\sqrt[4]{(\lambda^4 b^4+b^4)}}{\sqrt[4]{(\lambda^4 d^4+d^4)}} = \dfrac{\sqrt[4]{b^4}\cdot\sqrt[4]{(\lambda^4+1)}}{\sqrt[4]{d^4}\cdot\sqrt[4]{(\lambda^4+1)}} = \dfrac{\sqrt[4]{b^4}}{\sqrt[4]{d^4}} = \dfrac{b}{d}$

Hence $\dfrac{a}{c} = \dfrac{\sqrt[4]{(a^4+b^4)}}{\sqrt[4]{(c^4+d^4)}}$;

that is, $a : c :: \sqrt[4]{(a^4+b^4)} : \sqrt[4]{(c^4+d^4)}$.

Ex. 3. If $a : b = c : d = e : f$, prove that each of these ratios is equal to the ratio $a+c+e : b+d+f$.

Let $\dfrac{a}{b} = \lambda$, $\dfrac{c}{d} = \lambda$, $\dfrac{e}{f} = \lambda$.

Then $a = \lambda b$, $c = \lambda d$, $e = \lambda f$.

Now $\dfrac{a+c+e}{b+d+f} = \dfrac{\lambda b + \lambda d + \lambda f}{b+d+f} = \dfrac{\lambda(b+d+f)}{b+d+f} = \lambda$.

Hence $\dfrac{a+c+e}{b+d+f} = \dfrac{a}{b} = \dfrac{c}{d} = \dfrac{e}{f}$,

that is, $a+c+e : b+d+f = a : b = c : d = e : f$.

Ex. 4. If a, b, c are in continued proportion, show that
$$a^2+b^2 : b^2+c^2 = a : c.$$

Let $\dfrac{a}{b} = \lambda$. Then $\dfrac{b}{c} = \lambda$.

Hence $a = \lambda b$ and $b = \lambda c$.

Now $\dfrac{a^2+b^2}{b^2+c^2} = \dfrac{\lambda^2 b^2 + b^2}{b^2+c^2} = \dfrac{b^2(\lambda^2+1)}{\lambda^2 c^2 + c^2} = \dfrac{b^2(\lambda^2+1)}{c^2(\lambda^2+1)} = \dfrac{b^2}{c^2} = \dfrac{ac}{c^2} = \dfrac{a}{c}$.

Ex. 5. If $15a+b : 15c+d = 12a+b : 12c+d$, prove that
$$a : b = c : d.$$

Since $15a+b : 15c+d = 12a+b : 12c+d$,

and since product of extremes = product of means,

$$(15a+b)(12c+d) = (15c+d)(12a+b),$$

or, $\quad 180ac + 12bc + 15ad + bd = 180ac + 12ad + 15bc + bd,$

or, $\quad\quad\quad 12bc + 15ad = 12ad + 15bc,$

or, $\quad\quad\quad\quad 3ad = 3bc,$

or, $\quad\quad\quad\quad ad = bc.$

Whence, by Art. 355, $\quad a:b=c:d.$

Additional Examples will be found in page 137, to which we may add the following.

EXAMPLES.—CXXXi.

1. If $a:b=c:d$, show that $a+b:a=c+d:c$.

2. If $a:b=c:d$, show that $a^2-b^2:b^2=c^2-d^2:d^2$.

3. If $a_1:b_1=a_2:b_2$, show that $\dfrac{m_1 a_1 + m_2 a_2}{m_1 b_1 + m_2 b_2} = \dfrac{a_1}{b_1}$.

4. If $a:b::c:d$, show that
$$3a^2+ab+2b^2 : 3a^2-2b^2 :: 3c^2+cd+2d^2 : 3c^2-2d^2.$$

5. If $a:b=c:d$, show that
$$a^2+3ab+b^2 : c^2+3cd+d^2 = 2ab+3b^2 : 2cd+3d^2.$$

6. If $a:b=c:d=e:f$ then $a:b = mc-ne : md-nf$.

7. If $\dfrac{m}{n}a$, $\dfrac{m}{n}b$, any parts of a, b, be taken from a and b respectively, show that a, b, and the remainders form a proportion.

8. If $a:b=c:d=e:f$, show that
$$ac:bd = la^2+mc^2+ne^2 : lb^2+md^2+nf^2.$$

9. If $a_1:b_1=a_2:b_2=a_3:b_3$, show that
$$a_1^2+a_2^2+a_3^2 : b_1^2+b_2^2+b_3^2 :: a_1^2 : b_1^2.$$

10. If $a_1 : b_1 = a_2 : b_2 = a_3 : b_3$, show that
$$a_1 a_2 + a_2 a_3 + a_3 a_1 : b_1 b_2 + b_2 b_3 + b_3 b_1 = a_1^2 : b_1^2.$$

11. If $\dfrac{a^2 - ab + b^2}{a^2 + ab + b^2} = \dfrac{c^2 - cd + d^2}{c^2 + cd + d^2}$, show that either $\dfrac{a}{b} = \dfrac{c}{d}$ or $\dfrac{a}{b} = \dfrac{d}{c}$.

12. If $a^2 + b^2 : a^2 - b^2 = c^2 + d^2 : c^2 - d^2$, show that
$$a : b = c : d.$$

13. If $a : b = c : d$, show that
$$\frac{(a+c)(a^2+c^2)}{(a-c)(a^2-c^2)} = \frac{(b+d)(b^2+d^2)}{(b-d)(b^2-d^2)}.$$

14. If $a_1 : b_1 = a_2 : b_2$, show that
$$a_1 : b_1 = \sqrt{(a_1^2 + a_2^2)} : \sqrt{(b_1^2 + b_2^2)}.$$

On the Geometrical Treatment of Proportion.

360. The definition of Proportion (viz. the equality of ratios) is the same in Euclid as in Algebra. (Eucl. Book v. Def. 6 and 8.)

But the ways of testing whether two ratios are equal are quite different in Euclid and in Algebra.

The algebraic test is, as we have said, that the two fractions representing the ratios must be equal.

Euclid's test is given in Book v. Def. 5, where it stands thus:

"The first of four magnitudes is said to have the same ratio to the second which the third has to the fourth, when any equimultiples whatsoever of the first and third being taken and any equimultiples whatsoever of the second and fourth:

"If the multiple of the first be *less* than that of the second, the multiple of the third is also less than that of the fourth: or,

"If the multiple of the first be *equal* to that of the second, the multiple of the third is also equal to that of the fourth: or,

"If the multiple of the first be *greater* than that of the second, the multiple of the third is also greater than that of the fourth."

We shall now show, first, how to deduce Euclid's test of the equality of ratios from the algebraic test, and secondly, how to deduce the algebraic test from that employed by Euclid.

361.   I.   To show that if quantities be proportional according to the algebraical test they will also be proportional according to the geometrical test.

If a, b, c, d be proportional according to the algebraical test,

$$\frac{a}{b} = \frac{c}{d},$$

Multiply each side by $\frac{m}{n}$, and we get

$$\frac{ma}{nb} = \frac{mc}{nd},$$

Now, from the nature of fractions,
  if ma be less than nb, mc will also be less than nd, and
  if ma be equal to nb, mc will also be equal to nd, and
  if ma be greater than nb, mc will also be greater than nd.

Since then of the four quantities a, b, c, d equimultiples have been taken of the first and third, and equimultiples of the second and fourth, and it appears that when the multiple of the first is greater than, equal to, or less than the multiple of the second, the multiple of the third is also greater than, equal to, or less than the multiple of the fourth, it follows that a, b, c, d are proportionals according to the geometrical test.

362.   II.   To deduce the algebraic test of proportionality from that given by Euclid.

Let a, b, c, d be proportional according to Euclid.

Then if $\quad\dfrac{a}{b}$ is not equal to $\dfrac{c}{d},$

let $\quad\dfrac{a}{b+x}$ be equal to $\dfrac{c}{d}$ ....(1).

Take m and n such that

ma is greater than nb,

but less than $n(b+x)$...............(2).

Then, by Euclid's definition,

mc is greater than nd...............(3).

But since, by (1), $\dfrac{ma}{n(b+x)} = \dfrac{mc}{nd}$,

and, by (2), ma is less than $n(b+x)$,

it follows that mc is less than nd...............(4).

The results (3) and (4) therefore contradict each other.

Hence (1) cannot be true.

Therefore $\dfrac{a}{b}$ is equal to $\dfrac{c}{d}$.

We shall conclude this chapter with a mixed collection of Examples on Ratio and Proportion.

EXAMPLES.—CXXXII.

1. If $a-b : b-c :: b : c$, show that b is a mean proportional between a and c.

2. If $a : b :: c : d$, show that
$$a^2 + b^2 : \dfrac{a^3}{a+b} = c^2 + d^2 : \dfrac{c^3}{c+d}.$$
and $\quad a : b :: \sqrt[4]{(ma^4 + nc^4)} : \sqrt[4]{(mb^4 + nd^4)}$.

3. If $a : b :: c : d$, prove that
$$\dfrac{ma - nb}{ma + nb} = \dfrac{mc - nd}{mc + nd}.$$

4. If $\quad 5a + 3b : 7a + 3b :: 5b + 3c : 7b + 3c$, b is a mean proportional between a and c.

5. If 4 quantities be proportional, and the first be the greatest, the fourth is the least.

If $a+b, m+n, m-n, a-b$ be four such quantities, show that b is greater than n.

manner, and it also travels 15 miles an hour quicker. Supposing the rates of travelling uniform, what are they in miles per hour?

27. An article is sold at a loss of as much per cent. as it is worth in pounds. Show that it cannot be sold for more than £25.

XXIX. ON VARIATION.

363. If a sum of money is put out at interest at 5 per cent., the principal is 20 times as great as the annual interest, whatever the sum may be.

Hence if x be the principal, and y the interest,
$$x = 20y.$$

Now if we change x we must change y *in the same proportion,* for so long as the rate of interest remains the same, x will always be 20 times as great as y, and hence if x be doubled or trebled, y will also be doubled or trebled.

This is an instance of what is called DIRECT VARIATION, of which we may give the following definition.

DEF. One quantity y is said to vary directly as another quantity x, when y depends on x in such a manner that any increase or decrease made in the value of x produces a proportional increase or decrease in the value of y.

364. If $x = my$, where m is a constant quantity, that is, a quantity which is not altered by any change in the values of x and y,
$$y \text{ will vary directly as } x.$$

For any *increase* made in the value of x must produce a proportional *increase* in the value of y. Thus if x be doubled, y must also be doubled, to preserve the equality of x and my, since m cannot be changed.

365. Suppose a man can reap an acre of corn in a day.

Then 10 men can reap 60 acres in 6 days,

and 20 men can reap 60 acres in 3 days.

So that to do the same amount of work if we *double* the number of men we must *halve* the number of days.

This is an instance of what is called INVERSE VARIATION, of which we may give the following definition.

DEF. One quantity y is said to *vary inversely* as another quantity x, when y depends on x in such a manner that any *increase or decrease* made in the value of x produces a proportional *decrease or increase* in the value of y.

366. If $x = \dfrac{m}{y}$, where m is constant,

y will vary inversely as x.

For any *increase* made in the value of x must produce a proportional DECREASE in the value of y. Thus if x be doubled, y must be *halved*, to preserve the equality of x and $\dfrac{m}{y}$.

For
$$2x = \dfrac{2m}{y} = \dfrac{m}{\frac{y}{2}}.$$

367. If 1 man can reap 1 acre in 1 day,

5 men can reap 20 acres in 4 days,

and 10 men can reap 80 acres in 8 days.

That is, the number of acres reaped will depend on the product of the number of men into the number of days.

This is an example of *joint* variation, of which we may give the following definition.

DEF. One quantity x is said to vary *jointly* as two others y and z, when any change made in x produces a proportional change in the product of y and z.

368. One quantity x is said to vary directly as y and inversely as z when x varies as $\dfrac{y}{z}$.

manner, and it also travels 15 miles an hour quicker. Supposing the rates of travelling uniform, what are they in miles per hour?

27. An article is sold at a loss of as much per cent. as it is worth in pounds. Show that it cannot be sold for more than £25.

XXIX. ON VARIATION.

363. If a sum of money is put out at interest at 5 per cent., the principal is 20 times as great as the annual interest, whatever the sum may be.

Hence if x be the principal, and y the interest,
$$x = 20y.$$

Now if we change x we must change y *in the same proportion*, for so long as the rate of interest remains the same, x will always be 20 times as great as y, and hence if x be doubled or trebled, y will also be doubled or trebled.

This is an instance of what is called DIRECT VARIATION, of which we may give the following definition.

DEF. One quantity y is said to vary directly as another quantity x, when y depends on x in such a manner that any increase or decrease made in the value of x produces a proportional increase or decrease in the value of y.

364. If $x = my$, where m is a constant quantity, that is, a quantity which is not altered by any change in the values of x and y,

y will vary directly as x.

For any *increase* made in the value of x must produce a proportional *increase* in the value of y. Thus if x be doubled, y must also be doubled, to preserve the equality of x and my, since m cannot be changed.

ON VARIATION.

365. Suppose a man can reap an acre of corn in a day.

Then 10 men can reap 60 acres in 6 days,

and 20 men can reap 60 acres in 3 days.

So that to do the same amount of work if we *double* the number of men we must *halve* the number of days.

This is an instance of what is called INVERSE VARIATION, of which we may give the following definition.

DEF. One quantity y is said to *vary inversely* as another quantity x, when y depends on x in such a manner that any *increase or decrease* made in the value of x produces a proportional *decrease or increase* in the value of y.

366. If $x = \dfrac{m}{y}$, where m is constant,

y will vary inversely as x.

For any *increase* made in the value of x must produce a proportional DECREASE in the value of y. Thus if x be doubled, y must be *halved*, to preserve the equality of x and $\dfrac{m}{y}$.

For
$$2x = \frac{2m}{y} = \frac{m}{\frac{y}{2}}.$$

367. If 1 man can reap 1 acre in 1 day,

5 men can reap 20 acres in 4 days,

and 10 men can reap 80 acres in 8 days.

That is, the number of acres reaped will depend on the product of the number of men into the number of days.

This is an example of *joint* variation, of which we may give the following definition.

DEF. One quantity x is said to vary *jointly* as two others y and z, when any change made in x produces a proportional change in the product of y and z.

368. One quantity x is said to vary directly as y and inversely as z when x varies as $\dfrac{y}{z}$.

369. Theorem. If x varies as y when z is constant, and as z when y is constant, then when y and z are both variable,

$$x \text{ varies as } yz.$$

Let $$x = m \cdot yz.$$

Then we have to show that m is constant.

Now when z is constant,

$$x \text{ varies as } y;$$
$$\therefore mz \text{ is constant.}$$

Now z cannot involve y, since z is constant when y changes, and therefore m cannot involve y.

Similarly it may be shown that m cannot involve z;

$$\therefore m \text{ is constant,}$$

and x varies as yz.

370. The symbol $\propto$ is used to express variation; thus $x \propto y$ stands for the words x *varies as* y.

371. Variation is only an abbreviated form of expressing proportion.

Thus when we say that x varies as y, we mean that x bears to y the same ratio that any given value of x bears to the corresponding value of y, or

$$x : y = \text{a given value of } x : \text{the corresponding value of } y.$$

And similarly for the other kinds of variation, as will be seen from our examples.

Ex. 1. If $x \propto y$ and $y \propto z$, to show that $x \propto z$.

Let $$x = my, \text{ and } y = nz.$$

Then substituting this value of y in the first equation.

$$x = mnz;$$

and therefore, since mn is constant,

$$x \propto z.$$

Ex. 2. If $x \propto y$ and $x \propto z$, then will $x \propto \sqrt{(yz)}$.

Let $\qquad x = my$, and $x = nz$.

Then $\qquad x^2 = mnyz$;

$$\therefore x = \sqrt{(mn)} \cdot \sqrt{(yz)}.$$

Now $\sqrt{(mn)}$ is constant;

$$\therefore x \propto \sqrt{(yz)}.$$

Ex. 3. If y vary as x, and when $x=1$, $y=2$, what will be the value of y when $x=2$?

Here $y : x =$ a given value of y : corresponding value of x;

$$\therefore y : x = 2 : 1 :$$

$$\therefore y = 2x.$$

Hence, when $x=2$, $y=4$.

Ex. 4. If A vary inversely as B, and when $A=2$, $B=12$, what will B become when $A=9$?

Here $A : \dfrac{1}{B} =$ a given value of A : $\dfrac{1}{\text{corresponding value of } B}$;

$$\therefore A : \dfrac{1}{B} = 2 : \dfrac{1}{12};$$

$$\therefore \dfrac{A}{12} = \dfrac{2}{B}.$$

Hence, when $A=9$,

$$\dfrac{9}{12} = \dfrac{2}{B},$$

whence $\qquad B = \dfrac{24}{9} = \dfrac{8}{3} = 2\dfrac{2}{3}.$

Ex. 5. If A vary jointly as B and C, and when $A=6$, $B=6$, and $C=15$, find the value of A when $B=10$ and $C=3$.

Here

$A : BC =$ a given value of A : corresponding value of BC;

$$\therefore A : BC = 6 : 6 \times 15;$$

$$\therefore 90 A = 6 BC.$$

Hence, when $B = 10$ and $C = 3$,

$$90A = 6 \times 10 \times 3;$$
$$\therefore A = \frac{180}{90} = 2.$$

Ex. 6. If z vary as x directly and y inversely, and if when $z = 2$, $x = 3$ and $y = 4$, what is the value of z when $x = 15$ and $y = 8$?

Here $z : \dfrac{x}{y} =$ a given value of $z : \dfrac{\text{corresponding value of } x}{\text{corresponding value of } y}$;

$$\therefore z : \frac{x}{y} = 2 : \frac{3}{4};$$

$$\therefore \frac{3z}{4} = \frac{2x}{y}.$$

Hence, when $x = 15$ and $y = 8$,

$$\frac{3z}{4} = \frac{30}{8};$$

$$\therefore z = \frac{120}{24} = 5.$$

EXAMPLES.—cxxxiii.

1. If $A \propto \dfrac{1}{B}$ and $B \propto \dfrac{1}{C}$ then will $A \propto C$.

2. If $A \propto B$ then will $\dfrac{A}{P} \propto \dfrac{B}{P}$.

3. If $A \propto B$ and $C \propto D$ then will $AC \propto BD$.

4. If $x \propto y$, and when $x = 7$, $y = 5$, find the value of x when $y = 12$.

5. If $x \propto \dfrac{1}{y}$, and when $x = 10$, $y = 2$, find the value of y when $x = 4$.

6. If $x \propto yz$, and when $x=1$, $y=2$, $z=3$, find the value of y when $x=4$ and $z=2$.

7. If $x \propto \dfrac{y}{z}$, and when $x=6$, $y=4$, and $z=3$, find the value of x when $y=5$ and $z=7$.

8. If $3x+5y \propto 5x+3y$, and when $x=2$, $y=5$, find the value of $\dfrac{x}{y}$.

9. If $A \propto B$ and $B^3 \propto C^2$, express how A varies in respect of C.

10. If z vary conjointly as x and y, and $z=4$ when $x=1$ and $y=2$, what will be the value of x when $z=30$ and $y=3$?

11. If $A \propto B$, and when A is 8, B is 12; express A in terms of B.

12. If the square of x vary as the cube of y, and $x=3$ when $y=4$, find the equation between x and y.

13. If the square of x vary inversely as the cube of y, and $x=2$ when $y=3$, find the equation between x and y.

14. If the cube of x vary as the square of y and $x=3$ when $y=2$, find the equation between x and y.

15. If $x \propto z$ and $y \propto \dfrac{1}{z}$, show that $x \propto \dfrac{1}{y}$.

16. Show that in triangles of equal area the altitudes vary inversely as the bases.

17. Show that in parallelograms of equal area the altitudes vary inversely as the bases.

18. If $y = p+q+r$, where p is invariable, q varies as x, and r varies as x^2, find the relation between y and x, supposing that when $x=1$, $y=6$; when $x=2$, $y=11$; and when $x=3$, $y=18$.

19. The volume of a pyramid varies jointly as the area of its base and its altitude. A pyramid, the base of which is 9

feet square and the height of which is 10 feet, is found to contain 10 cubic yards. What must be the height of a pyramid upon a base 3 feet square in order that it may contain 2 cubic yards?

20. The amount of glass in a window, the panes of which are in every respect equal, varies as the number, length, and breadth of the panes jointly. Show that if their number varies as the square of their breadth inversely, and their length varies as their breadth inversely, the whole area of glass varies as the square of the length of the panes.

XXX. ON ARITHMETICAL PROGRESSION.

372. An **Arithmetical Progression** is a series of numbers which increase or decrease *by a constant difference*.

Thus, the following series are ARITHMETICAL PROGRESSIONS:

$$2, 4, 6, 8, 10;$$
$$9, 7, 5, 3, 1.$$

The Constant Difference being 2 in the first series and -2 in the second.

373. In Algebra we express an Arithmetical Progression thus: taking a to represent the first term and d to represent the constant difference, we shall have as a series of numbers in Arithmetical Progression

$$a, a+d, a+2d, a+3d,$$

and so on.

We observe that the terms of the series differ only in the *coefficient* of d, and that each coefficient of d is always less by 1 than the number of the term in which that particular coefficient stands. Thus

the coefficient of d in the 3rd term is 2,
.... in the 4th 3,
.... in the 5th .... 4.

Consequently the coefficient of d in the nth term will be $n-1$.

Therefore the nth term of the series will be $a + (n-1)d$.

374. If the series be
$$a, a+d, a+2d, \ldots\ldots\ldots$$
and z the last term, the term next before z will clearly be $z-d$, and the term next before it will be $z-2d$, and so on.

Hence, the series written backwards will be
$$z, z-d, z-2d, \ldots\ldots a+2d, a+d, a.$$

375. *To find the sum of a series of numbers in Arithmetical Progression.*

Let a denote the first term.
... d .... the constant difference.
... z .... the last term.
... n .... the number of terms.
... s .... the sum of the n terms.

Then $s = a + (a+d) + (a+2d) + \ldots\ldots + (z-2d) + (z-d) + z.$

Also $s = z + (z-d) + (z-2d) + \ldots\ldots + (a+2d) + (a+d) + a,$

the series in the second case being the same as in the first, but written in the reverse order.

Therefore, by adding the two series together, we get
$$2s = (a+z) + (a+z) + (a+z) + \ldots\ldots + (a+z) + (a+z) + (a+z);$$
and since on the right-hand side of this equation we have a series of n numbers each equal to $a + z$, we get
$$2s = n(a+z);$$
$$\therefore s = \frac{n}{2}(a+z).$$

This result may be put in another form, because in the place of z we may put $a + (n-1)d$, by Article 373.

Hence $$s = \frac{n}{2}\{a + a + (n-1)d\},$$

that is, $$= \frac{n}{2}\{2a + (n-1)d\}.$$

376. We have now obtained the following results:

$$z = a + (n-1)d \quad \ldots\ldots\ldots\ldots (A),$$

$$s = \frac{n}{2}(a+z) \quad \ldots\ldots\ldots\ldots (B),$$

$$s = \frac{n}{2}\{2a + (n-1)d\} \quad \ldots\ldots\ldots (C).$$

From one or more of these equations we have in Examples to determine the values of a, d, n, s or z. We shall now proceed to give instances of such Examples.

Ex. 1. Find the LAST TERM of the series
7, 10, 13, ... to 20 terms.

Taking the equation $z = a + (n-1)d$, for a put 7 and for n put 20, and we get

$$z = 7 + (20-1)d,$$
or, $\quad z = 7 + 19d.$

Now d is always found by *taking the first term from the second*, and in this case,

$$d = 10 - 7 = 3,$$
$$\therefore z = 7 + 19 \times 3 = 7 + 57 = 64.$$

Ex. 2. Find the last term of the series
12, 8, 4, ... to 11 terms.

In the equation $\quad z = a + (n-1)d,$
put $\quad\quad\quad\quad\quad a = 12$ and $n = 11.$
Then $\quad\quad\quad\quad z = 12 + 10d.$
Now $\quad\quad\quad\quad d = 8 - 12 = -4.$
Hence $\quad\quad\quad z = 12 - 40 = -28.$

EXAMPLES.—CXXXiv.

Find the last term of each of the following series

1. 2, 5, 8 ...... to 17 terms.
2. 4, 8, 12 ...... to 50 terms.

3. $7, \dfrac{29}{4}, \dfrac{15}{2}$ ...... to 16 terms.

4. $\dfrac{1}{2}, -1, -\dfrac{5}{2}$ ...... to 23 terms.

5. $\dfrac{5}{6}, \dfrac{1}{2}, \dfrac{1}{6}$ ...... to 12 terms.

6. $-12, -8, -4$ ...... to 14 terms.

7. $-3, 5, 13$ ...... to 16 terms.

8. $\dfrac{n-1}{n}, \dfrac{n-2}{n}, \dfrac{n-3}{n}$ ...... to n terms.

9. $(x+y)^2, x^2+y^2, (x-y)^2$ ...... to n terms.

10. $\dfrac{a-b}{a+b}, \dfrac{4a-3b}{a+b}, \dfrac{7a-5b}{a+b}$ ...... to n terms.

377. Ex. 1. Find the sum of the series

$$3, 5, 7 \text{ ...... to 12 terms.}$$

In the equation $\quad s = \dfrac{n}{2}\{2a + (n-1)d\}$

put 3 for a and 12 for n, and we get

$$s = \dfrac{12}{2}\{6 + 11d\}.$$

Now $d = 5 - 3 = 2$, and so

$$s = \dfrac{12}{2}\{6 + 22\} = 6 \times 28 = 168.$$

Ex. 2. Find the sum of the series

$$10, 7, 4 \text{ ...... to 10 terms.}$$

$$s = \dfrac{n}{2}\{2a + (n-1)d\};$$

put 10 for a and 10 for n, then

$$s = \dfrac{10}{2}\{20 + 9d\}.$$

Now $d = 7 - 10 = -3$, and therefore

$$s = \frac{10}{2}\{20 - 27\} = 5 \times (-7) = -35.$$

EXAMPLES.—CXXXV.

Find the sum of the following series:

1. $1, 2, 3 \ldots\ldots$ to 100 terms.
2. $2, 4, 6 \ldots\ldots$ to 50 terms.
3. $3, 7, 11 \ldots\ldots$ to 20 terms.
4. $\frac{1}{4}, \frac{1}{2}, \frac{3}{4} \ldots\ldots$ to 15 terms.
5. $-9, -7, -5 \ldots\ldots$ to 12 terms.
6. $\frac{5}{6}, \frac{1}{2}, \frac{1}{6} \ldots\ldots$ to 17 terms.
7. $1, 2, 3 \ldots\ldots$ to n terms.
8. $1, 4, 7 \ldots\ldots$ to n terms.
9. $1, 8, 15 \ldots\ldots$ to n terms.
10. $\frac{n-1}{n}, \frac{n-2}{n}, \frac{n-3}{n} \ldots\ldots$ to n terms.

378. **Ex.** What is the CONSTANT DIFFERENCE when the first term is 24 and the tenth term is -12?

Taking the equation (A),

$$z = a + (n - 1)d,$$

and regarding the tenth as the last term, we get

$$-12 = 24 + (10 - 1)d,$$

or
$$-36 = 9d,$$

whence we obtain $\quad d = -4.$

EXAMPLES.—CXXXVI.

What is the Constant Difference in the following cases?

1. When the first term is 100 and the twentieth is -14.
2. .... fifty-first is $-x$.
3. .... $-\frac{1}{2}$ .... forty-ninth is $5\frac{1}{2}$.
4. .... $-\frac{3}{4}$ .... twenty-fifth is $-21\frac{3}{4}$.
5. .... -10 .... sixth is -20.
6. .... 150 .... ninety-first is 0.

379. **Ex.** What is the FIRST TERM when the 40th term is 28 and the 43rd term is 32?

Taking equation (A),
$$z = a + (n-1)d,$$
and regarding the last term to be the 40th, we get
$$28 = a + 39d \dots\dots\dots\dots\dots(1).$$
Again, regarding the last term to be the 43rd, we get
$$32 = a + 42d \dots\dots\dots\dots\dots(2).$$
From equations (1) and (2) we may find the value of a to be -24.

EXAMPLES.—CXXXVII.

1. What is the first term when

   (1) The 59th term is 70 and the 66th term is 84;

   (2) The 20th term is $93 - 35b$ and the 21st is $98 - 37b$;

   (3) The second term is $\frac{1}{2}$ and the 55th is $5·8$;

   (4) The second term is 4 and the 87th is -20?

2. The sum of the 3rd and 8th terms of a series is 31, and the sum of the 5th and 10th terms is 43. Find the sum of 10 terms.

3. The sum of the 1st and 3rd terms of a series is 0, and the sum of the 2nd and 7th terms is 40. Find the sum of 7 terms.

4. If 24 and 33 be the fourth and fifth terms of a series, what is the 100th term?

5. Of how many terms does an Arithmetical Progression consist, whose difference is 3, first term 5 and last term 302?

6. Supposing that a body falls through a space of $16\frac{1}{12}$ feet in the first second of its fall, and in each succeeding second $32\frac{1}{6}$ feet more than in the next preceding one, how far will a body fall in 20 seconds?

7. What debt can be discharged in a year by weekly payments in arithmetical progression; the first payment being 1 shilling and the last £5. 3s.?

8. Find the 41st term and the sum of 41 terms in each of the following series:

(1) $-5, 4, 13 \ldots$

(2) $4a^2, 0, -4a^2 \ldots$

(3) $1+x, 5+3x, 9+5x \ldots$

(4) $-4\frac{1}{2}, -1\cdot 4 \ldots$

(5) $\frac{1}{4}, \frac{9}{20} \ldots$

9. To how many terms do the following series extend, and what is the sum of all the terms?

(1) $1002 \ldots 10, 2.$

(2) $-6, 2 \ldots, 186.$

(3) $2\frac{1}{2}x$, $\cdot 8x$ ...... $-72\cdot 3x$.

(4) $\frac{1}{2}$, $\frac{1}{4}$ ...... -24

(5) $m-1$ ...... $137(1-m)$, $139(1-m)$.

(6) $x+254$, ... $x+2$, $x-2$.

380. *To insert 3 arithmetic means between 2 and 10.*

The number of terms will be 5.

Taking the equation $z = a + (n-1)d$,

we have $\qquad 10 = 2 + (5-1)d$.

Whence $\qquad 8 = 4d$; $\therefore d = 2$.

Hence the series will be

$\qquad$ 2, 4, 6, 8, 10.

EXAMPLES.—CXXXVIII.

1. Insert 4 arithmetic means between 3 and 18.
2. Insert 5 arithmetic means between 2 and -2.
3. Insert 3 arithmetic means between 3 and $\frac{2}{3}$.
4. Insert 4 arithmetic means between $\frac{1}{2}$ and $\frac{1}{3}$.

381. *To insert 3 arithmetic means between a and b.*

The number of terms in the series will be 5, since there are to be 3 terms in addition to the first term a and the last term b.

Taking the equation $z = a + (n-1)d$,

we have to find d, having given

$\qquad a$, $z = b$ and $n = 5$.

Hence $b = a + (5-1)d$,

or, $4d = b-a$, $\therefore d = \dfrac{b-a}{4}$.

Hence the series will be

$$a,\ a + \dfrac{b-a}{4},\ a + \dfrac{b-a}{2},\ a + \dfrac{3(b-a)}{4},\ b,$$

that is, $a,\ \dfrac{3a+b}{4},\ \dfrac{a+b}{2},\ \dfrac{a+3b}{4},\ b.$

EXAMPLES.—CXXXIX.

1. Insert 3 arithmetic means between m and n.
2. Insert 4 arithmetic means between $m+1$ and $m-1$.
3. Insert 4 arithmetic means between n^2 and n^2+1.
4. Insert 3 arithmetic means between x^2+y^2 and x^2-y^2.

382. We shall now give the general form of the proposition "*To insert* m *arithmetic means between* a *and* b."

The number of terms in the series will be $m+2$.

Then taking the equation $z = a + (n-1)d$,
we have in this case $b = a + (m+2-1)d$,
or, $b = a + (m+1)d$.

Hence $d = \dfrac{b-a}{m+1}$,

and the form of the series will be

$$a,\ a + \dfrac{b-a}{m+1},\ a + \dfrac{2b-2a}{m+1},\ \ldots\ldots,\ b - \dfrac{2b-2a}{m+1},\ b - \dfrac{b-a}{m+1},\ b,$$

that is,

$$a,\ \dfrac{am+b}{m+1},\ \dfrac{am-a+2b}{m+1},\ \ldots\ldots,\ \dfrac{bm-b+2a}{m+1},\ \dfrac{bm+a}{m+1},\ b.$$

XXXI. ON GEOMETRICAL PROGRESSION.

383. A **Geometrical** Progression is a series of numbers which increase or decrease *by a constant factor*.

Thus the following series are GEOMETRICAL PROGRESSIONS,

$$2, 4, 8, 16, 32, 64;$$

$$12, 3, \frac{3}{4}, \frac{3}{16}, \frac{3}{64};$$

$$4, -\frac{1}{2}, \frac{1}{16}, -\frac{1}{128}, \frac{1}{1024}.$$

The Constant Factors being 2 in the first series, $\frac{1}{4}$ in the second, and $-\frac{1}{8}$ in the third.

NOTE. That which we shall call the Constant Factor is usually called the Common Ratio.

384. In Algebra we express a Geometrical Progression thus: taking a to represent the first term and f to represent the Constant Factor, we shall have as a series of numbers in Geometrical Progression

$$a, af, af^2, af^3, \text{ and so on.}$$

We observe that the terms of the series differ only in the *index* of f, and that each index of f is always less by 1 than the number of the term in which that particular index stands.

Thus the index of f in the 3rd term is 2,
in the 4th 3,
in the 5th .... 4.

Consequently the index of f in the nth term will be $n-1$.

Therefore the nth term of the series will be af^{n-1}.

[S.A.] 8

Hence if z be the last term,
$$z = af^{n-1}.$$

385. If the series contain n terms, a being the first term and f the Constant Factor,

the last term will be af^{n-1},

the last term but *one* will be af^{n-2},

the last term but *two* will be af^{n-3}.

Now $af^{n-1} \times f = af^{n-1} \times f^1 = af^{n-1+1} = af^n$,

$af^{n-2} \times f = af^{n-2} \times f^1 = af^{n-2+1} = af^{n-1}$,

$af^{n-3} \times f = af^{n-3} \times f^1 = af^{n-3+1} = af^{n-2}$.

386. We may now proceed *to find the sum of a series of numbers in Geometrical Progression*.

Let a denote the first term,

    f the constant factor,

    n the number of terms,

    s the sum of the n terms.

Then $s = a + af + af^2 + \ldots + af^{n-3} + af^{n-2} + af^{n-1}$.

Now multiply both sides of this equation by f, then
$$fs = af + af^2 + af^3 + \ldots + af^{n-2} + af^{n-1} + af^n.$$

Hence, subtracting the first equation from the second,
$$fs - s = af^n - a.$$
$$\therefore s(f-1) = a(f^n - 1);$$
$$\therefore s = \frac{a(f^n - 1)}{f - 1}.$$

NOTE. The proposition just proved presents a difficulty to a beginner, which we shall endeavour to explain. When we multiply the series of n *terms*
$$a + af + af^2 + \ldots\ldots + af^{n-3} + af^{n-2} + af^{n-1}$$

by f, we shall obtain another series

$$af + af^2 + af^3 + \ldots\ldots + af^{n-2} + af^{n-1} + af^n,$$

which also contains n *terms*.

Though we cannot fill up the gap in each series completely, we see that the terms in the two series must be the same, except the *first term* in the former series, and the *last term* in the latter. Hence, when we subtract, all the terms will disappear except these two.

387. From the formulæ:

$$z = af^{n-1} \ldots\ldots\ldots\ldots\ldots\ldots\ldots\text{(A)},$$
$$s = \frac{a(f^n - 1)}{f - 1} \ldots\ldots\ldots\ldots\ldots\text{(B)},$$

prove the following:

(α) $\quad s = \dfrac{fz - a}{f - 1}.$ $\qquad$ (γ) $\quad a = fz - (f - 1)s.$

(β) $\quad a = \dfrac{z}{f^{n-1}}.$ $\qquad$ (δ) $\quad f = \dfrac{s - a}{s - z}.$

388. **Ex.** Find the LAST TERM of the series

$$3,\ 6,\ 12\ \ldots\ldots \text{ to 9 terms.}$$

The Constant Factor is $\dfrac{6}{3}$, that is, 2.

In the formula

$$z = af^{n-1},$$

putting 3 for a, 2 for f, and 9 for n, we get

$$z = 3 \times 2^8 = 3 \times 256 = 768.$$

EXAMPLES.—cxl.

Find the last term of the following series

1. $1,\ 2,\ 4\ \ldots\ldots$ to 7 terms.
2. $4,\ 12,\ 36\ \ldots\ldots$ to 10 terms.
3. $5,\ 20,\ 80\ \ldots\ldots$ to 9 terms.

4. 8, 4, 2 ...... to 15 terms.

5. 2, 6, 18 ...... to 9 terms.

6. $\dfrac{1}{64}, \dfrac{1}{16}, \dfrac{1}{4}$ ...... to 11 terms.

7. $-\dfrac{2}{3}, \dfrac{1}{3}, -\dfrac{1}{6}$ ...... to 7 terms.

389. **Ex.** Find the sum of the series

$$6, 3, \dfrac{3}{2} \ldots\ldots \text{ to 8 terms.}$$

Generally, $\quad s = \dfrac{a(f^n - 1)}{f - 1}$

and here $\quad a = 6, f = \dfrac{1}{2}, n = 8,$

$$\therefore s = \dfrac{6\left(\dfrac{1^8}{2^8} - 1\right)}{\dfrac{1}{2} - 1} = \dfrac{6\left(\dfrac{1}{256} - 1\right)}{-\dfrac{1}{2}}$$

$$= \dfrac{\dfrac{6}{256} - 6}{-\dfrac{1}{2}} = \dfrac{6 - \dfrac{6}{256}}{\dfrac{1}{2}} = \dfrac{765}{64}.$$

EXAMPLES.—CXli.

Find the sum of the following series:

1. 2, 4, 8 ...... to 15 terms.

2. 1, 3, 9 ...... to 6 terms.

3. a, ax^2, ax^4 ...... to 13 terms.

4. $a, \dfrac{a}{x}, \dfrac{a}{x^2}$ ...... to 9 terms.

5. $a^2 - x^2, a - x, \dfrac{a-x}{a+x}$ ...... to 7 terms.

390. To find the sum of an Infinite Series in Geometrical Progression, when the Constant Factor is a proper fraction.

If f be a proper fraction and n very large,

f^n is a very small number.

Hence if the number of terms be *infinite*, f^n is so small that we may neglect it in the expression

$$s = \frac{a(f^n - 1)}{f - 1},$$

and we get

$$s = \frac{-a}{f - 1}$$

$$= \frac{a}{1 - f}.$$

391. Ex. 1. Find the sum of the series $\frac{4}{3} + 1 + \frac{3}{4} + \ldots$ to infinity.

Here
$$f = 1 \div \frac{4}{3} = \frac{3}{4};$$

$$\therefore s = \frac{a}{1-f} = \frac{\frac{4}{3}}{1 - \frac{3}{4}} = \frac{16}{3} = 5\frac{1}{3}.$$

Ex. 2. Sum to infinity the series $\frac{3}{2} - \frac{2}{3} + \frac{8}{27} - \ldots$

Here
$$f = -\frac{2}{3} \div \frac{3}{2} = -\frac{4}{9};$$

$$\therefore s = \frac{a}{1-f} = \frac{\frac{3}{2}}{1 - \left(-\frac{4}{9}\right)} = \frac{\frac{3}{2}}{1 + \frac{4}{9}} = \frac{27}{26}.$$

EXAMPLES.—cxlii.

Find the sum of the following infinite series:

1. $1, \dfrac{1}{2}, \dfrac{1}{4}, \ldots\ldots$
2. $1, \dfrac{1}{4}, \dfrac{1}{16}, \ldots\ldots$
3. $3, \dfrac{1}{3}, \dfrac{1}{27}, \ldots\ldots$
4. $\dfrac{2}{3}, \dfrac{1}{3}, \dfrac{1}{6}, \ldots\ldots$
5. $\dfrac{3}{4}, \dfrac{1}{4}, \ldots\ldots$
6. $\dfrac{1}{2}, -\dfrac{1}{3}, \ldots\ldots$
7. $8, \dfrac{2}{3}, \ldots\ldots$
8. $1\dfrac{1}{2}, \cdot 5, \ldots\ldots$
9. $4^3, 2^4, \ldots\ldots$
10. $2x^3, -\cdot 25x, \ldots\ldots$
11. $a, b, \ldots\ldots$
12. $\dfrac{1}{10}, \dfrac{1}{10^2}, \ldots\ldots$
13. $x, -y, \ldots\ldots$
14. $\dfrac{86}{100}, \dfrac{86}{10000}, \ldots\ldots$
15. $\cdot 54444, \ldots\ldots$
16. $\cdot 83636, \ldots\ldots$

392. *To insert 3 geometric means between* 10 *and* 160.

Taking the equation $\qquad z = af^{n-1}$,

we put 10 for a, 160 for z, and 5 for n, and we obtain

$$160 = 10 \cdot f^{5-1};$$

$$\therefore 16 = f^4.$$

Now $\qquad 16 = 2 \times 2 \times 2 \times 2 = 2^4;$

$$\therefore 2^4 = f^4.$$

Hence $f = 2$, and the series will be

$$10,\ 20,\ 40,\ 80,\ 160.$$

EXAMPLES.—cxliii.

1. Insert 3 geometric means between 3 and 243.
2. Insert 4 geometric means between 1 and 1024.
3. Insert 3 geometric means between 1 and 16.
4. Insert 4 geometric means between $\dfrac{1}{2}$ and $\dfrac{243}{64}$.

393. *To insert m geometric means between a and b.*

The number of terms in the series will be $m+2$.

In the formula $z = af^{n-1}$,

putting b for z, and $m+2$ for n, we get

$$b = af^{m+2-1},$$
or, $$b = af^{m+1};$$
$$\therefore f^{m+1} = \dfrac{b}{a},$$
or, $$f = \dfrac{b^{\frac{1}{m+1}}}{a^{\frac{1}{m+1}}}.$$

Hence the series will be,

$$a,\ a \times \dfrac{b^{\frac{1}{m+1}}}{a^{\frac{1}{m+1}}},\ a \times \dfrac{b^{\frac{2}{m+1}}}{a^{\frac{2}{m+1}}},\ \ldots\ldots,\ b \div \dfrac{b^{\frac{2}{m+1}}}{a^{\frac{2}{m+1}}},\ b \div \dfrac{b^{\frac{1}{m+1}}}{a^{\frac{1}{m+1}}},\ b,$$

that is,

$$a,\ (a^m \cdot b)^{\frac{1}{m+1}},\ (a^{m-1} \cdot b^2)^{\frac{1}{m+1}},\ \ldots\ldots,\ (a^2 \cdot b^{m-1})^{\frac{1}{m+1}},\ (a \cdot b^m)^{\frac{1}{m+1}},\ b.$$

394. We shall now give some mixed Examples on Arithmetical and Geometrical Progression.

EXAMPLES.—cxliv.

1. Sum the following series:

   (1) $8 + 15 + 22 + \ldots\ldots$ to 12 terms.

   (2) $116 + 108 + 100 + \ldots\ldots$ to 10 terms.

(3) $3 + \dfrac{1}{2} + \dfrac{1}{12} + \ldots$ to infinity.

(4) $2 - \dfrac{1}{4} + \dfrac{1}{32} - \ldots$ to infinity.

(5) $\dfrac{1}{2} - \dfrac{2}{3} - \dfrac{11}{6} - \ldots$ to 13 terms.

(6) $\dfrac{1}{2} - \dfrac{1}{3} + \dfrac{2}{9} - \ldots$ to 6 terms.

(7) $\dfrac{1}{2} - 1 - \dfrac{5}{2} - \ldots$ to 29 terms.

(8) $\dfrac{5}{7} + 1 + 1\dfrac{2}{7} + \ldots$ to 8 terms.

(9) $\dfrac{1}{3} + \dfrac{2}{9} + \dfrac{4}{27} + \ldots$ to infinity.

(10) $\dfrac{3}{5} - \dfrac{14}{10} - \dfrac{51}{15} - \ldots$ to 10 terms.

(11) $\sqrt{\dfrac{3}{5}} - \sqrt{6} + 2\sqrt{(15)} - \ldots$ to 8 terms.

(12) $-\dfrac{7}{5} + \dfrac{7}{2} - \dfrac{35}{4} + \ldots$ to 5 terms.

2. If the continued product of 5 terms in Geometrical Progression be 32, show that the middle term is 2.

3. If a, b, c are in arithmetic progression, and a, b', c are in geometrical progression, show that $\dfrac{b}{b'} = \dfrac{a+c}{2\sqrt{(ac)}}$.

4. Show that the arithmetical mean between a and b is greater than the geometrical mean.

5. The sum of the first three terms of an arithmetic series is 12, and the sixth term is 12 also. Find the sum of the first 6 terms.

6. What is necessary that a, b, c may be in geometric progression?

ON GEOMETRICAL PROGRESSION.

7. If $2n$, x and $\frac{1}{2n}$ are in geometric progression, what is x?

8. If $2n$, y and $\frac{1}{2n}$ are in arithmetic progression, what is y?

9. The sum of a geometric progression whose first term is 1, constant factor 3, and number of terms 4, is equal to the sum of an arithmetic progression, whose first term is 4 and constant difference 4; how many terms are there in the arithmetic progression?

10. The first $(7+n)$ natural numbers when added together make 153. Find n.

11. Prove that the sum of any number of terms of the series 1, 3, 5, …… is the square of the number of terms.

12. If the sum of a series of 5 terms in arithmetic progression be 95, show that the middle term is 19.

13. There is an arithmetical progression whose first term is $3\frac{1}{3}$, the constant difference is $1\frac{4}{9}$, and the sum of the terms is 22. Required the number of terms.

14. The 3 digits of a certain number are in arithmetical progression; if the number be divided by the sum of the digits in the units' and tens' place, the quotient is 107. If 396 be subtracted from the number, its digits will be inverted. Required the number.

15. If the $(p+q)^{th}$ term of a geometric progression be m, and the $(p-q)^{th}$ term be n, show that the pth term is $\sqrt{(mn)}$.

16. The difference between two numbers is 48, and the arithmetic mean exceeds the geometric by 18. Find the numbers.

17. Place three arithmetic means between 1 and 11.

18. The first term of an increasing arithmetic series is ·034, the constant difference ·0004, and the sum 2·748. Find the number of terms.

19. Place nine arithmetic means between 1 and −1.

20. Prove that every term of the series 1, 2, 4, ... is greater by unity than the sum of all that precede it.

21. Show that if a series of mp terms forming a geometrical progression whose constant factor is r be divided into sets of p consecutive terms, the sums of the sets will form a geometrical progression whose constant factor is r^p.

22. Find five numbers in arithmetical progression, such that their sum is 55, and the sum of their squares 765.

23. In a geometrical progression of 5 terms the difference of the extremes is to the difference of the 2nd and 4th terms as 10 to 3, and the sum of the 2nd and 4th terms equals twice the product of the 1st and 2nd. Find the series.

24. Show that the amounts of a sum of money put out at Compound Interest form a series in geometrical progression.

25. A certain number consists of three digits in geometrical progression. The sum of the digits is 13, and if 792 be added to the number, the digits will be inverted. Find the number.

26. The population of a county increases in 4 years from 10000 to 14641; what is the rate of increase?

XXXII. ON HARMONICAL PROGRESSION.

395. A **Harmonical** Progression is a series of numbers of which the reciprocals form an Arithmetical Progression.

Thus the series of numbers $a, b, c, d,$ ...... is a HARMONICAL PROGRESSION, if the series $\frac{1}{a}, \frac{1}{b}, \frac{1}{c}, \frac{1}{d},$ is an Arithmetical Progression.

If a, b, c be in Harmonical Progression, b is called the *Harmonical Mean* between a and c.

NOTE. There is no way of finding a general expression for the sum of a Harmonical Series, but many problems with reference to such a series may be solved by inverting the terms and treating the reciprocals as an Arithmetical Series.

396. *If* a, b, c *be in Harmonical Progression, to show that*
$$a : c :: a-b : b-c.$$

Since $\dfrac{1}{a}, \dfrac{1}{b}, \dfrac{1}{c}$ are in Arithmetical Progression,

$$\frac{1}{c}-\frac{1}{b}=\frac{1}{b}-\frac{1}{a},$$

or
$$\frac{b-c}{bc}=\frac{a-b}{ab},$$

or
$$\frac{ab}{bc}=\frac{a-b}{b-c},$$

or
$$\frac{a}{c}=\frac{a-b}{b-c}.$$

397. *To insert* m *harmonic means between* a *and* b.

First to insert m arithmetic means between $\dfrac{1}{a}$ and $\dfrac{1}{b}$.

Proceeding as in Art. 357, we have
$$\frac{1}{b}=\frac{1}{a}+(m+1)\,d,$$

or
$$a=b+(m+1).\,abd$$

$$\therefore d=\frac{a-b}{ab\,(m+1)}.$$

Hence the arithmetic series will be
$$\frac{1}{a},\ \frac{1}{a}+\frac{a-b}{ab(m+1)},\ \frac{1}{a}+\frac{2(a-b)}{ab(m+1)},\ \ldots\ \frac{1}{a}+\frac{m(a-b)}{ab(m+1)},\ \frac{1}{b},$$

or,
$$\frac{1}{a},\ \frac{bm+a}{ab(m+1)},\ \frac{bm+2a-b}{ab(m+1)},\ \ldots\ \frac{am+b}{ab(m+1)},\ \frac{1}{b}.$$

Therefore the Harmonic Series is
$$a,\ \frac{ab(m+1)}{bm+a},\ \frac{ab(m+1)}{bm+2a-b},\ \ldots\ \frac{ab(m+1)}{am+b},\ b.$$

398. Given a and b the first two terms of a series in Harmonical Progression, to find the nth term.

$\dfrac{1}{a}, \dfrac{1}{b}$ are the first two terms of an Arithmetical Series of which the common difference is $\dfrac{1}{b} - \dfrac{1}{a}$.

The nth term of this Arithmetical Series is

$$\frac{1}{a} + (n-1)\left(\frac{1}{b} - \frac{1}{a}\right)$$

$$= \frac{1}{a} + \frac{(n-1)(a-b)}{ab} = \frac{b + na - a - nb + b}{ab}$$

$$= \frac{(na - a) - (nb - 2b)}{ab} = \frac{(n-1)a - (n-2)b}{ab},$$

∴ the nth term of the Harmonical Series is

$$\frac{ab}{(n-1)a - (n-2)b}.$$

399. Let a and c be any two numbers,

b the Harmonical Mean between them.

Then $\quad\quad \dfrac{1}{b} - \dfrac{1}{a} = \dfrac{1}{c} - \dfrac{1}{b}$,

or $\quad\quad \dfrac{2}{b} = \dfrac{a+c}{ac}$;

$$\therefore b = \frac{2ac}{a+c}.$$

400. The following results should be remembered.

Arithmetical Mean between a and $c = \dfrac{a+c}{2}$.

Geometrical Mean between a and $c = \sqrt{ac}$.

Harmonical Mean between a and $c = \dfrac{2ac}{a+c}$.

Hence if we denote the Means by the letters A, G, H respectively,

$$A \times H = \frac{a+c}{2} \times \frac{2ac}{a+c}$$

$$= ac$$

$$= G^2;$$

that is, G is a mean proportional between A and H.

401. To show that A, G, H are in descending order of magnitude.

Since $(\sqrt{a} - \sqrt{c})^2$ must be a positive quantity.

$(\sqrt{a} - \sqrt{c})^2$ is greater than 0,

or $\quad a - 2\sqrt{ac} + c$ greater than 0,

or $\quad a + c$ greater than $2\sqrt{ac}$,

or $\quad \dfrac{a+c}{2}$ greater than $\sqrt{ac}$;

that is, A is greater than G.

Also, since $a+c$ is greater than $2\sqrt{ac}$,

$\sqrt{ac}\,(a+c)$ is greater than $2ac$;

$\therefore \sqrt{ac}$ is greater than $\dfrac{2ac}{a+c}$;

i.e. G is greater than H.

Examples.—cxlv.

1. Insert two harmonic means between 6 and 24.

2. ...... four.. 2 and 3.

3. ...... three .... $\dfrac{1}{3}$ and $\dfrac{3}{2}$.

4. ...... four.. $\dfrac{1}{3}$ and $\dfrac{1}{18}$.

5. Insert five harmonic means between -1 and 2^{-1}.

6. ...... five .... $\frac{1}{2}$ and $-\frac{1}{2}$.

7. ...... six .... 3 and $\frac{6}{23}$.

8. ...... n .... $2x$ and $3y$.

9. The sum of three terms of a harmonical series is $\frac{11}{12}$, and the first term is $\frac{1}{2}$: find the series, and continue it both ways.

10. The arithmetical mean between two numbers exceeds the geometrical by 13, and the geometrical exceeds the harmonical by 12. What are the numbers?

11. There are four numbers a, b, c, d, the first three in arithmetical, the last three in harmonical progression; show that $a : b = c : d$.

12. If x is the harmonic mean between m and n, show that
$$\frac{1}{x-m} + \frac{1}{x-n} = \frac{1}{m} + \frac{1}{n}.$$

13. The sum of three terms of a harmonic series is 11, and the sum of their squares is 49; find the numbers.

14. If x, y, z be the pth, q^{th}, and r^{th} terms of a H.P., show that $(r-q)yz + (p-r)xz + (q-p)xy = 0$.

15. If the H.M. between each pair of the numbers, a, b, c be in A.P., then b^2, a^2, c^2 will be in H.P.: and if the H.M. be in H.P., b, a, c will be in H.P.

16. Show that $\frac{c+2a}{c-b} + \frac{c+2b}{c-a} = 4, > 7, \text{ or } > 10$, according as c is the A., G. or H. mean between a and b.

XXXIII. PERMUTATIONS.

402. The different arrangements in respect of order of succession which can be made of a given number of things are called **Permutations**.

Thus if from a box of letters I select *two*, P and Q, I can make *two* permutations of them, placing P first on the left and then on the right of Q, thus:

$$P, Q \text{ and } Q, P.$$

If I now take *three* letters, P, Q and R, I can make *six* permutations of them, thus:

P, Q, R ; P, R, Q, two in which P stands first.
Q, P, R ; Q, R, P, ... Q ....
R, P, Q ; R, Q, P, ... R ....

403. In the Examples just given *all* the things in each case are taken together; but we may be required to find how many permutations can be made out of a number of things, when *a certain number only* of them are taken at a time.

Thus the permutations that can be formed out of the letters P, Q, and R taken *two at a time* are six in number, thus:

$$P, Q\ ;\ P, R\ ;\ Q, P\ ;\ Q, R\ ;\ R, P\ ;\ R, Q.$$

404. *To find the number of permutations of* n *different things taken* r *at a time.*

Let $a, b, c, d \ldots$ stand for n different things.

First to find the number of permutations of the n things taken *two at a time*.

If a be placed before each of the other things $b, c, d \ldots$ of which the number is $n-1$, we shall have $n-1$ permutations in which a stands first, thus

$$ab,\ ac,\ ad,\ \ldots\ldots$$

If b be placed before each of the other things, $a, c, d \ldots$ we shall have $n-1$ permutations in which b stands first, thus:

$$ba, bc, bd, \ldots\ldots$$

Similarly there will be $n-1$ permutations in which c stands first: and so of the rest. In this way we get every possible permutation of the n things taken two at a time.

Hence there will be $n \cdot (n-1)$ permutations of n things taken *two* at a time.

Next to find the number of permutations of the n things taken *three* at a time.

Leaving a out, we can form $(n-1) \cdot (n-2)$ permutations of the remaining $(n-1)$ things taken *two* at a time, and if we place a before each of these permutations we shall have $(n-1) \cdot (n-2)$ permutations of the n things taken *three* at a time in which a stands first.

Similarly there will be $(n-1) \cdot (n-2)$ permutations of the n things taken *three* at a time in which b stands first: and so for the rest.

Hence the whole number of permutations of the n things taken *three* at a time will be $n \cdot (n-1) \cdot (n-2)$, the factors of the formula decreasing each by 1, and *the figure in the last factor being 1 less than the number taken at a time.*

We now assume that the formula holds good for the number of permutations of n things taken $r-1$ at a time, and we shall proceed to show that it will hold good for the number of permutations of n things taken r at a time.

The number of permutations of the n things taken $r-1$ at a time will be

$$n \cdot (n-1) \cdot (n-2) \ldots\ldots [n-\{(r-1)-1\}],$$

that is $\quad n \cdot (n-1) \cdot (n-2) \ldots\ldots (n-r+2)$.

Leaving a out we can form $(n-1) \cdot (n-2) \ldots\ldots (n-1-r+2)$ permutations of the $(n-1)$ remaining things taken $r-1$ at a time.

Putting a before each of these, we shall have

$$(n-1) \cdot (n-2) \ldots\ldots (n-r+1)$$

permutations of the n things taken r at a time in which a stands first.

So again we shall have $(n-1).(n-2)\ldots(n-r+1)$ permutations of the n things taken r at a time in which b stands first; and so on.

Hence the whole number of permutations of the n things taken r at a time will be

$$n.(n-1).(n-2)\ldots(n-r+1).$$

If then the formula holds good when the n things are taken $r-1$ at a time, it will hold good when they are taken r at a time.

But we have shown it to hold when they are taken 3 at a time; hence it will hold when they are taken 4 at a time, and so on: therefore it is true for all integral values of r.*

405. If the n things be taken *all* together, $r=n$, and the formula gives

$$n.(n-1).(n-2)\ldots(n-n+1);$$

that is, $\qquad n.(n-1).(n-2)\ldots 1$

as the number of permutations that can be formed of n different things taken *all* together.

For brevity the formula

$$n.(n-1).(n-2)\ldots 1,$$

which is the same as $\quad 1.2.3\ldots n,$

is written $\lfloor n$. This symbol is called *factorial n*.

Similarly $\qquad \lfloor r$ is put for $1.2.3\ldots r$;

$\qquad\qquad\quad \lfloor r-1 \ldots$ for $1.2.3\ldots (r-1).$

Obs. $\qquad \lfloor n = n.\lfloor n-1 = n.(n-1).\lfloor n-2 = \&c.$

406. *To find the number of permutations of n things taken all together when certain of the things are alike.*

Let the n things be represented by the letters $a, b, c, d\ldots$ and suppose that $\quad a$ recurs p times,

$\qquad\qquad\qquad\qquad b \ldots q$ times,

$\qquad\qquad\qquad\qquad c \ldots r$ times,

and so on.

---

* Another proof of this Theorem may be seen in Art. 475.

[S.A.]   T

Let P represent the whole number of permutations.

Then if all the p letters a were changed into p other letters, different from each other and from all the rest of the n letters, the places of these p letters in *any one* permutation could now be interchanged, each interchange giving rise to a new permutation, and thus from each single permutation we could form $1 . 2 \ldots\ldots p$ permutations in all, and the whole number of permutations would be $(1 . 2 \ldots p) P$, that is $\lfloor p . P$.

Similarly if in addition the q letters b were changed into q letters different from each other and from all the rest of the n letters, the whole number of permutations would be

$$\lfloor q . \lfloor p . P ;$$

and if the r letters c were also similarly changed, the whole number of permutations would be

$$\lfloor r . \lfloor q . \lfloor p . P ;$$

and so on, if more were alike.

But when the p, q, and r, &c., letters have thus been changed, we shall have n letters all different, and the number of permutations that can be formed of them is $\lfloor n$ (Art. 405).

Hence $\quad P . \lfloor p . \lfloor q . \lfloor r \ldots\ldots = \lfloor n ;$

$$\therefore P = \frac{\lfloor n}{\lfloor p . \lfloor q . \lfloor r} \ldots\ldots$$

EXAMPLES.—cxlvi.

1. How many permutations can be formed out of 12 things taken 2 at a time?

2. How many permutations can be formed out of 16 things taken 3 at a time?

3. How many permutations can be formed out of 20 things taken 4 at a time?

4. How many changes can be rung with 5 bells out of 8?

5. How many permutations can be made of the letters in the word *Examination* taken all together?

6. In how many ways can 8 men be placed side by side?

7. In how many ways can 10 men be placed side by side?

8. Three flags are required to make a signal. How many signals can be given by 20 flags of 5 different colours, there being 4 of each colour?

9. How many different permutations can be formed out of the letters in *Algebra* taken all together?

10. The number of things : number of permutations of the things taken 3 at a time $= 1 : 20$. How many things are there?

11. The number of permutations of m things taken 3 at a time : the number of permutations of $m+2$ things taken 3 at a time $= 1 : 3$. Find m.

12. In the permutations of a, b, c, d, e, f, g taken all together, find how many begin with cd.

13. Find the number of permutations of the letters of the product $a^2 b^3 c^4$ written at full length.

14. Find the number of permutations that can be formed out of the letters in each of the following words: *Conceit, Talavera, Calcutta, Proposition, Mississippi.*

XXXIV. COMBINATIONS.

407. The **Combinations** of a number of things are the different collections that can be formed out of them by taking a certain number at a time, without regard to the order in which the things stand in each collection.

Thus the combinations of a, b, c, d taken *two* at a time are ab, ac, ad, bc, bd, cd.

Here from each combination we could make *two* permutations: thus ab, ba; ac, ca; and so on: for ab, ba are the same combination, and so are ac, ca.

Similarly the combinations of a, b, c, d taken *three* at a time are abc, abd, acd, bcd.

Here from each combination we could make *six* permutations; thus $abc, acb, bac, bca, cab, cba$: and so on.

And, generally, in accordance with Art. 405, any combination of n things may be made into $1.2.3 \ldots n$ permutations.

408. *To find the number of combinations of* n *different things taken* r *at a time.*

Let C_r denote the number of combinations required.

Since each combination contains r things it can be made into $\lfloor r$ permutations (Art. 405);

$\therefore$ the whole number of permutations $= \lfloor r \cdot C_r$.

But also (from Art. 404) the whole number of permutations of n things taken r at a time

$$= n(n-1) \ldots \ldots (n-r+1);$$
$$\therefore \lfloor r \cdot C_r = n(n-1) \ldots \ldots (n-r+1);$$
$$\therefore C_r = \frac{n(n-1) \ldots \ldots (n-r+1)}{\lfloor r}.$$

409. *To show that the number of combinations of* n *things taken* r *at a time is the same as the number taken* $n-r$ *at a time.*

$$C_r = \frac{n.(n-1) \ldots \ldots (n-r+1)}{1.2.3 \ldots \ldots r};$$

and

$$C_{n-r} = \frac{n.(n-1) \ldots \ldots \{n-(n-)+1\}}{1.2.3 \ldots \ldots (n-r)}$$

$$= \frac{n.(n-1) \ldots \ldots (r+1)}{1.2.3 \ldots \ldots (n-r)}.$$

Hence

$$\frac{C_r}{C_{n-r}} = \frac{n.(n-1) \ldots \ldots (n-r+1)}{1.2.3 \ldots \ldots r} \times \frac{1.2.3 \ldots \ldots (n-r)}{n.(n-1) \ldots \ldots (r+1)}$$

$$= \frac{n.(n-1) \ldots \ldots (n-r+1).(n-r) \ldots \ldots 3.2.1}{1.2.3 \ldots \ldots r.(r+1) \ldots \ldots (n-1).n}$$

$$= \frac{\lfloor n}{\lfloor n}$$

$$= 1.$$

That is, $\qquad C_r = C_{n-r}$.

410. Making $r = 1, 2, 3 \ldots\ldots r-1, r, r+1$ in order,

$$C_1 = n, \quad C_2 = \frac{n}{1} \cdot \frac{n-1}{2}, \quad C_3 = \frac{n}{1} \cdot \frac{n-1}{2} \cdot \frac{n-2}{3},$$

....

$$C_{r-1} = \frac{n \cdot (n-1) \ldots\ldots (n-r+2)}{1 \cdot 2 \ldots\ldots (r-1)}$$

$$C_r = \frac{n \cdot (n-1) \ldots\ldots (n-r+2) \cdot (n-r+1)}{1 \cdot 2 \ldots\ldots (r-1) \cdot r}$$

$$C_{r+1} = \frac{n \cdot (n-1) \ldots\ldots (n-r+1) \cdot (n-r)}{1 \cdot 2 \ldots\ldots r \cdot (r+1)}$$

....

$$C_n = 1.$$

Hence the general expression for the factor connecting C_r, one of the set of numbers $C_1, C_2 \ldots\ldots C_{r+1} \ldots\ldots C_n$, with C_{r-1}, that which stands next before it, is $\frac{n-r+1}{r}$, that is,

$$C_r = \frac{n-r+1}{r} \cdot C_{r-1}.$$

With regard to this factor $\frac{n-r+1}{r}$, we observe

(1) It is always positive, because $n+1$ is greater than r.

(2) Its value continually decreases, for

$$\frac{n-r+1}{r} = \frac{n+1}{r} - 1,$$

which decreases as r increases.

(3) Though $\frac{n-r+1}{r}$ continually decreases, yet for several successive values of r it is greater than unity, and therefore each of the corresponding terms is greater than the preceding.

(4) When r is such that $\frac{n-r+1}{r}$ is less than unity the corresponding term is less than the preceding.

(5) If n and r be such that $\dfrac{n-r+1}{r} = 1$, C_r and C_{r-1} are a pair of equal terms, each greater than any preceding or subsequent term.

Hence up to a certain term (or pair of terms) the terms increase, and after that decrease: this term (or pair of terms) is the greatest of the series, and it is the object of the next Article to determine what value of r gives this greatest term (or pair of terms).

411. *To find the value of r for which the number of combinations of n things taken r together is the greatest.*

$$C_{r-1} = \frac{n \cdot (n-1) \ldots (n-r+2)}{1 \cdot 2 \ldots (r-1)}$$

$$C_r = \frac{n \cdot (n-1) \ldots (n-r+2)}{1 \cdot 2 \ldots (r-1)} \cdot \frac{(n-r+1)}{r}$$

$$C_{r+1} = \frac{n \cdot (n-1) \ldots (n-r+1)}{1 \cdot 2 \ldots r} \cdot \frac{n-r}{r+1}.$$

Hence, if C_r denote the number of combinations required, $\dfrac{C_r}{C_{r-1}}$ and $\dfrac{C_r}{C_{r+1}}$ must neither of them be less than 1.

But $\qquad \dfrac{C_r}{C_{r-1}} = \dfrac{n-r+1}{r},$

and $\qquad \dfrac{C_r}{C_{r+1}} = \dfrac{r+1}{n-r}.$

Hence $\dfrac{n-r+1}{r}$ is not less than 1 and $\dfrac{r+1}{n-r}$ is not less than 1,

or, $\qquad n-r+1$ is not less than r and $r+1$ not less than $n-r$,

or, $\qquad n+1$ is not less than $2r$ and $2r$ not less than $n-1$;

∴ $2r$ is not greater than $n+1$ and not less than $n-1$.

Hence $2r$ can have only three values, $n-1$, n, $n+1$.

Now $2r$ must be an even number, and therefore

(1) If n be odd, $n-1$ and $n+1$ being both even numbers $2r$ may be equal to $n-1$ or $n+1$;

COMBINATIONS.

$$\therefore r - \frac{n}{2} = \frac{1}{2} \text{ or } r = \frac{n+1}{2}.$$

(2) If n be even, $n-1$ and $n+1$ being both odd numbers, $2r$ can only be equal to n;

$$\therefore r = \frac{n}{2}.$$

Ex. 1. Of eight things how many must be taken together that the number of combinations may be the greatest possible?

Here $n = 8$, an even number, therefore the number to be taken is 4, which will give $\dfrac{8 \times 7 \times 6 \times 5}{1 \times 2 \times 3 \times 4}$ or 70 combinations.

Ex. 2. If the number of things be 9, then the number to be taken is $\dfrac{9-1}{2}$ or $\dfrac{9+1}{2}$, that is 4 or 5, which will give respectively

$$\frac{9 \times 8 \times 7 \times 6}{1 \times 2 \times 3 \times 4}, \text{ or 126 combinations, and}$$

$$\frac{9 \times 8 \times 7 \times 6 \times 5}{1 \times 2 \times 3 \times 4 \times 5}, \text{ or 126 combinations.}$$

EXAMPLES.—cxlvii.

1. Out of 100 soldiers how many different parties of 4 can be chosen?

2. How many combinations can be made of 6 things taken 5 at a time?

3. Of the combinations of the first 10 letters of the alphabet taken 5 together, in how many will *a* occur?

4. How many words can be formed, consisting of 3 consonants and one vowel, in a language containing 19 consonants and 5 vowels?

5. The number of combinations of n things taken 4 at a time : the number taken 2 at a time $= 15 : 2$. Find n.

6. The number of combinations of n things, taken 5 at

a time, is $3\frac{3}{5}$ times the number of combinations taken 3 at a time. Find n.

7. Out of 17 consonants and 5 vowels, how many words can be formed, each containing 2 vowels and 3 consonants?

8. Out of 12 consonants and 5 vowels how many words can be formed, each containing 6 consonants and 3 vowels?

9. The number of permutations of n things, 3 at a time, is 6 times the number of combinations, 4 at a time. Find n.

10. How many different sums may be formed with a guinea, a half-guinea, a crown, a half-crown, a shilling, and a sixpence?

11. At a game of cards, 3 being dealt to each person, any one can have 425 times as many hands as there are cards in the pack. How many cards are there?

12. There are 12 soldiers and 16 sailors. How many different parties of 6 can be made, each party consisting of 3 soldiers and 3 sailors?

13. On how many nights can a different patrol of 5 men be draughted from a corps of 36? On how many of these would any one man be taken?

XXXV. THE BINOMIAL THEOREM. POSITIVE INTEGRAL INDEX.

412. THE **Binomial Theorem**, first explained by Newton, is a method of raising a binomial expression to any power without going through the process of actual multiplication.

413. *To investigate the Binomial Theorem for a Positive Integral Index.*

By actual multiplication we can show that

$$(x+a_1)(x+a_2) = x^2 + (a_1+a_2)x + a_1a_2$$

$$(x+a_1)(x+a_2)(x+a_3) = x^3 + (a_1+a_2+a_3)x^2$$
$$+ (a_1a_2 + a_1a_3 + a_2a_3)x + a_1a_2a_3$$

$$(x+a_1)(x+a_2)(x+a_3)(x+a_4) = x^4 + (a_1+a_2+a_3+a_4)x^3$$
$$+ (a_1a_2 + a_1a_3 + a_1a_4 + a_2a_3 + a_2a_4 + a_3a_4)x^2$$
$$+ (a_1a_2a_3 + a_1a_2a_4 + a_1a_3a_4 + a_2a_3a_4)x + a_1a_2a_3a_4.$$

In these results we observe the following laws:

I. Each product is composed of a descending series of powers of x. The index of x in the first term is the same as the number of factors, and the indices of x decrease by unity in each succeeding term.

II. The number of terms is greater by 1 than the number of factors.

III. The coefficient of the *first* term is unity.

of the *second* the sum of $a_1, a_2, a_3 \ldots$

of the *third* the sum of the products of
$a_1, a_2, a_3 \ldots$ *taken two at a time.*

of the *fourth* the sum of the products of
$a_1, a_2, a_3 \ldots$ *taken three at a time.*

and the *last* term is the product of all the quantities

$$a_1, a_2, a_3 \ldots$$

Suppose now this law to hold for $n-1$ factors, so that

$$(x+a_1)(x+a_2)(x+a_3)\ldots(x+a_{n-1})$$
$$= x^{n-1} + S_1 \cdot x^{n-2} + S_2 \cdot x^{n-3} + S_3 \cdot x^{n-4} + \ldots + S_{n-1},$$

where $S_1 = a_1 + a_2 + a_3 + \ldots + a_{n-1}$,

that is, the sum of $a_1, a_2, a_3 \ldots a_{n-1}$,

$$S_2 = a_1a_2 + a_1a_3 + a_2a_3 + \ldots + a_1a_{n-1} + a_2a_{n-1} + \ldots$$

that is, the sum of the products of $a_1, a_2, a_3 \ldots a_{n-1}$, taken two at a time,

$S_3 = a_1 a_2 a_3 + a_1 a_2 a_4 + \ldots + a_1 a_2 a_{n-1} + a_1 a_3 a_{n-1} + \ldots$

that is, the sum of the products of $a_1, a_2 \ldots a_{n-1}$, taken three at a time,

....

$S_{n-1} = a_1 a_2 a_3 \ldots a_{n-1}$,

that is, the product of $a_1, a_2, a_3 \ldots a_{n-1}$.

Now multiply both sides by $x + a_n$.

Then

$(x + a_1)(x + a_2) \ldots (x + a_{n-1})(x + a_n)$
$= x^n + S_1 x^{n-1} + S_2 x^{n-2} + S_3 x^{n-3} + \ldots$
$\quad + a_n x^{n-1} + a_n S_1 x^{n-2} + a_n S_2 x^{n-3} + \ldots + a_n S_{n-1}$
$= x^n + (S_1 + a_n) x^{n-1} + (S_2 + a_n S_1) x^{n-2}$
$\quad + (S_3 + a_n S_2) x^{n-3} + \ldots + a_n S_{n-1}.$

Now $S_1 + a_n = a_1 + a_2 + a_3 + \ldots + a_{n-1} + a_n$,

that is, the sum of $a_1, a_2, a_3 \ldots a_n$,

$S_2 + a_n S_1 = S_2 + a_n (a_1 + a_2 + \ldots + a_{n-1})$,

that is, the sum of the products of $a_1, a_2 \ldots a_n$, taken two at a time,

$S_3 + a_n S_2 = S_3 + a_n (a_1 a_2 + a_1 a_3 + \ldots)$,

that is, the sum of the products of $a_1, a_2 \ldots a_n$, taken three at a time,

....

$a_n S_{n-1} = a_1 a_2 a_3 \ldots a_{n-1} a_n$,

that is, the product of $a_1, a_2, a_3 \ldots a_n$.

If then the law holds good for $n-1$ factors, it will hold good for n factors: and as we have shown that it holds good up to 4 factors it will hold for 5 factors: and hence for 6 factors: and so on for any number.

Now let each of the n quantities $a_1, a_2, a_3 \ldots a_n$ be equal to a, and let us write our result thus:

$(x+a_1)(x+a_2)\ldots(x+a_n) = x^n + A_1 \cdot x^{n-1} + A_2 \cdot x^{n-2} + \ldots + A_n.$

The left-hand side becomes

$(x+a)(x+a)\ldots(x+a)$ to n factors, that is, $(x+a)^n$.

And on the right-hand side

$A_1 = a + a + a + \ldots$ to n terms $= na$,

$A_2 = a^2 + a^2 + a^2 + \ldots$ to as many terms as are equal to the number of combinations of n things taken *two* at a time, that is $\dfrac{n \cdot (n-1)}{1 \cdot 2}$;

$$\therefore A_2 = \frac{n \cdot (n-1)}{1 \cdot 2} \cdot a^2,$$

$A_3 = a^3 + a^3 + a^3 + \ldots$ to as many terms as are equal to the number of combinations of n things taken *three* at a time, that is $\dfrac{n \cdot (n-1) \cdot (n-2)}{1 \cdot 2 \cdot 3}$;

$$\therefore A_3 = \frac{n \cdot (n-1) \cdot (n-2)}{1 \cdot 2 \cdot 3} \cdot a^3,$$

. . . .

$A_n = a \cdot a \cdot a \ldots$ to n factors $= a^n$.

Hence we obtain as our final result

$(x+a)^n = x^n + nax^{n-1} + \dfrac{n \cdot (n-1)}{1 \cdot 2} a^2 x^{n-2}$

$\qquad + \dfrac{n \cdot (n-1) \cdot (n-2)}{1 \cdot 2 \cdot 3} \cdot a^3 x^{n-3} + \ldots + a^n.$

414. Ex. Expand $(x+a)^6$.

Here the number of terms will be *seven*, and we have

$(x+a)^6 = x^6 + 6ax^5 + \dfrac{6 \cdot 5}{1 \cdot 2} a^2 x^4 + \dfrac{6 \cdot 5 \cdot 4}{1 \cdot 2 \cdot 3} a^3 x^3$

$\qquad + \dfrac{6 \cdot 5 \cdot 4 \cdot 3}{1 \cdot 2 \cdot 3 \cdot 4} a^4 x^2 + \dfrac{6 \cdot 5 \cdot 4 \cdot 3 \cdot 2}{1 \cdot 2 \cdot 3 \cdot 4 \cdot 5} a^5 x + a^6$

$= x^6 + 6ax^5 + 15a^2x^4 + 20a^3x^3 + 15a^4x^2 + 6a^5x + a^6.$

NOTE. The coefficients of terms equidistant from the end and from the beginning are the same. The general proof of this will be given in Art. 420.

Hence in the Example just given when the coefficients of *four* terms had been found those of the other three might have been written down at once.

EXAMPLES.—cxlviii.

Expand the following expressions:

1. $(a+x)^4$.   2. $(b+c)^6$.   3. $(a+b)^7$.
4. $(x+y)^8$.   5. $(5+4a)^4$.   6. $(a^2+bc)^5$.

415. Since
$$(x+a)^n = x^n + nax^{n-1} + \frac{n \cdot (n-1)}{1 \cdot 2} \cdot a^2 x^{n-2} + \ldots + a^n,$$
if we put $x=1$, we shall have
$$(1+a)^n = 1 + na + \frac{n \cdot (n-1)}{1 \cdot 2} \cdot a^2 + \ldots + a^n.$$

416. Every binomial may be reduced to such a form that the part to be expanded may have 1 for its first term.

Thus since
$$x + a = x\left(1 + \frac{a}{x}\right),$$
$$(x+a)^n = x^n \left(1 + \frac{a}{x}\right)^n;$$
and we may then expand $\left(1 + \frac{a}{x}\right)^n$ and multiply each term of the result by x^n.

Ex. Expand $(2x+3y)^5$.

$$(2x+3y)^5 = (2x)^5 \cdot \left(1 + \frac{3y}{2x}\right)^5$$
$$= 32x^5 \cdot \left\{ 1 + 5 \cdot \frac{3y}{2x} + \frac{5 \cdot 4}{1 \cdot 2} \cdot \left(\frac{3y}{2x}\right)^2 + \frac{5 \cdot 4 \cdot 3}{1 \cdot 2 \cdot 3} \cdot \left(\frac{3y}{2x}\right)^3 \right.$$
$$\left. + \frac{5 \cdot 4 \cdot 3 \cdot 2}{1 \cdot 2 \cdot 3 \cdot 4} \cdot \left(\frac{3y}{2x}\right)^4 + \left(\frac{3y}{2x}\right)^5 \right\}$$

$$= 32x^5 \left\{ 1 + \frac{15y}{2x} + \frac{90y^2}{4x^2} + \frac{270y^3}{8x^3} + \frac{405y^4}{16x^4} + \frac{243y^5}{32x^5} \right\}$$

$$= 32x^5 + 240x^4y + 720x^3y^2 + 1080x^2y^3 + 810xy^4 + 243y^5.$$

417. The expansion of $(x-a)^n$ will be precisely the same as that of $(x+a)^n$, except that the sign of terms in which the odd powers of a enter, that is the second, fourth, sixth, and other even terms, will be *negative*.

Thus $(x-a)^n = x^n - nax^{n-1} + \dfrac{n \cdot (n-1)}{1 \cdot 2} \cdot a^2 x^{n-2}$

$$- \frac{n \cdot (n-1) \cdot (n-2)}{1 \cdot 2 \cdot 3} \cdot a^3 x^{n-3} + \ldots$$

for $\quad (x-a)^n = \{x + (-a)\}^n$

$$= x^n + n(-a)x^{n-1} + \frac{n \cdot (n-1)}{1 \cdot 2}(-a)^2 x^{n-2} + \&c.$$

$$= x^n - nax^{n-1} + \frac{n \cdot (n-1)}{1 \cdot 2} a^2 x^{n-2} + \&c.$$

Ex. Expand $(a-c)^5$.

$$(a-c)^5 = a^5 - 5a^4c + \frac{5 \cdot 4}{1 \cdot 2}a^3c^2 - \frac{5 \cdot 4 \cdot 3}{1 \cdot 2 \cdot 3}a^2c^3 + \frac{5 \cdot 4 \cdot 3 \cdot 2}{1 \cdot 2 \cdot 3 \cdot 4}ac^4 - c^5.$$

$$= a^5 - 5a^4c + 10a^3c^2 - 10a^2c^3 + 5ac^4 - c^5$$

EXAMPLES.—cxlix.

Expand the following expressions:

1. $(a-x)^6$.
2. $(b-c)^7$.
3. $(2x-3y)^5$.
4. $(1-2x)^5$.
5. $(1-x)^{10}$.
6. $(a^3-b^2)^8$.

418. A trinomial, as $a+b+c$, may be raised to any power by the Binomial Theorem, if we regard two terms as one, thus:

$$(a+b+c)^n = (a+b)^n + n \cdot (a+b)^{n-1} \cdot c$$

$$+ \frac{n \cdot (n-1)}{1 \cdot 2} \cdot (a+b)^{n-2} \cdot c^2 + \ldots$$

Ex. Expand $(1 + x + x^2)^3$.

$$(1 + x + x^2)^3 = (1+x)^3 + 3(1+x)^2 \cdot x^2 + \frac{3 \cdot 2}{1 \cdot 2}(1+x) \cdot x^4 + x^6$$

$$= (1 + 3x + 3x^2 + x^3) + 3(1 + 2x + x^2)x^2$$
$$+ 3(1+x)x^4 + x^6$$

$$= 1 + 3x + 3x^2 + x^3 + 3x^2 + 6x^3 + 3x^4 + 3x^4$$
$$+ 3x^5 + x^6$$

$$= 1 + 3x + 6x^2 + 7x^3 + 6x^4 + 3x^5 + x^6.$$

EXAMPLES.—cl.

Expand the following expressions:

1. $(a + 2b - c)^3$. 2. $(1 - 2x + 3x^2)^3$. 3. $(x^3 - x^2 + x)^3$.

4. $(3x^{\frac{4}{3}} + 2x^{\frac{1}{6}} + 1)^3$. 5. $\left(x + 1 - \frac{1}{x}\right)^3$. 6. $(a^{\frac{1}{4}} + b^{\frac{1}{4}} - c^{\frac{1}{4}})^3$.

419. *To find the rth or general term of the expansion of* $(x + a)^n$.

We have to determine *three* things to enable us to write down the rth term of the expansion of $(x + a)^n$.

      1. The index of x in that term.
      2. The index of a in that term.
      3. The coefficient of that term.

Now the index of x, decreasing by 1 in each term, is in the rth term $n - r + 1$; and the index of a, increasing by 1 in each term, is in the rth term $r - 1$.

For example, in the *third* term

    the index of x is $n - 3 + 1$, that is, $n - 2$;
    the index of a is $3 - 1$, that is, 2.

In assigning its proper coefficient to the rth term we have to determine the *last* factor in the denominator and also in the numerator of the fraction

$$\frac{n \cdot (n-1) \cdot (n-2) \cdot (n-3) \ldots\ldots}{1 \cdot 2 \cdot 3 \cdot 4 \ldots\ldots}.$$

Now the *last* factor of the denominator is less by 1 than the number of the term to which it belongs. Thus in the 3rd term the last factor of the denominator is 2, and in the rth term the last factor of the denominator is $r-1$.

The *last* factor of the numerator is formed by subtracting from n the number of the term to which it belongs and adding 2 to the result.

Thus in the 3rd term the last factor of the numerator is
$$n-3+2, \text{ that is } n-1;$$
in the 4th .... $n-4+2$, that is $n-2$;
and so in the rth .... $n-r+2$.

Observe also that the factors of the numerator *decrease* by unity, and the factors of the denominator *increase* by unity, so that the coefficient of the rth term is

$$\frac{n \cdot (n-1) \cdot (n-2)\ldots\ldots(n-r+2)}{1 \cdot 2 \cdot 3 \ldots\ldots (r-1)}.$$

Collecting our results, we write the rth term of the expansion of $(x+a)^n$ thus:

$$\frac{n \cdot (n-1) \cdot (n-2)\ldots\ldots(n-r+2)}{1 \cdot 2 \cdot 3 \ldots\ldots (r-1)} \cdot a^{r-1} \cdot x^{n-r+1}.$$

Obs. The index of a is the same as the last factor in the denominator. The sum of the indices of a and x is n.

EXAMPLES.—cli.

Find

1. The 8th term of $(1+x)^{11}$.
2. The 5th term of $(a^2-b^2)^{12}$.
3. The 4th term of $(a-b)^{100}$.
4. The 9th term of $(2ab-cd)^{14}$.
5. The middle term of $(a-b)^{16}$.
6. The middle term of $(a^{\frac{1}{8}}+b^{\frac{1}{8}})^8$.
7. The two middle terms of $(a-b)^{19}$.
8. The two middle terms of $(a+x)^{13}$.

9. Show that the coefficient of the middle term of
$(a+x)^{4n}$ is $2^{2n} \times \dfrac{1.3.5\ldots\ldots(4n-1)}{1.2.3\ldots\ldots 2n}$.

10. Show that the coefficient of the middle term of
$(a+x)^{4n+2}$ is $2^{n+1} \times \dfrac{(2n+3)(2n+5)\ldots\ldots(4n-1)(4n+1)}{1.2\ldots\ldots n}$.

420. *To show that the coefficient of the rth term from the beginning of the expansion of $(x+a)^n$ is identical with the coefficient of the rth term from the end.*

Since the number of terms in the expansion is $n+1$, there are $n+1-r$ terms before the rth term from the end, and therefore the rth term from the end is the $(n-r+2)^{th}$ term from the beginning.

Thus in the expansion of $(x+a)^5$, that is,
$$x^5 + 5ax^4 + 10a^2x^3 + 10a^3x^2 + 5a^4x + a^5,$$
the 3rd term from the end is the $(5-3+2)^{th}$, that is the 4th term from the beginning.

Now if we denote the coefficient of the rth term by C_r, and the coefficient of the $(n-r+2)^{th}$ term by C_{n-r+2}, we have
$$C_r = \frac{n.(n-1)\ldots\ldots(n-r+2)}{1.2\ldots\ldots(r-1)},$$
$$C_{n-r+2} = \frac{n.(n-1)\ldots\ldots\{n-(n-r+2)+2\}}{1.2\ldots\ldots(n-r+2-1)}$$
$$= \frac{n.(n-1)\ldots\ldots r}{1.2\ldots\ldots(n-r+1)}.$$

Hence
$$\frac{C_r}{C_{n-r+2}} = \frac{n.(n-1)\ldots\ldots(n-r+2)}{1.2\ldots\ldots(r-1)} \times \frac{1.2\ldots\ldots(n-r+1)}{n.(n-1)\ldots\ldots r}$$
$$= \frac{n.(n-1)\ldots\ldots(n-r+2).(n-r+1)\ldots\ldots 2.1}{1.2\ldots\ldots(r-1).r\ldots\ldots(n-1).n}$$
$$= \frac{\lfloor n}{\lfloor n} = 1, \text{ which proves the proposition.}$$

421. *To find the greatest term in the expansion of $(x+a)^n$, n being a positive integer.*

The rth term of the expansion $(x+a)^n$ is

$$\frac{n \cdot (n-1) \ldots\ldots (n-r+2)}{1 \cdot 2 \ldots\ldots (r-1)} \cdot a^{r-1} \cdot x^{n-r+1}.$$

The $(r+1)^{\text{th}}$ term of the expansion $(x+a)^n$ is

$$\frac{n \cdot (n-1) \ldots\ldots (n-r+2) \cdot (n-r+1)}{1 \cdot 2 \ldots\ldots (r-1) \cdot r} \cdot a^r \cdot x^{n-r}.$$

Hence it follows that we obtain the $(r+1)^{\text{th}}$ term by multiplying the rth term by

$$\frac{n-r+1}{r} \cdot \frac{a}{x}.$$

When this multiplier is first less than 1, the rth term is the greatest in the expansion.

Now $\quad \dfrac{n-r+1}{r} \cdot \dfrac{a}{x}$ is first less than 1

when $\quad na - ra + a$ is first less than rx,

or $\quad na + a$ first less than $rx + ra$,

or $\quad r(x+a)$ first greater than $a(n+1)$,

or $\quad r$ first greater than $\dfrac{a(n+1)}{x+a}$.

If r be equal to $\dfrac{a(n+1)}{x+a}$, then $\dfrac{n-r+1}{r} \cdot \dfrac{a}{x} = 1$, and the $(r+1)^{\text{th}}$ term is equal to the rth, and each is greater than any other term.

Ex. Find the greatest term in the expansion of $(4+a)^7$, when $a = \dfrac{3}{2}$.

Here $\quad \dfrac{a(n+1)}{x+a} = \dfrac{\frac{3}{2}(7+1)}{4+\frac{3}{2}} = \dfrac{12}{11} = \dfrac{24}{11} = 2\dfrac{2}{11}.$

The first whole number greater than $2\frac{2}{11}$ is 3, therefore the greatest term of the expansion is the 3rd.

[S.A.]

422. *To find the sum of all the coefficients in the expansion of $(1+x)^n$.*

Since $(1+x)^n = 1 + nx + \dfrac{n \cdot (n-1)}{1 \cdot 2} x^2 + \ldots\ldots$

$$+ \dfrac{n \cdot (n-1)}{1 \cdot 2} x^{n-2} + nx^{n-1} + x^n \ldots\ldots$$

putting $\quad x = 1$, we get

$$2^n = 1 + n + \dfrac{n \cdot (n-1)}{1 \cdot 2} + \ldots\ldots + \dfrac{n \cdot (n-1)}{1 \cdot 2} + n + 1;$$

or, $\quad 2^n =$ the sum of all the coefficients.

423. *To show that the sum of the coefficients of the odd terms in the expansion of $(1+x)^n$ is equal to the sum of the coefficients of the even terms.*

Since

$$(1+x)^n = 1 + nx + \dfrac{n \cdot (n-1)}{1 \cdot 2} x^2 + \dfrac{n \cdot (n-1) \cdot (n-2)}{1 \cdot 2 \cdot 3} x^3 + \ldots\ldots$$

putting $\quad x = -1$, we get

$$(1-1)^n = 1 - n + \dfrac{n \cdot (n-1)}{1 \cdot 2} - \dfrac{n \cdot (n-1) \cdot (n-2)}{1 \cdot 2 \cdot 3} + \ldots\ldots$$

or, $\quad 0 = \left\{ 1 + \dfrac{n \cdot (n-1)}{1 \cdot 2} + \ldots\ldots \right\}$

$$- \left\{ n + \dfrac{n \cdot (n-1) \cdot (n-2)}{1 \cdot 2 \cdot 3} + \ldots\ldots \right\}$$

= sum of coefficients of odd terms − sum of coefficients of even terms;

∴ sum of coefficients of odd terms = sum of coefficients of even terms.

Hence, by the preceding Article,

$$\text{sum of coefficients of odd terms} = \dfrac{2^n}{2} = 2^{n-1};$$

$$\text{sum of coefficients of even terms} = \dfrac{2^n}{2} = 2^{n-1}.$$

XXXVI. THE BINOMIAL THEOREM.
FRACTIONAL AND NEGATIVE INDICES.

424. We have shown that when m is a positive integer,
$$(1+x)^m = 1 + mx + \frac{m \cdot (m-1)}{1 \cdot 2} x^2 + \ldots$$

We have now to show that this equation holds good when m is a positive fraction, as $\frac{3}{2}$, a negative integer, as -3, or a negative fraction, as $-\frac{3}{4}$.

We shall give the proof devised by Euler.

425. If m be a positive integer we know that
$$(1+x)^m = 1 + mx + \frac{m \cdot (m-1)}{1 \cdot 2} x^2 + \frac{m \cdot (m-1) \cdot (m-2)}{1 \cdot 2 \cdot 3} x^3 + \ldots$$

Let us agree to represent a series of the form
$$1 + mx + \frac{m \cdot (m-1)}{1 \cdot 2} x^2 + \ldots$$
by the symbol $f(m)$, *whatever the value of* m *may be*.

Then we know that when m is a positive integer
$$(1+x)^m = f(m) \ ;$$
and we have to show that, also, when m is fractional or negative
$$(1+x)^m = f(m).$$

Since
$$f(m) = 1 + mx + \frac{m \cdot (m-1)}{1 \cdot 2} x^2 + \ldots$$
$$f(n) = 1 + nx + \frac{n \cdot (n-1)}{1 \cdot 2} x^2 + \ldots$$

If we multiply together the two series, we shall obtain an expression of the form

$$1 + ax + bx^2 + cx^3 + dx^4 + \ldots\ldots$$

that is, a series of ascending powers of x in which the coefficients a, b, c …… are formed by various combinations of m and n.

To determine the mode in which a and b are formed, let us commence the multiplication of the two series and continue it as far as terms involving x^2, thus

$$f(m) = 1 + mx + \frac{m \cdot (m-1)}{1 \cdot 2} x^2 + \ldots\ldots$$

$$f(n) = 1 + nx + \frac{n \cdot (n-1)}{1 \cdot 2} x^2 + \ldots\ldots$$

$$f(m) \times f(n) = 1 + mx + \frac{m \cdot (m-1)}{1 \cdot 2} x^2 + \ldots\ldots$$

$$+ nx + mnx^2 + \ldots\ldots$$

$$+ \frac{n \cdot (n-1)}{1 \cdot 2} x^2 + \ldots\ldots$$

$$1 + (m+n) \cdot x + \left\{ \frac{m \cdot (m-1)}{1 \cdot 2} + mn + \frac{n \cdot (n-1)}{1 \cdot 2} \right\} x^2 + \ldots\ldots$$

Comparing this product with the assumed expression

$$1 + ax + bx^2 + cx^3 + dx^4 + \ldots\ldots$$

we see that $\quad a = m + n,$

and $\quad b = \dfrac{m \cdot (m-1)}{1 \cdot 2} + mn + \dfrac{n \cdot (n-1)}{1 \cdot 2}$

$\quad\quad = \dfrac{m^2 - m + 2mn + n^2 - n}{1 \cdot 2}$

$\quad\quad = \dfrac{(m+n) \cdot (m+n-1)}{1 \cdot 2}.$

Similarly we could show *by actual multiplication* that

$$c = \frac{(m+n).(m+n-1).(m+n-2)}{1.2.3},$$

$$d = \frac{(m+n).(m+n-1).(m+n-2).(m+n-3)}{1.2.3.4}.$$

Thus we might determine the successive coefficients to any extent, but we may ascertain *the law of their formation* by the following considerations.

The *forms* of the coefficients, that is, the way in which m and n are involved in them, do not depend in any way on the *values* of m and n, but will be precisely the same whether m and n be positive integers or any numbers whatsoever.

If then we can determine the law of their formation when m and n are positive integers, we shall know the law of their formation for all values of m and n.

Now when m and n are *positive integers*,

$$f(m) = (1+x)^m,$$
$$f(n) = (1+x)^n;$$
$$\therefore f(m) \times f(n) = (1+x)^m \times (1+x)^n$$
$$= (1+x)^{m+n}$$
$$= 1 + (m+n)x + \frac{(m+n).(m+n-1)}{1.2}x^2 + \ldots$$
$$= f(m+n).$$

Hence we conclude that *whatever be the values* of m and n

$$f(m) \times f(n) = f(m+n).$$

Hence $f(m+n+p) = f(m).f(n+p)$
$$= f(m).f(n).f(p),$$

and so generally

$$f(m+n+p+\ldots) = f(m).f(n).f(p)\ldots$$

Now let $m = n = p = \ldots = \frac{h}{k}$, h and k being positive integers, then

$$f\left(\frac{h}{k} + \frac{h}{k} + \frac{h}{k} + \ldots \text{ to } k \text{ terms}\right)$$
$$= f\left(\frac{h}{k}\right) \cdot f\left(\frac{h}{k}\right) \cdot f\left(\frac{h}{k}\right) \ldots \text{ to } k \text{ factors}.$$

or, $$f(h) = \left\{f\left(\frac{h}{k}\right)\right\}^k,$$

or, $$(1+x)^h = \left\{f\left(\frac{h}{k}\right)\right\}^k;$$

$$\therefore (1+x)^{\frac{h}{k}} = f\left(\frac{h}{k}\right)$$

$$= 1 + \frac{h}{k}x + \frac{\frac{h}{k}\cdot\left(\frac{h}{k}-1\right)}{1 \cdot 2}x^2 + \ldots$$

which proves the theorem for a positive fractional index.

Again, since $f(m) \cdot f(n) = f(m+n)$ for all values of m and n, let $n = -m$, then

$$f(m) \cdot f(-m) = f(m-m)$$
$$= f(0).$$

Now the series $$1 + mx + \frac{m \cdot (m-1)}{1 \cdot 2}x^2 + \ldots$$

becomes 1 when $m = 0$, that is, $f(0) = 1$;

$$\therefore f(m) \cdot f(-m) = 1;$$

$$\therefore f(-m) = \frac{1}{f(m)} = \frac{1}{(1+x)^m} = (1+x)^{-m};$$

$$\therefore (1+x)^{-m} = f(-m)$$

$$= 1 + (-m)x + \frac{-m(-m-1)}{1 \cdot 2}x^2 + \ldots$$

which proves the theorem for a *negative* index, integral or fractional.

426. Ex. Expand $(a+x)^{\frac{1}{2}}$ to four terms.

$$(a+x)^{\frac{1}{2}} = a^{\frac{1}{2}} + \frac{1}{2} \cdot a^{\frac{1}{2}-1} \cdot x + \frac{\frac{1}{2} \cdot (\frac{1}{2}-1)}{1 \cdot 2} \cdot a^{\frac{1}{2}-2} \cdot x^2 +$$

$$\frac{\frac{1}{2} \cdot (\frac{1}{2}-1)(\frac{1}{2}-2)}{1 \cdot 2 \cdot 3} \cdot a^{\frac{1}{2}-3} \cdot x^3 \ldots$$

$$= a^{\frac{1}{2}} + \frac{1}{2} \cdot a^{-\frac{1}{2}} \cdot x + \frac{-\frac{1}{4}}{2} \cdot a^{-\frac{3}{2}} \cdot x^2 + \frac{\frac{3}{8}}{6} \cdot a^{-\frac{5}{2}} \cdot x^3 \ldots$$

$$= a^{\frac{1}{2}} + \frac{x}{2a^{\frac{1}{2}}} - \frac{x^2}{8a^{\frac{3}{2}}} + \frac{x^3}{16a^{\frac{5}{2}}} \ldots$$

Or we might proceed thus, as is explained in Art. 416.

$$(a+x)^{\frac{1}{2}} = a^{\frac{1}{2}} \left(1 + \frac{x}{a}\right)^{\frac{1}{2}}$$

$$= a^{\frac{1}{2}} \left\{ 1 + \frac{1}{2} \cdot \frac{x}{a} + \frac{\frac{1}{2} \cdot (\frac{1}{2}-1)}{1 \cdot 2} \cdot \frac{x^2}{a^2} + \frac{\frac{1}{2}(\frac{1}{2}-1)(\frac{1}{2}-2)}{1 \cdot 2 \cdot 3} \cdot \frac{x^3}{a^3} \ldots \right\}$$

$$= a^{\frac{1}{2}} \left\{ 1 + \frac{x}{2a} - \frac{x^2}{8a^2} + \frac{x^3}{16a^3} \ldots \right\}$$

$$= a^{\frac{1}{2}} + \frac{x}{2a^{\frac{1}{2}}} - \frac{x^2}{8a^{\frac{3}{2}}} + \frac{x^3}{16a^{\frac{5}{2}}} \ldots$$

EXAMPLES.—clii.

Expand the following expressions:

1. $(1+x)^{\frac{1}{2}}$ to five terms.
2. $(1+a)^{\frac{2}{3}}$ to four terms.
3. $(a+x)^{\frac{1}{3}}$ to five terms.
4. $(1+2x)^{\frac{1}{2}}$ to five terms.
5. $\left(a + \frac{4x}{3}\right)^{\frac{3}{4}}$ to four terms.
6. $(a^{\frac{1}{4}} + x^{\frac{1}{4}})^{\frac{4}{5}}$ to four terms.
7. $(1-x^2)^{\frac{1}{2}}$ to five terms.
8. $(1-a^2)^{\frac{7}{3}}$ to four terms.
9. $(1-3x)^{\frac{3}{4}}$ to four terms.
10. $\left(x^2 - \frac{2y}{3}\right)^{\frac{3}{2}}$ to four terms.
11. $(1-x)^{\frac{5}{6}}$ to four terms.
12. $\left(\frac{2x}{3} - \frac{3y}{2}\right)^{\frac{2}{3}}$ to three terms.

THE BINOMIAL THEOREM.

427. *To expand* $(1+x)^{-n}$.

$$(1+x)^{-n} = 1 + (-n) \cdot x + \frac{-n \cdot (-n-1)}{1 \cdot 2} x^2$$
$$+ \frac{-n \cdot (-n-1) \cdot (-n-2)}{1 \cdot 2 \cdot 3} x^3 + \ldots$$
$$= 1 - nx + \frac{n(n+1)}{1 \cdot 2} x^2 - \frac{n(n+1)(n+2)}{1 \cdot 2 \cdot 3} \cdot x^3 + \ldots$$

the terms being alternately positive and negative.

Ex. Expand $(1+x)^{-3}$ to five terms.

$$(1+x)^{-3} = 1 - 3x + \frac{3 \cdot 4}{1 \cdot 2} x^2 - \frac{3 \cdot 4 \cdot 5}{1 \cdot 2 \cdot 3} x^3 + \frac{3 \cdot 4 \cdot 5 \cdot 6}{1 \cdot 2 \cdot 3 \cdot 4} \cdot x^4 - \ldots$$
$$= 1 - 3x + 6x^2 - 10x^3 + 15x^4 - \ldots$$

428. *To expand* $(1-x)^{-n}$.

$$(1-x)^{-n} = 1 - (-n) \cdot x + \frac{-n \cdot (-n-1)}{1 \cdot 2} \cdot x^2$$
$$- \frac{-n(-n-1)(-n-2)}{1 \cdot 2 \cdot 3} x^3 + \ldots$$
$$= 1 + nx + \frac{n \cdot (n+1)}{1 \cdot 2} \cdot x^2 + \frac{n \cdot (n+1)(n+2)}{1 \cdot 2 \cdot 3} x^3 + \ldots$$

the terms being all positive.

Ex. Expand $(1-x)^{-3}$ to five terms.

$$(1-x)^{-3} = 1 + 3x + \frac{3 \cdot 4}{1 \cdot 2} x^2 + \frac{3 \cdot 4 \cdot 5}{1 \cdot 2 \cdot 3} x^3 + \frac{3 \cdot 4 \cdot 5 \cdot 6}{1 \cdot 2 \cdot 3 \cdot 4} x^4 + \ldots$$
$$= 1 + 3x + 6x^2 + 10x^3 + 15x^4 + \ldots$$

EXAMPLES.—cliii.

Expand

1. $(1+a)^{-2}$ to five terms.
2. $(1-3x)^{-1}$ to five terms.
3. $\left(1-\frac{x}{4}\right)^{-1}$ to four terms.
4. $\left(1-\frac{x}{2}\right)^{-2}$ to five terms.
5. $(a^2 - 2x)^{-5}$ to five terms.
6. $(a^{\frac{1}{3}} - x^{\frac{1}{3}})^{-6}$ to four terms.

EXAMPLES.—cliv.

Expand

1. $(1+x^2)^{-\frac{1}{2}}$ to five terms.
2. $(1-x^2)^{-\frac{3}{2}}$ to five terms.
3. $(x^5+z^6)^{-\frac{2}{5}}$ to four terms.
4. $(1+2x)^{-\frac{1}{2}}$ to five terms.
5. $(a^2+x^2)^{-\frac{1}{2}}$ to four terms.
6. $(a^3+x^3)^{-\frac{1}{3}}$ to four terms.

430. *Observations on the general expression for the term involving x^r in the expansions $(1+x)^n$ and $(1-x)^n$.*

The general expression for the term involving x^r, that is the $(r+1)^{th}$ term, in the expansion of $(1+x)^n$ is

$$\frac{n \cdot (n-1) \ldots (n-r+1)}{1 \cdot 2 \ldots \ldots r} \cdot x^r.$$

From this we must deduce the form in all cases.

Thus the $(r+1)^{th}$ term of the expansion of $(1-x)^n$ is found by changing x into $(-x)$, and therefore it is

$$\frac{n \cdot (n-1) \ldots (n-r+1)}{1 \cdot 2 \ldots \ldots r} \cdot (-x)^r$$

or,

$$(-1)^r \frac{n \cdot (n-1) \ldots (n-r+1)}{1 \cdot 2 \ldots \ldots r} \cdot x^r.$$

If n be negative and $= -m$, the $(r+1)^{th}$ term of the expansion of $(1+x)^n$ is

$$\frac{(-m)(-m-1)\ldots-m-r+1)}{1.2\ldots\ldots\ldots r}x^r,$$

or,

$$\frac{(-1)^r \cdot \{m\cdot(m+1)\ldots(m+r-1)\}}{1.2\ldots\ldots\ldots r}x^r.$$

If n be negative and $= -m$, the $(r+1)^{th}$ term of the expansion of $(1+x)^n$ is

$$\frac{(-1)^r \cdot \{m\cdot(m+1)\ldots(m+r-1)\}}{1.2\ldots\ldots\ldots r}\cdot(-x)^r.$$

or,

$$\frac{m\cdot(m+1)\ldots(m+r-1)}{1.2\ldots\ldots\ldots r}\cdot x^r.$$

EXAMPLES.—clv.

Find the rth terms of the following expansions:

1. $(1+x)^7$.   2. $(1-x)^{12}$.   3. $(a-x)^8$.   4. $(5x+2y)^6$.

5. $(1+x)^{-2}$.   6. $(1-3x)^{-4}$.   7. $(1-x)^{-\frac{1}{2}}$.   8. $(a+x)^{\frac{1}{3}}$.

9. $(1-2x)^{-\frac{7}{2}}$.   10. $(a^2-x^2)^{-\frac{3}{4}}$.

11. Find the $(r+1)^{th}$ term of $(1-x)^{-3}$.

12. Find the $(r+1)^{th}$ term of $(1-4x)^{-\frac{1}{2}}$.

13. Find the $(r+1)^{th}$ term of $(1+x)^{2r}$.

14. Show that the coefficient of x^{r+1} in $(1+x)^{n+1}$ is the sum of the coefficients of x^r and x^{r+1} in $(1+x)^n$.

15. What is the fourth term of $\left(a-\frac{1}{x}\right)^{-\frac{1}{2}}$?

16. What is the fifth term of $(a^2-b^2)^{\frac{3}{2}}$?

17. What is the ninth term of $(a^2+2x^2)^{\frac{1}{2}}$?

18. What is the tenth term of $(a+b)^{-m}$?

19. What is the seventh term of $(a+b)^{\frac{1}{m}}$?

431. The following are examples of the application of the Binomial Theorem to the approximation to roots of numbers.

(1) To approximate to the square root of 104.

$$\sqrt{104} = \sqrt{(100+4)} = 10\left(1 + \frac{4}{100}\right)^{\frac{1}{2}}$$

$$= 10\left\{1 + \frac{1}{2}\cdot\frac{4}{100} + \frac{\frac{1}{2}\left(\frac{1}{2}-1\right)}{1.2}\cdot\left(\frac{4}{100}\right)^2 \right.$$

$$\left. + \frac{\frac{1}{2}\left(\frac{1}{2}-1\right)\left(\frac{1}{2}-2\right)}{1.2.3}\cdot\left(\frac{4}{100}\right)^3 + \ldots\right\}$$

$$= 10\left\{1 + \frac{2}{100} - \frac{2}{10000} + \frac{4}{1000000} - \ldots\right\}$$

$$= 10.19804 \text{ nearly.}$$

(2) To approximate to the fifth root of 2.

$$\sqrt[5]{2} = (1+1)^{\frac{1}{5}}$$

$$= 1 + \frac{1}{5} + \frac{1}{2}\cdot\frac{1}{5}\left(\frac{1}{5}-1\right) + \frac{1}{6}\cdot\frac{1}{5}\cdot\left(\frac{1}{5}-1\right)\left(\frac{1}{5}-2\right) + \ldots$$

$$= 1 + \frac{1}{5} - \frac{2}{25} + \frac{3}{250} - \frac{21}{2500} + \ldots$$

$$= 1 + \frac{3}{25} + \frac{9}{2500} \text{ nearly}$$

$$= 1.1236 \text{ nearly.}$$

(3) To approximate to the cube root of 25.

$$\sqrt[3]{25} = \sqrt[3]{(27-2)} = 3\left\{1 - \frac{2}{27}\right\}^{\frac{1}{3}}.$$

Here we take the cube next *above* 25, so as to make the second term of the binomial as small as possible, and then proceed as before.

EXAMPLES.—clvi.

Approximate to the following roots:

1. $\sqrt[3]{31}$.   2. $\sqrt[7]{108}$.   3. $\sqrt[5]{260}$.   4. $\sqrt[5]{31}$.

XXXVII. SCALES OF NOTATION.

432. The symbols employed in our common system of Arithmetical Notation are the nine digits and zero. These digits when written consecutively acquire local values from their positions with respect to the place of units, the value of every digit increasing *ten-fold* as we advance towards the left hand, and hence the number *ten* is called the RADIX of the Scale.

If we agree to represent the number ten by the letter t, a number, expressed according to the conventions of Arithmetical Notation by 3245, would assume the form

$$3t^3 + 2t^2 + 4t + 5$$

if expressed according to the conventions of Algebra.

433. Let us now suppose that some other number, as *five*, is the radix of a scale of notation, then a number expressed in this scale arithmetically by 2341 will, if *five* be represented by f, assume the form

$$2f^3 + 3f^2 + 4f + 1$$

if expressed algebraically.

And, generally, if r be the radix of a scale of notation, a number expressed arithmetically in that scale by 6789 will, when expressed algebraically, since the value of each digit increases *r-fold* as we advance towards the left hand, be represented by

$$6r^3 + 7r^2 + 8r + 9.$$

434. The number which denotes the radix of any scale will be represented in that scale by 10.

Thus in the scale whose radix is five, the number **five will** be represented by 10.

In the same scale seven, being equal to five + two, will therefore be represented by 12.

Hence the series of natural numbers as far as *twenty-five* will be represented in the scale whose radix is five thus :

1, 2, 3, 4, 10, 11, 12, 13, 14, 20, 21, 22, 23, 24, 30, 31, 32, 33, 34, 40, 41, 42, 43, 44, 100.

435. In the scale whose radix is *eleven* we shall require a new symbol to express the number ten, for in that scale the number eleven is represented by 10. If we agree to express ten in this scale by the symbol t, the series of natural numbers as far as twenty-three will be represented in this scale thus :

1, 2, 3, 4, 5, 6, 7, 8, 9, t, 10, 11, 12, 13, 14, 15, 16, 17, 18, 19, 1t, 20, 21

436. In the scale whose radix is *twelve* we shall require another new symbol to express the number eleven. If we agree to express this number by the symbol e, the natural numbers from nine to thirteen will be represented in the scale whose radix is twelve thus :

9, t, e, 10, 11.

Again, the natural numbers from twenty to twenty-five will be represented thus :

18, 19, 1t, 1e, 20, 21.

437. The scale of notation of which the radix is *two*, is called the Binary Scale.

The names given to the scales, up to that of which the radix is twelve, are Ternary, Quaternary, Quinary, Senary, Septenary, Octonary, Nonary, Denary, Undenary and Duodenary.

438. To perform the operations of Addition, Subtraction, Multiplication, and Division in a scale of notation whose index is r, we proceed in the same way as we do for numbers expressed in the common scale, with this difference only, that r must be used where ten would be used in the common scale : which will be understood better by the following examples.

Ex. 1. Find the sum of 4325 and 5234 in the senary scale.

$$4325$$
$$5234$$
the sum $= 14003$

which is obtained by adding the numbers in vertical lines, carrying 1 for every six contained in the several results, and setting down the excesses above it.

Thus 4 units and 5 units make nine units, that is, six units together with 3 units, so we set down 3 and carry 1 to the next column.

Ex. 2. Find the difference between 62345 and 53466 in the septenary scale.

$$62345$$
$$53466$$
the difference $= 5546$

which is obtained by the following process. We cannot take six units from five units, we therefore add *seven* units to the five units, making 12 units, and take six units from twelve units, and then we add 1 to the lower figure in the second column, and so on.

Ex. 3. Multiply 2471 by 358 in the duodenary scale.

$$2\;4\;7\;1$$
$$3\;5\;8$$
$$1\;7\;0\;8\;8$$
$$e\;t\;e\;5$$
$$7\;1\;9\;3$$
$$8\;3\;3\;3\;1\;8$$

Ex. 4. Divide 367286 by 8 in the nonary scale.

$$8\;)\;367286$$
$$42033$$

The following is the process. We ask how often 8 is contained in 36, which in the nonary scale represents *thirty-three* units; the answer is 4 and 1 over. We then ask how often 8 is contained in 17, which in the nonary scale represents *sixteen* units; the answer is 2 and no remainder. And so for the other digits.

Ex. 5. Divide 1184323 by 589 in the duodenary scale.

```
  589) 1184323 (2483
        e56
        ----
        22t3
        1te0
        ----
        3e32
        39t0
        ----
        1523
        1523
        ----
```

Ex. 6. Extract the square root of 10534521 in the senary scale.

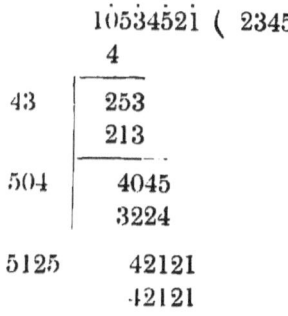

EXAMPLES.—clvii.

1. Add 23561, 42513, 645325 in the septenary scale.
2. Add 3074852, 4635628, 1247653 in the nonary scale.
3. Subtract 267862 from 358423 in the nonary scale.
4. Subtract 124321 from 211010 in the quinary scale.
5. Multiply 57264 by 675 in the octonary scale.
6. Multiply 1456 by 6541 in the septenary scale.
7. Divide 243012 by 5 in the senary scale.
8. Divide 3756025 by 6 in the octonary scale.
9. Extract the square root of 25400544 in the senary scale.
10. Extract the square root of 568988t1 in the duodenary scale.

439. *To transform a given integral number from one scale to another.*

Let N be the given integer expressed in the first scale,

r the radix of the new scale in which the number is to be expressed,

$a, b, c \ldots\ldots m, p, q$ the digits, $n+1$ in number, expressing the number in the new scale;

so that the number in the new scale will be expressed thus:
$$ar^n + br^{n-1} + cr^{n-2} + \ldots\ldots + mr^2 + pr + q.$$

We have now from the equation
$$N = ar^n + br^{n-1} + cr^{n-2} + \ldots\ldots + mr^2 + pr + q$$
to determine the values of $a, b, c \ldots\ldots m, p, q$.

Divide N by r, the remainder is q. Let A be the quotient; then
$$A = ar^{n-1} + br^{n-2} + cr^{n-3} + \ldots\ldots + mr + p.$$

Divide A by r, the remainder is p. Let B be the quotient; then
$$B = ar^{n-2} + br^{n-3} + cr^{n-4} + \ldots\ldots + m.$$

Hence the first digit to the right of the number expressed in the new scale is q, the first remainder;

second .... p, the second remainder;
third .... m, the third remainder;
and thus all the digits may be determined.

Ex. 1. Transform 235791 from the common scale to the scale whose radix is 6.

6	235791
6	39298 remainder 3
6	6549 remainder 4
6	1091 remainder 3
6	181 remainder 5
6	30 remainder 1
6	5 remainder 0
	0 remainder 5

The number required is therefore 5015343.

The digits by which a number can be expressed in a scale whose radix is r will be $1, 2, 3\ldots\ldots r-1$, because these, with 0, are the only remainders which can arise from a division in which the divisor is r.

Ex. 2. Express 3598 in the scale whose radix is 12.

```
12 | 3598
12 |  299  remainder t
12 |   24  remainder e
12 |    2  remainder 0
   |    0  remainder 2
```

∴ the number required is $20et$.

440. The method of transforming a given integer from one scale to another is of course applicable to cases in which both scales are other than the common scale. We must, however, be careful to perform the operation of division in accordance with the principles explained in Art. 438, Ex. 4.

Ex. Transform 142532 from the scale whose radix is 6 to the scale whose radix is 5.

```
5 | 142532
5 |  20330  remainder 2
5 |   2303  remainder 3
5 |    300  remainder 3
5 |     33  remainder 3
5 |      4  remainder 1
  |      0  remainder 4
```

The required number is therefore 413332.

EXAMPLES.—clviii.

Express

1. 1828 in the septenary scale.
2. 1820 in the senary scale.
3. 43751 in the duodenary scale.

[S.A.]

4. 3700 in the quinary scale.
5. 7631 in the binary scale.
6. 215855 in the duodenary scale.
7. 790158 in the septenary scale.

Transform

8. 34002 from the quinary to the quaternary scale.
9. 8978 from the undenary to the duodenary scale.
10. 3256 from the septenary to the duodenary scale.
11. 37704 from the nonary to the octonary scale.
12. 5056 from the septenary to the quaternary scale.
13. 654321 from the duodenary to the septenary scale.
14. 2304 from the quinary to the undenary scale.

441. In any scale the positive integral powers of the number which denotes the radix of the scale are expressed by 10, 100, 1000 ......

Thus twenty-five, which is the square of five, is expressed in the scale whose radix is five by 100: one hundred and twenty-five will be expressed by 1000, and so on.

Generally, the nth power of the number denoting the radix in any scale is expressed by 1 followed by n cyphers.

The highest number that can be expressed by p digits in a scale whose radix is r is expressed by $r^p - 1$.

Thus the highest number that can be expressed by 4 digits in the scale whose radix is five is

$10^4 - 1$, or $10000 - 1$, that is 4444.

The least number that can be expressed by p digits in a scale whose radix is r is expressed by r^{p-1}.

Thus the least number that can be expressed by 4 digits in the scale whose radix is five is

10^{4-1} or 10^3, that is 1000.

442. In a scale whose radix is r, the sum of the digits of an integer divided by $(r-1)$ will leave the same remainder as the integer leaves when divided by $r-1$.

Let N be the number, and suppose
$$N = ar^n + br^{n-1} + cr^{n-2} + \ldots + mr^2 + pr + q.$$
Then
$$N = a(r^n - 1) + b(r^{n-1} - 1) + c(r^{n-2} - 1) + \ldots + m(r^2 - 1) + p(r-1)$$
$$+ \{a + b + c + \ldots + m + p + q\}.$$

Now all the expressions $r^n - 1, r^{n-1} - 1 \ldots r^2 - 1, r - 1$ are divisible by $r - 1$;

$$\therefore \frac{N}{r-1} = \text{an integer} + \frac{a + b + c + \ldots m + p + q}{r-1};$$

which proves the proposition, for since the quotients differ by an integer, their fractional parts must be the same, that is, the remainders after division are the same.

NOTE. From this proposition is derived the test of the accuracy of the result of Multiplication in Arithmetic by *casting out the nines*.

For let $\qquad A = 9m + a,$

and $\qquad\qquad B = 9n + b\;;$

then $\qquad\qquad AB = 9(9mn + an + bm) + ab\;;$

that is, AB and ab when divided by 9 will leave the same remainder.

Radical Fractions.

443. As the local value of each digit in a scale whose radix is r increases r-fold as we advance from right to left, so does the local value of each decrease in the same proportion as we advance from left to right.

If then we affix a line of digits to the right of the units' place, each one of these having from its position a value one-r^{th} part of the value it would have if it were one place further to the left, we shall have on the right hand of the units' place a series of Fractions of which the denominators are successively $r, r^2, r^3, \ldots\ldots$, while the numerators may be any numbers between $r-1$ and zero. These are called **Radical Fractions**.

In our common system of notation the word *Radical* is replaced by *Decimal*, because *ten* is the radix of the scale.

Now adopting the ordinary system of notation, and marking the place of units by putting a dot · to the right of it, we have the following results :

In the denary scale
$$246\cdot 4789 = 2 \times 10^2 + 4 \times 10 + 6 + \frac{4}{10} + \frac{7}{10^2} + \frac{8}{10^3} + \frac{9}{10^4};$$
in the quinary scale
$$324\cdot 4213 = 3 \times 10^2 + 2 \times 10 + 4 + \frac{4}{10} + \frac{2}{10^2} + \frac{1}{10^3} + \frac{3}{10^4},$$
remembering that *in this scale* 10 stands for *five* and not for *ten* (Art. 434).

444. *To show that in any scale a radical fraction is a proper fraction.*

Suppose the fraction to contain n digits, $a, b, c \ldots\ldots$

Then, since $r-1$ is the *highest* value that each of the digits can have,

$\dfrac{a}{r} + \dfrac{b}{r^2} + \ldots$ is not greater than $(r-1)\left(\dfrac{1}{r} + \dfrac{1}{r^2} + \ldots \text{ to } n \text{ terms}\right)$

not greater than $(r-1)\left\{\dfrac{1}{r} \cdot \dfrac{\left(\frac{1}{r}\right)^n - 1}{\frac{1}{r} - 1}\right\}$;

not greater than $(r-1)\left\{\dfrac{r^n - 1}{r^n(r-1)}\right\}$;

not greater than $\dfrac{r^n - 1}{r^n}$;

not greater than $1 - \dfrac{1}{r^n}$.

Hence the given fraction is less than 1, and is therefore a proper fraction.

445. *To transform a fraction expressed in a given scale into a radical fraction in any other scale.*

Let F be the given fraction expressed in the first scale,

r the radix of the new scale in which the fraction is to be expressed,

$a, b, c \ldots$ the digits expressing the fraction in the new scale, so that

$$F = \frac{a}{r} + \frac{b}{r^2} + \frac{c}{r^3} + \ldots$$

from which equation the values of $a, b, c \ldots$ are to be determined.

Multiplying both sides of the equation by r,

$$Fr = a + \frac{b}{r} + \frac{c}{r^2} + \ldots$$

Now $\frac{b}{r} + \frac{c}{r^2} + \ldots$ is a proper fraction by Art. 444.

Hence the integral part of Fr will $= a$, *the first digit of the new fraction*, and the fractional part of Fr will

$$= \frac{b}{r} + \frac{c}{r^2} + \ldots$$

Giving to this fractional part of Fr the symbol F_1 we have

$$F_1 = \frac{b}{r} + \frac{c}{r^2} + \ldots$$

Multiplying both sides of the equation by r,

$$F_1 r = b + \frac{c}{r} + \ldots$$

Hence the integral part of $F_1 r = b$, *the second digit of the fraction*, and thus, by a similar process, all the digits of the new fraction may be found.

Ex. 1. Express $\dfrac{3}{7}$ as a radical fraction in the quinary scale:

$$\dfrac{3}{7} \times 5 = \dfrac{15}{7} = 2 + \dfrac{1}{7},$$

$$\dfrac{1}{7} \times 5 = \dfrac{5}{7} = 0 + \dfrac{5}{7},$$

$$\dfrac{5}{7} \times 5 = \dfrac{25}{7} = 3 + \dfrac{4}{7},$$

$$\dfrac{4}{7} \times 5 = \dfrac{20}{7} = 2 + \dfrac{6}{7},$$

$$\dfrac{6}{7} \times 5 = \dfrac{30}{7} = 4 + \dfrac{2}{7},$$

$$\dfrac{2}{7} \times 5 = \dfrac{10}{7} = 1 + \dfrac{3}{7};$$

therefore fraction is $\cdot \dot{2}0324\dot{1}$ recurring.

Ex. 2. Express ·84375 in the octonary scale:

$$\begin{array}{r} \cdot 84375 \\ 8 \\ \hline 6 \cdot 75000 \\ 8 \\ \hline 6 \cdot 00000 \end{array}$$

The fraction required is ·66.

Ex. 3. Transform ·42765 from the nonary to the senary scale.

$$\begin{array}{r} \cdot 42765 \\ 6 \\ \hline 2 \cdot 78133 \\ 6 \\ \hline 5 \cdot 23820 \\ 6 \\ \hline 1 \cdot 55430 \\ 6 \\ \hline 3 \cdot 65800 \end{array}$$

The fraction required is ·2513 ...

SCALES OF NOTATION. 327

Ex. 4. Transform $e124 \cdot t275$ from the duodenary to the quaternary scale:

```
4 | e124                    ·t275
  |                            4
4 | 2937 – remainder 0      ─────
  |                         3·4t58
4 |  83t – remainder 3          4
  |                         ─────
4 |  20e – remainder 2      1·75t8
  |                             4
4 |   62 – remainder 3      ─────
  |                         2·5e68
4 |   16 – remainder 2          4
  |                         ─────
4 |    4 – remainder 2      1·et28
  |
4 |    1 – remainder 0
  |
  |    0 – remainder 1
```

Number required is $10223230 \cdot 3121 \ldots$

EXAMPLES.—clix.

1. Express $\dfrac{25}{36}$ in the senary scale.

2. Express $\dfrac{3}{11}$ in the septenary scale.

3. Express $23 \cdot 125$ in the nonary scale.

4. Express $1820 \cdot 3375$ in the senary scale.

5. In what scale is 17486 written 212542?

6. In what scale is 511173 written 1746305?

7. Show that a number in the Common scale is divisible:

   (1) by 3 if the sum of its digits is divisible by 3.

   (2) by 4 if the last two digits be divisible by 4.

   (3) by 8 if the last three digits be divisible by 8.

   (4) by 5 if the number ends with 5 or 0.

(5) by 11 if the difference between the sum of the digits in the odd places and the sum of those in the even places be divisible by 11.

8. If N be a number in the scale whose radix is r, and n be the number resulting when the digits of N are reversed, show that $N-n$ is divisible by $r-1$.

XXXVIII. ON LOGARITHMS.

446. DEF. The **Logarithm** of a number to a given base is the index of the power to which the base must be raised to give the number.

Thus if $m = a^x$, x is called the logarithm of m to the base a.

For instance, if the base of a system of Logarithms be 2,

3 is the logarithm of the number 8,

because $8 = 2^3$:

and if the base be 5, then

3 is the logarithm of the number 125,

because $125 = 5^3$.

447. The logarithm of a number m to the base a is written thus, $\log_a m$; and so, if $m = a^x$,

$$x = \log_a m.$$

Hence it follows that $m = a^{\log_a m}$.

448. Since $1 = a^0$, the logarithm of unity to any base is zero.

Since $a = a^1$, the logarithm of the base of any system is unity.

449. We now proceed to describe that which is called the Common System of logarithms.

The base of the system is 10.

By a *system* of logarithms to the base 10, we mean a succession of values of x which satisfy the equation

$$m = 10^x$$

for all positive values of m, integral or fractional.

Such a system is formed by the series of logarithms of the natural numbers from 1 to 100000, which constitute the logarithms registered in our ordinary tables, and which are therefore called *tabular logarithms*.

450. Now
$$1 = 10^0,$$
$$10 = 10^1,$$
$$100 = 10^2,$$
$$1000 = 10^3,$$

and so on.

Hence the logarithm of 1 is 0,
of 10 is 1,
of 100 is 2,
of 1000 is 3,

and so on.

Hence for all numbers between 1 and 10 the logarithm is a decimal less than 1,

between 10 and 100 the logarithm is a decimal between 1 and 2,

between 100 and 1000 .... a decimal between 2 and 3, and so on.

451. The logarithms of the natural numbers from 1 to 12 stand thus in the tables :

No.	Log	No.	Log
1	0·0000000	7	0·8450980
2	0·3010300	8	0·9030900
3	0·4771213	9	0·9542425
4	0·6020600	10	1·0000000
5	0·6989700	11	1·0413927
6	0·7781513	12	1·0791812

The logarithms are calculated to seven places of decimals

452. The integral parts of the logarithms of numbers higher than 10 are called the *characteristics* of those logarithms, and the decimal parts of the logarithms are called the *mantissæ*.

Thus 1 is the characteristic,
·0791812 the mantissa,
of the logarithm of 12.

453. The logarithms for 100 and the numbers that succeed it (and in some tables those that precede 100) have no characteristic prefixed, because it can be supplied by the reader, being 2 for all numbers between 100 and 1000, 3 for all between 1000 and 10000, and so on. Thus in the Tables we shall find

No.	Log
100	0000000
101	0043214
102	0086002
103	0128372
104	0170333
105	0211893

which we read thus:

the logarithm of 100 is 2,
of 101 is 2·0043214,
of 102 is 2·0086002; and so on.

454. Logarithms are of great use in making arithmetical computations more easy, for by means of a Table of Logarithms the operation

of Multiplication is changed into that of Addition,
... Division .... Subtraction,
... Involution .... Multiplication,
... Evolution .... Division,

as we shall show in the next four Articles.

455. *The logarithm of a product is equal to the sum of the logarithms of its factors.*

Let
$$m = a^x,$$
and
$$n = a^y.$$
Then
$$mn = a^{x+y};$$
$$\therefore \log_a mn = x + y$$
$$= \log_a m + \log_a n.$$

Hence it follows that
$$\log_a mnp = \log_a m + \log_a n + \log_a p,$$
and similarly it may be shown that the Theorem holds good for any number of factors.

Thus the operation of Multiplication is changed into that of Addition.

Suppose, for instance, we want to find the product of 246 and 357, we add the logarithms of the factors, and the sum is the logarithm of the product: thus

$$\log 246 = 2\cdot 3909351$$
$$\log 357 = 2\cdot 5526682$$
$$\text{their sum} = 4\cdot 9436033$$

which is the logarithm of 87822, the product required.

NOTE. We do not write $\log_{10} 246$, for so long as we are treating of logarithms to the particular base 10, we may omit the suffix.

456. *The logarithm of a quotient is equal to the logarithm of the dividend diminished by the logarithm of the divisor.*

Let
$$m = a^x,$$
and
$$n = a^y.$$
Then
$$\frac{m}{n} = a^{x-y};$$
$$\therefore \log_a \frac{m}{n} = x - y$$
$$= \log_a m - \log_a n.$$

Thus the operation of Division is changed into that of Subtraction.

If, for example, we are required to divide 371·49 by 52·376, we proceed thus,

$$\log 371{\cdot}49 = 2{\cdot}5699471$$
$$\log 52{\cdot}376 = 1{\cdot}7191323$$
$$\text{their difference} = {\cdot}8508148$$

which is the logarithm of 7·092752, the quotient required.

457. *The logarithm of any power of a number is equal to the product of the logarithm of the number and the index denoting the power.*

Let $\quad m = a^x.$

Then $\quad m^r = a^{rx};$

$\therefore \log_a m^r = rx$

$\qquad = r \cdot \log_a m.$

Thus the operation of Involution is changed into Multiplication.

Suppose, for instance, we have to find the fourth power of 13, we may proceed thus,

$$\log 13 = 1{\cdot}1139434$$
$$\underline{\qquad\qquad 4}$$
$$4{\cdot}4557736$$

which is the logarithm of 28561, the number required.

458. *The logarithm of any root of a number is equal to the quotient arising from the division of the logarithm of the number by the number denoting the root.*

Let $\quad m = a^x.$

Then $\quad m^{\frac{1}{r}} = a^{\frac{x}{r}};$

$\therefore \log_a m^{\frac{1}{r}} = \dfrac{x}{r}$

$\qquad = \dfrac{1}{r} \cdot \log_a m.$

Thus the operation of Evolution is changed into Division.

ON LOGARITHMS. 333

If, for example, we have to find the fifth root of 16807, we proceed thus,

$$5 \mid \underline{4\cdot2254902}, \text{ the log of } 16807$$
$$\cdot8450980$$

which is the logarithm of 7, the root required.

459. The common system of Logarithms has this advantage over all others for numerical calculations, that its base is the same as the radix of the common scale of notation.

Hence it is that the same mantissa serves for all numbers which have the same significant digits and differ only in the position of the place of units relatively to those digits.

For, since log 60 = log   10 + log 6 = 1 + log 6,
        log 600 = log 100 + log 6 = 2 + log 6,
        log 6000 = log 1000 + log 6 = 3 + log 6,

it is clear that if we know the logarithm of any number, as 6, we also know the logarithms of the numbers resulting from multiplying that number by the powers of 10.

So again, if we know that

$$\log 1\cdot7692 \text{ is } \cdot247783,$$

we also know that

$$\log 17\cdot692 \text{ is } 1\cdot247783,$$
$$\log 176\cdot92 \text{ is } 2\cdot247783,$$
$$\log 1769\cdot2 \text{ is } 3\cdot247783,$$
$$\log 17692 \text{ is } 4\cdot247783,$$
$$\log 176920 \text{ is } 5\cdot247783.$$

460. We must now treat of the logarithms of numbers less than unity.

Since $\qquad 1 = 10^0,$

$$\cdot1 = \frac{1}{10} = 10^{-1},$$

$$\cdot01 = \frac{1}{100} = 10^{-2},$$

the logarithm of a number

.... between 1 and ·1 lies between 0 and −1,

.... between ·1 and ·01 .... −1 and −2,

.... between ·01 and ·001 .... −2 and −3,

and so on.

Hence the logarithms of all numbers less than unity are negative.

We do not require a separate table for these logarithms, for we can deduce them from the logarithms of numbers greater than unity by the following process:

$$\log ·6 = \log \frac{6}{10} = \log 6 - \log 10 = \log 6 - 1,$$

$$\log ·06 = \log \frac{6}{100} = \log 6 - \log 100 = \log 6 - 2,$$

$$\log ·006 = \log \frac{6}{1000} = \log 6 - \log 1000 = \log 6 - 3.$$

Now the logarithm of 6 is ·7781513.

Hence

$\log ·6 \;\;\; = -1 + ·7781513$, which is written $\bar{1}·7781513$,

$\log ·06 \;\; = -2 + ·7781513$, which is written $\bar{2}·7781513$,

$\log ·006 = -3 + ·7781513$, which is written $\bar{3}·7781513$,

the characteristics only being negative and the mantissæ positive.

461. Thus the same mantissæ serve for the logarithms of all numbers, *whether greater or less than unity*, which have the same significant digits, and differ only in the position of the place of units relatively to those digits.

It is best to regard the Table as a register of the logarithms of numbers which have *one* significant digit before the decimal point.

For instance, when we read in the tables 144 | 1583625, we interpret the entry thus

    log 1·44 is ·1583625.

We then obtain the following rules for the characteristic to be attached in each case.

I. If the decimal point be shifted one, two, three ... n places to the right, prefix as a characteristic 1, 2, 3 ... n.

II. If the decimal point be shifted one, two, three ... n places to the left, prefix as a characteristic $\bar{1}, \bar{2}, \bar{3} ... \bar{n}$.

Thus   log   1·44 is   ·1583625,

∴ log   14·4 is 1·1583625,

log   144 is 2·1583625,

log   1440 is 3·1583625,

log   ·144 is $\bar{1}$·1583625,

log   ·0144 is $\bar{2}$·1583625,

log   ·00144 is $\bar{3}$·1583625.

462. In calculations with negative characteristics we follow the rules of algebra. Thus,

(1) If we have to add the logarithms $\bar{3}$·64628 and 2·4236., we first add the mantissæ, and the result is 1·06995, and then add the characteristics, and this result is $\bar{1}$.

The final result is $\bar{1}$ + 1·06995, that is, ·06995.

(2) To subtract $\bar{5}$·6249372 from $\bar{3}$·2456973, we may arrange the numbers thus,

         $-3 + ·2456973$
         $-5 + ·6249372$
         $1 + ·6207601$

the 1 carried on from the last subtraction in the decimal places changing −5 into −4, and then −4 subtracted from −3 giving 1 as a result.

Hence the resulting logarithm is 1·6207601.

(3) To multiply $\bar{3}\cdot7482569$ by 5.

$$\frac{\bar{3}\cdot7482569}{\overline{12}\cdot7412845} \quad 5$$

the 3 carried on from the last multiplication of the decimal places being added to -15, and thus giving -12 as a result.

(4) To divide $\overline{14}\cdot2456736$ by 4.

Increase the negative characteristic so that it may be exactly divisible by 4, making a proper compensation, thus,

$$\overline{14}\cdot2456736 = \overline{16} + 2\cdot2456736.$$

Then $\quad \dfrac{\overline{14}\cdot2456736}{4} = \dfrac{\overline{16}+2\cdot2456736}{4} = \bar{4} + \cdot5614184;$

and so the result is $\bar{4}\cdot5614184$.

Examples.—clx.

1. Add $\bar{3}\cdot1651553$, $\bar{4}\cdot7505855$, $6\cdot6879746$, $\bar{2}\cdot6150026$.
2. Add $4\cdot6843785$, $\bar{5}\cdot6650657$, $3\cdot8905196$, $3\cdot4675284$.
3. Add $2\cdot5324716$, $3\cdot6650657$, $\bar{5}\cdot8905196$, $\cdot3156215$.
4. From $2\cdot483269$ take $\bar{3}\cdot742891$.
5. From $\bar{2}\cdot352678$ take $\bar{5}\cdot428619$.
6. From $5\cdot349162$ take $\bar{3}\cdot624329$.
7. Multiply $\bar{2}\cdot4596721$ by 3.
8. Multiply $\bar{7}\cdot429683$ by 6.
9. Multiply $\bar{9}\cdot2843617$ by 7.
10. Divide $\bar{6}\cdot3725409$ by 3.
11. Divide $\overline{14}\cdot432962$ by 6.
12. Divide $\bar{4}\cdot53627188$ by 9.

463. We shall now explain how a system of logarithms calculated to a base a may be transformed into another system of which the base is b.

Let m be a number of which the logarithm in the first system is x and in the second y.

Then $\qquad m = a^x$,

and $\qquad m = b^y$.

Hence $\qquad b^y = a^x$,

$$\therefore b = a^{\frac{x}{y}};$$

$$\therefore \frac{x}{y} = \log_a b;$$

$$\therefore \frac{y}{x} = \frac{1}{\log_a b};$$

$$\therefore y = \frac{1}{\log_a b} x.$$

Hence if we multiply the logarithm of any number in the system of which the base is a by $\dfrac{1}{\log_a b}$, we shall obtain the logarithm of the same number in the system of which the base is b.

This constant multiplier $\dfrac{1}{\log_a b}$ is called THE MODULUS *of the system of which the base is b* with reference to the system of which the base is a.

464. The common system of logarithms is used in all numerical calculations, but there is another system, which we must notice, employed by the discoverer of logarithms, Napier, and hence called THE NAPIERIAN SYSTEM.

The base of this system, denoted by the symbol e, is the number which is the sum of the series

$$2 + \frac{1}{2} + \frac{1}{2.3} + \frac{1}{2.3.4} + \ldots \text{ ad inf.},$$

of which sum the first eight digits are $2\cdot7182818$.

465. Our common logarithms are formed from the Logarithms of the Napierian System by multiplying each of the

latter by a common multiplier called The Modulus of the Common System

This modulus is, in accordance with the conclusion of Art. 463, $\dfrac{1}{\log_e 10}$.

That is, if l and N be the logarithms of the same number in the common and Napierian systems respectively,

$$l = \dfrac{1}{\log_e 10} \cdot N.$$

Now $\log_e 10$ is $2\cdot 30258509$;

$$\therefore \dfrac{1}{\log_e 10} \text{ is } \dfrac{1}{2\cdot 30258509} \text{ or } \cdot 43429448,$$

and so the modulus of the common system is $\cdot 43429448$.

466. To prove that $\log_a b \times \log_b a = 1$.

Let $\qquad x = \log_a b$.

Then $\qquad b = a^x$;

$\qquad \therefore b^{\frac{1}{x}} = a$;

$\qquad \therefore \dfrac{1}{x} = \log_b a$.

Thus $\qquad \log_a b \times \log_b a = x \times \dfrac{1}{x}$

$\qquad\qquad = 1.$

467. The following are simple examples of the method of applying the principles explained in this Chapter.

Ex. 1. Given $\log 2 = \cdot 3010300$, $\log 3 = \cdot 4771213$ and $\log 7 = \cdot 8450980$, find $\log 42$.

Since $\qquad 42 = 2 \times 3 \times 7$

$\qquad \log 42 = \log 2 + \log 3 + \log 7$

$\qquad\qquad = \cdot 3010300 + \cdot 4771213 + \cdot 8450980$

$\qquad\qquad = 1 \cdot 6232493.$

Ex. 2. Given $\log 2 = \cdot 3010300$ and $\log 3 = \cdot 4771213$, find the logarithms of 64, 81 and 96.

$$\log 64 = \log 2^6 = 6 \log 2$$
$$\log 2 = \cdot 3010300$$
$$6$$
$$\therefore \log 64 = 1 \cdot 8061800$$

$$\log 81 = \log 3^4 = 4 \log 3$$
$$\log 3 = \cdot 4771213$$
$$4$$
$$\therefore \log 81 = 1 \cdot 9084852$$

$$\log 96 = \log (32 \times 3) = \log 32 + \log 3,$$
and $$\log 32 = \log 2^5 = 5 \log 2;$$
$$\therefore \log 96 = 5 \log 2 + \log 3 = 1 \cdot 5051500 + \cdot 4771213 = 1 \cdot 9822713.$$

Ex. 3. Given $\log 5 = \cdot 6989700$, find the logarithm of $\sqrt[7]{(6 \cdot 25)}$.

$$\log (6 \cdot 25)^{\frac{1}{7}} = \frac{1}{7} \log 6 \cdot 25 = \frac{1}{7} \log \frac{625}{100} = \frac{1}{7} (\log 625 - \log 100)$$
$$= \frac{1}{7} (\log 5^4 - 2) = \frac{1}{7} (4 \log 5 - 2)$$
$$= \frac{1}{7} (2 \cdot 7958800 - 2) = \cdot 1136657.$$

EXAMPLES.—clxi.

1. Given $\log 2 = \cdot 3010300$, find $\log 128$, $\log 125$ and $\log 2500$.

2. Given $\log 2 = \cdot 3010300$ and $\log 7 = \cdot 8450980$, find the logarithms of 50, ·005 and 196.

3. Given $\log 2 = \cdot 3010300$, and $\log 3 = \cdot 4771213$, find the logarithms of 6, 27, 54 and 576.

4. Given $\log 2 = \cdot 3010300$, $\log 3 = \cdot 4771213$, $\log 7 = \cdot 8450980$, find $\log 60$, $\log \cdot 03$, $\log 1 \cdot 05$, and $\log \cdot 0000432$.

5. Given $\log 2 = \cdot 3010300$, $\log 18 = 1\cdot 2552725$ and $\log 21 = 1\cdot 3222193$, find $\log \cdot 00075$ and $\log 31\cdot 5$.

6. Given $\log 5 = \cdot 6989700$, find the logarithms of 2, $\cdot 064$, and $\left(\dfrac{2^{60}}{5^{20}}\right)^{\frac{1}{14}}$.

7. Given $\log 2 = \cdot 3010300$, find the logarithms of 5, $\cdot 125$, and $\left(\dfrac{5^{90}}{2^{40}}\right)^{\frac{1}{15}}$.

8. What are the logarithms of $\cdot 01$, 1 and 100 to the base 10? What to the base $\cdot 01$?

9. What is the characteristic of $\log 1593$, (1) to base 10, (2) to base 12?

10. Given $\dfrac{4^x}{2^{x+y}} = 8$, and $x = 3y$, find x and y.

11. Given $\log 4 = \cdot 6020600$, $\log 1\cdot 04 = \cdot 0170333$:

(a) Find the logarithms of 2, 25, $83\cdot 2$, $(\cdot 625)^{\frac{1}{100}}$.

(b) How many digits are there in the integral part of $(1\cdot 04)^{6000}$?

12. Given $\log 25 = 1\cdot 3979400$, $\log 1\cdot 03 = \cdot 0128372$:

(a) Find the logarithms of 5, 4, $51\cdot 5$, $(\cdot 064)^{\frac{1}{100}}$.

(b) How many digits are there in the integral part of $(1\cdot 03)^{600}$?

13. Having given $\log 3 = \cdot 4771213$, $\log 7 = \cdot 8450980$, $\log 11 = 1\cdot 0413927$:

find the logarithms of 7623, $\dfrac{77}{300}$ and $\dfrac{3}{539}$.

14. Solve the equations:

(1) $4096^x = \dfrac{8}{64^x}$.

(2) $\left(\dfrac{1}{\cdot 4}\right)^x = 6\cdot 25$.

(3) $a^x \cdot b^x = m$.

(4) $a^{mx} b^{nx} = c$.

(5) $a^{3x} \cdot b^{4-x} = c^{2x-1}$.

(6) $a^x b^m = c^{1-3x}$.

468. We have explained in Arts. 459—461 the advantages of the Common System of Logarithms, which may be stated in a more general form thus:

Let A be any sequence of figures (such as 2·35916), having *one* digit in the integral part.

Then any number N having the same sequence of figures (such as 235·916 or ·0023591G) is of the form $A \times 10^n$, where n is an integer, positive or negative.

Therefore $\log_{10} N = \log_{10}(A \times 10^n) = \log_{10} A + n$.

Now A lies between 10^0 and 10^1, and therefore $\log A$ lies between 0 and 1, and is therefore a proper fraction.

But $\log_{10} N$ and $\log_{10} A$ differ only by the integer n;

$\therefore \log_{10} A$ is the fractional part of $\log_{10} N$.

Hence *the logarithms of all numbers having* THE SAME SEQUENCE OF FIGURES *have the same mantissa*.

Therefore one register serves for the mantissa of logarithms of all such numbers. This renders the tables more comprehensive.

Again, considering all numbers which have the same sequence of figures, the number containing *two* digits in the integral part $= 10 \cdot A$, and therefore the characteristic of its logarithm is 1.

Similarly the number containing m digits in the integral part $= 10^m \cdot A$, and therefore the characteristic of its logarithm is m.

Also numbers which have no digit in the integral part and one cypher after the decimal point are equal to $A \cdot 10^{-1}$ and $A \cdot 10^{-2}$ respectively, and therefore the characteristics of their logarithms are -1 and -2 respectively.

Similarly the number having m *cyphers* following the decimal point $= A \cdot 10^{-(m+1)}$;

$\therefore$ *the characteristic of its logarithm is* $-(m+1)$.

Hence *we see that the characteristics of the logarithms of all numbers can be determined by inspection and therefore need not be registered. This renders the tables less bulky.*

469. The method of using Tables of Logarithms does not fall within the scope of this treatise, but an account of it may be found in the Author's work on ELEMENTARY TRIGONOMETRY.

470. We proceed to give a short explanation of the way in which Logarithms are applied to the solution of questions relating to Compound Interest.

471. Suppose r to represent the interest on £1 for a year, then the interest on P pounds for a year is represented by Pr, and the amount of P pounds for a year is represented by $P + Pr$.

472. *To find the amount of a given sum for any time at compound interest.*

Let P be the original principal,

    r the interest on £1 for a year,

    n the number of years.

Then if P_1, P_2, $P_3 \dots P_n$ be the amounts at the end of 1, 2, 3 ... n years,

$$P_1 = P + Pr = P(1+r),$$
$$P_2 = P_1 + P_1 r = P_1(1+r) = P(1+r)^2$$
$$P_3 = P_2 + P_2 r = P_2(1+r) = P(1+r)^3,$$
$$\dots\dots\dots\dots\dots\dots\dots\dots\dots\dots\dots\dots$$
$$P_n = P(1+r)^n.$$

473. Now suppose P_n, P and r to be given: then by the aid of Logarithms we can find n, for

$$\log P_n = \log \{P(1+r)^n\}$$
$$= \log P + n \log (1+r);$$
$$\therefore n = \frac{\log P_n - \log P}{\log (1+r)}.$$

ON LOGARITHMS. 343

474. If the interest be payable at intervals other than a year, the formula $P_n = P(1+r)^n$ is applicable to the solution of the question, it being observed that r represents the interest on £1 for the period on which the interest is calculated, half-yearly, quarterly, or for any other period, and n represents the number of such periods.

For example, to find the interest on P pounds for 4 years at compound interest, reckoned quarterly, at 5 per cent. per annum.

Here
$$r = \frac{1}{4} \text{ of } \frac{5}{100} = \frac{1\cdot 25}{100} = \cdot 0125,$$
$$n = 4 \times 4 = 16;$$
$$\therefore P_n = P(1 + \cdot 0125)^{16}.$$

EXAMPLES.—clxii.

N.B.—The Logarithms required may be found from the extracts from the Tables given in pages 329, 330.

1. In how many years will a sum of money double itself at 4 per cent. compound interest?

2. In how many years will a sum of money double itself at 3 per cent. compound interest?

3. In how many years will a sum of money double itself at 10 per cent. compound interest?

4. In how many years will a sum of money treble itself at 5 per cent. compound interest?

5. If £P at compound interest, rate r, double itself in n years, and at rate $2r$ in m years: show that $m : n$ is greater than $1 : 2$.

6. In how many years will £1000 amount to £1800 at 5 per cent. compound interest?

7. In how many years will £P double itself at 6 per cent. per ann. compound interest payable half-yearly?

APPENDIX.

475. The following is another method of proving the principal theorem in Permutations, to which reference is made in the note on page 289.

To prove that the number of permutations of n *things taken* r *at a time is* n . (n – 1)......(n – r + 1).

Let there be n things a, b, c, d ......

If n things be taken 1 at a time, the number of permutations is of course n.

Now take any one of them, as a, then $n-1$ are left, and any one of these may be put after a to form a permutation, 2 at a time, in which a stands first: and hence since there are n things which may begin and each of these n may have $n-1$ put after it, there are altogether $n(n-1)$ permutations of n things taken 2 at a time.

Take any one of these, as ab, then there are $n-2$ left, and any one of these may be put after ab, to form a permutation, 3 at a time, in which ab stands first: and hence since there are $n(n-1)$ things which may begin, and each of these $n(n-1)$ may have $n-2$ put after it, there are altogether $n(n-1)(n-2)$ permutations of n things taken 3 at a time.

If we take any one of these as abc, there are $n-3$ left, and so the number of permutations of n things taken 4 at a time is $n . (n-1)(n-2)(n-3)$.

So we see that to find the number of permutations, taken r at a time, we must multiply the number of permutations, taken $r-1$ at a time, by the number formed by subtracting $r-1$ from n, since this will be the number of endings any one of these permutations may have.

Hence the number of permutations of n things taken 5 at a time is

$n(n-1)(n-2)(n-3) \times (n-4)$, or $n(n-1)(n-2)(n-3)(n-4)$;

and since each time we multiply by an additional factor the number of factors is equal to the number of things taken at a time, it follows that the number of permutations of n things taken r at a time is the product of the factors

$$n . (n-1)(n-2)......(n-r+1).$$

ANSWERS

i. (Page 10.)

1. $5a + 7b + 12c$.
2. $a + 3b + 2c$.
3. $2a + 2b + 2c$.
4. $8a + 2b + 2c$.
5. $2x - 7a + 3b - 2$.
6. 0.
7. $12b + 3c$.

ii. (Page 10.)

1. $2a$.
2. $2a + 5x$.
3. $3a - 3x$.
4. $8x + 5y$.
5. $4a + b + 2c$.
6. $2a$.
7. 4.
8. $13x - y - 6z$.
9. $10a - 7b - x$.

iii. (Page 10.)

1. $2b$.
2. $x + 2y$.
3. $a + 5c + d$.
4. $2y + 2z$.
5. $2r$.
6. $2b + 2c$.
7. $a - 3b - c$.
8. $3y + z$.

iv. (Page 11.)

1. $4a - b$.
2. $4b$.
3. $a + b - 4c$.
4. $2b$.
5. $14x + 2$.
6. $2x + a$.
7. $6x - a$.
8. a.
9. $2a - b$.
10. $2a$.
11. c.
12. $x + 3a$.
13. $29a - 27b + 6c$.

v. (Page 16.)

Addition.

1. $7a - 2b$.
2. $-10b + 6c$.
3. $-11x - 8y - 6z$.
4. $-6b - 5c + 3d$.
5. $2a$.
6. $-2x - 2a + b + 4y$.
7. $7a + 4b - 4c$.
8. $7a - b + 7c$.
9. $-6y + 2z$.

[S.A.]

Subtraction.

1. $2a + 2b$. 2. $a - c$. 3. $2a - 2b + 2c$.
4. $8x - 17y + 5$. 5. $7a - 16b + 20c$. 6. $5a - 3b - 8x$.
7. $-3a + 3b - 4c$. 8. $2v + 2c - 15$. 9. $11x - 7y + 4z$.
10. $6a - b + 5c$. 11. $12p - 9q + 2r$.

vi. (Page 20.)

1. $3xy$. 2. $12xy$. 3. $12x^2y^2$. 4. $3a^2bc^2$.
5. a^7. 6. a^8. 7. $12a^5b^3$. 8. $35a^6bc^4$.
9. $180a^4b^5c^4$. 10. $28a^7bc^{10}$. 11. $3a^{11}$. 12. $20a^4b^3cy$.
13. $76x^4y^4z^3$. 14. $51ab^4c^2yz$. 15. $48x^8y^{10}z^0$.
16. $12a^2bcxy$. 17. $8a^{14}b^5c^2$. 18. $9m^5n^3p^3$.
19. $abcx^2y^2z^4$. 20. $33a^{20}b^{10}m^2x$.

vii. (Page 22.)

1. $a^2 + ab - ac$. 2. $2a^2 + 6ab - 8ac$. 3. $a^4 + 3a^3 + 4a^2$.
4. $9a^5 - 15a^4 - 18a^3 + 21a^2$. 5. $a^3b - 2a^2b^2 + ab^3$.
6. $3a^5b - 9a^4b^3 + 3a^2b^4$. 7. $8m^3n + 9m^2n^2 + 10mn^3$.
8. $18a^6b + 8a^5b^2 - 6a^4b^3 + 8a^3b^4$. 9. $x^4y^4 - x^3y^3 + x^2y^2 - 7xy$.
10. $m^3n - 3m^2n^2 + 3mn^3 - n^4$. 11. $144a^5b^4 - 72a^4b^5 + 60a^3b^6$.
12. $104x^4y - 136x^3y^2 + 40x^2y^3 - 8xy^4$.

viii. (Page 27.)

1. $x^2 + 12x + 27$. 2. $x^2 + 8x - 105$. 3. $x^2 - 2x - 120$.
4. $x^2 - 15x + 56$. 5. $a^2 - 8a + 15$. 6. $y^2 + 7y - 78$.
7. $x^4 + x^2 - 20$. 8. $x^4 - 12x^3 + 50x^2 - 84x + 45$.
9. $x^4 - 31x^2 + 9$. 10. $a^6 - 3a^5 - 3a^4 + 13a^3 - 6a^2 - 6a + 4$.
11. $2x^4 - x^2 + 2x - 1$. 12. $x^4 + x^2y^2 + y^4$. 13. $x^3 - y^3$.
14. $a^6 - x^6$. 15. $x^5 - 5x^3 + 5x^2 - 1$.
16. $x^4 - 81y^4$. 17. $a^4 - 16b^4$. 18. $16a^4 - b^4$.

19. $a^5 - 4a^4b + 4a^3b^2 + 4a^2b^3 - 17ab^4 - 12b^5$.
20. $a^5 + 5a^4b + a^3b^2 - 10a^2b^3 + 12ab^4 - 9b^5$.
21. $a^4 + 4a^2x^2 + 16x^4$.   22. $81a^4 + 9a^2x^2 + x^4$.
23. $x^8 + 4a^2x^4 + 16a^4$.   24. $a^3 + b^3 + c^3 - 3abc$.
25. $x^5 + x^4y - 9x^3y^2 - 20x^2y^3 + 2xy^4 + 15y^5$.
26. $a^2b^2 + c^2d^2 - a^2c^2 - b^2d^2$.   27. $x^8 - a^8$.
28. $x^3 - ax^2 + bx^2 - cx^2 - abx + acx - bcx + abc$.
29. 1.   30. $x^6 - y^6$.   31. $a^{16} - x^{16}$.   32. -47.
33. 2.   34. -14.   35. $ab + ac + bc$.   36. -60.
37. 2.   38. m^2.

ix. (Page 28.)

1. $-a^2b$.   2. $-a^5$.   3. $-a^3b^3$.   4. $12a^3b^3$.
5. $-30x^4y^3$.   6. $-a^3 + a^2b - ab^2$.   7. $-6a^5 - 8a^4 + 10a^3$.
8. $a^4 + 2a^3 + 2a^2 + a$.   9. $-6x^3y + x^2y^2 + 7xy^3 - 12y^4$.
10. $5m^3 + m^2n - 13mn^2 + 7n^3$.   11. $-13r^3 - 22r^2 + 96r + 135$.
12. $-7x^4 + x^3z + 8x^2z^2 + 9xz^2 + 9z^3$.
13. $x^6 + x^3y^3$.   14. $x^4 + 2x^3y + 2x^2y^2 + 2xy^3 + y^4$.

x. (Page 32.)

1. $x^2 + 2ax + a^2$.   2. $x^2 - 2ax + a^2$.   3. $x^2 + 4x + 4$.
4. $x^2 - 6x + 9$.   5. $x^4 + 2x^2y^2 + y^4$.   6. $x^4 - 2x^2y^2 + y^4$.
7. $a^6 + 2a^3b^3 + b^6$.   8. $a^6 - 2a^3b^3 + b^6$.
9. $x^2 + y^2 + z^2 + 2xy + 2xz + 2yz$.
10. $x^2 + y^2 + z^2 - 2xy + 2xz - 2yz$.
11. $m^2 + n^2 + p^2 + r^2 + 2mn - 2mp - 2mr - 2np - 2nr + 2pr$.
12. $x^4 + 4x^3 - 2x^2 - 12x + 9$.   13. $x^4 - 12x^3 + 50x^2 - 84x + 49$.
14. $4x^4 - 28x^3 + 85x^2 - 126x + 81$.
15. $x^4 + y^4 + z^4 + 2x^2y^2 - 2x^2z^2 - 2y^2z^2$.

16. $x^8 - 8x^6y^2 + 18x^4y^4 - 8x^2y^6 + y^8$.
17. $a^6 + b^6 + c^6 + 2a^3b^3 + 2a^3c^3 + 2b^3c^3$.
18. $x^6 + y^6 + z^6 - 2x^3y^3 - 2x^3z^3 + 2y^3z^3$.
19. $x^2 + 4y^2 + 9z^2 + 4xy - 6xz - 12yz$.
20. $x^4 + 4y^4 + 25z^4 - 4x^2y^2 + 10x^2z^2 - 20y^2z^2$.
21. $x^3 + 3ax^2 + 3a^2x + a^3$. 22. $x^3 - 3ax^2 + 3a^2x - a^3$.
23. $x^3 + 3x^2 + 3x + 1$. 24. $x^3 - 3x^2 + 3x - 1$.
25. $x^3 + 6x^2 + 12x + 8$. 26. $a^6 - 3a^4b^2 + 3a^2b^4 - b^6$.
27. $a^3 + 3a^2b + 3ab^2 + b^3 + c^3 + 3a^2c + 6abc + 3b^2c + 3ac^2 + 3bc^2$.
28. $a^3 - 3a^2b + 3ab^2 - b^3 - c^3 - 3a^2c + 6abc - 3b^2c + 3ac^2 - 3bc^2$.
29. $m^4 - 2m^2n^2 + n^4$. 30. $m^4 + 2m^3n - 2mn^3 - n^4$.

xi. (Page 34.)

1. x^3. 2. x^8. 3. x^3y. 4. x^4yz^5. 5. $6bc$. 6. $8c^2$.
7. $16a^2b^6c^6$. 8. $121m^8n^8p^8$. 9. $12a^3xy^4$. 10. $8a^4bc^2$.

xii. (Page 35.)

1. $x^2 + 2x + 1$. 2. $y^3 - y^2 + y - 1$. 3. $a^2 + 2ab + 3b^2$.
4. $x^4 + mpx^2 + m^2p^2$. 5. $4ay - 7x + x^2$. 6. $8x^3y^5 - 4x^2y^2 - 2y$.
7. $27m^6n^6 - 18m^3n^4 + 9mp$. 8. $3x^2y^2 - 2xy^3 - y^4$.
9. $13a^2b - 9ab^2 + 7b$. 10. $19b^3c^2 + 12b^2c^3 - 7bc^4$.

xiii. (Page 36.)

1. -8. 2. $15a^5$. 3. $-21x^3y^6$.
4. $-6m^2n$. 5. $16a^3b$. 6. $a^2x^2 + ax + 1$.
7. $-2a^2 + 3a - x^2$. 8. $2 + 6a^2b - 8a^4b^6$.
9. $-12x^2 + 9xy - 8y^2$. 10. $-x^3 + b^3x^7z^2 + by^4$.

xiv. (Page 38.)

1. $x + 5$. 2. $x - 10$. 3. $x + 4$. 4. $x + 12$.
5. $x^2 + 7x + 12$. 6. $x^2 - 1$. 7. $x^2 + x + 1$.

ANSWERS. 349

8. $x^3 - 3x^2 + 3x + 1$. 9. $x^2 - 2x - 1$. 10. $x^2 - 2x + 1$.
11. $x^2 - x + 1$. 12. $x^3 - 2x^2 + 8$. 13. $x^2 + 3y^2$.
14. $a^3 + 3a^2b + 3ab^2 + b^3$. 15. $a^4 - 4a^3b + 6a^2b^2 - 4ab^3 + b^4$.
16. $x^2 - 6x + 5$. 17. $a^3 - 2a^2b + 3ab^2 + 4b^3$.
18. $2ax^2 - 3a^2x + a^3$. 19. $x^2 - x + 1$. 20. $x^2 - a^2$.
21. $x + 2y$. 22. $x^4 - x^3y + x^2y^2 - xy^3 + y^4$.
23. $x^5 + x^4y + x^3y^2 + x^2y^3 + xy^4 + y^5$. 24. $a + b - c$.
25. $-b + 2b^2 - b^3$. 26. $a - b + c - d$.
27. $x^2 - xy - xz + y^2 - yz + z^2$. 28. $x^{12} - x^9y^2 + x^6y^4 - x^3y^6 + y^8$.
29. $p + 2q - r$. 30. $a^4 - a^3b + a^2b^2 - ab^3 + b^4$.
31. $x^4 + x^3y + x^2y^2 + xy^3 + y^4$. 32. $2x^3 - 3x^2 + 2x$.
33. $a^4 + 3a^3 + 9a^2 + 27a + 81$. 34. $k^7 + k^4 + k$.
35. $x^2 - 9x - 10$. 36. $24x^2 - 2ax - 35a^2$.
37. $6x^2 - 7x + 8$. 38. $8x^3 + 12ax^2 - 18a^2x - 27a^3$.
39. $27x^3 - 36ax^2 + 48a^2x - 64a^3$. 40. $2a + 3b$.
41. $x + 2a$. 42. $a^2 - 4b^2$. 43. $x^2 - 3x - y$.
44. $x^2 - 3xy - 2y^2$. 45. $x^3 + 3x^2y + 9xy^2 + 27y^3$.
46. $a^3 + 2a^2b + 4ab^2 + 8b^3$. 47. $27a^3 - 18a^2b + 12ab^2 - 8b^3$.
48. $8x^3 - 12x^2y + 18xy^2 - 27y^3$. 49. $3a + 2b + c$.
50. $a^2 - 2ax + 4x^2$. 51. $x^2 + xy + y^2$. 52. $16x^2 - 4xy + y^2$.
53. $x^2 + xy - y^2$. 54. $ax^2 + 4a^2x + 2a^3$. 55. $a - x$.
56. $x - y - z$. 57. $3x^2 - x + 2$. 58. $4 - 6x + 8x^2 - 10x^3$.
59. $x + y$. 60. $ax + by - ab - xy$. 61. $bx + ay$.
62. $x^2 - ax + b^2$.

XV. (Page 40.)

1. $x^2 + ax + b$. 2. $y^2 - (l + m)y + lm$. 3. $x^2 + cx + d$.
4. $x^2 + ax - b$. 5. $x^2 - (b + d)x + bd$.

XVI. (Page 42.)

1. $m - n$, $m^2 - mn + n^2$, $m^4 - m^3n + m^2n^2 - mn^3 + n^4$,
$m^6 - m^5n + $ &c., $m^8 - m^7n + $ &c.

2. $m+n$, m^2+mn+n^2, m^3+m^2n+ &c., m^5+m^4n+ &c.,
m^6+m^5n+ &c.

3. $a-1$, a^2-a+1, a^4-a^3+ &c., a^6-a^5+ &c., a^7-a^6+ &c.

4. $y+1$, y^2+y+1, y^4+y^3+ &c., y^6+y^5+ &c., y^8+y^7+ &c.

xvii. (Page 43.)

1. $5x(x-3)$. 2. $3x(x^2+6x-2)$. 3. $7(7y^2-2y+1)$.
4. $4xy(x^2-3xy+2y^2)$. 5. $x(x^3-ax^2+bx+c)$.
6. $3x^3y^2(x^2y-7x+9y^2)$. 7. $27a^3b^6(2+4a^3b^2-9a^5b^3)$.
8. $45x^4y^7(x^3y^3-2x-8y)$.

xviii. (Page 44.)

1. $(x-a)(x-b)$. 2. $(a-x)(b+x)$. 3. $(b-y)(c+y)$.
4. $(a+m)(b+n)$. 5. $(ax+y)(bx-y)$. 6. $(ab+cd)(x-y)$.
7. $(cx+my)(dx-ny)$. 8. $(ac-bd)(bx-dy)$.

xix. (Page 45.)

1. $(x+5)(x+6)$. 2. $(x+5)(x+12)$. 3. $(y+12)(y+1)$.
4. $(y+11)(y+10)$. 5. $(m+20)(m+15)$. 6. $(m+6)(m+17)$.
7. $(a+8b)(a+b)$. 8. $(x+4m)(x+9m)$. 9. $(y+3n)(y+16n)$.
10. $(z+4p)(z+25p)$. 11. $(c^2+2)(c^2+3)$.
12. $(x^3+1)(x^3+3)$. 13. $(xy+2)(xy+16)$.
14. $(x^4y^2+3)(x^4y^2+4)$. 15. $(m^5+8)(m^5+2)$.
16. $(n+20q)(n+7q)$.

xx. (Page 45.)

1. $(x-5)(x-2)$. 2. $(x-19)(x-10)$.
3. $(y-11)(y-12)$. 4. $(y-20)(y-10)$.
5. $(n-23)(n-20)$. 6. $(n-56)(n-1)$.
7. $(x^3-4)(x^3-3)$. 8. $(ab-26)(ab-1)$.
9. $(b^2c^3-5)(b^2c^3-6)$. 10. $(xyz-11)(xyz-2)$.

xxi. (Page 46.)

1. $(x+12)(x-5)$. 2. $(x+15)(x-3)$. 3. $(a+12)(a-1)$.
4. $(a+20)(a-7)$. 5. $(b+25)(b-12)$. 6. $(b+30)(b-5)$.
7. $(x^4+4)(x^4-1)$. 8. $(xy+14)(xy-11)$.
9. $(m^5+20)(m^5-5)$. 10. $(n+30)(n-13)$.

xxii. (Page 46.)

1. $(x-11)(x+6)$. 2. $(x-9)(x+2)$. 3. $(m-12)(m+3)$.
4. $(n-15)(n+4)$. 5. $(y-14)(y+1)$. 6. $(z-20)(z+5)$.
7. $(x^5-10)(x^5+1)$. 8. $(cd-30)(cd+6)$.
9. $(m^3n-2)(m^3n+1)$. 10. $(p^4q^2-12)(p^4q^2+7)$.

xxiii. (Page 47.)

1. $(x-3)(x-12)$. 2. $(x+9)(x-5)$.
3. $(ab-18)(ab+2)$. 4. $(x^4-5m)(x^4+2m)$.
5. $(y^3+10)(y^3-9)$. 6. $(x^2+10)(x^2-11)$.
7. $x(x^2+3ax+4a^2)$. 8. $(x+m)(x+n)$.
9. $(y^3-3)(y^3-1)$. 10. $(xy-ab)(x-c)$.
11. $(x+a)(x-b)$. 12. $(x-c)(x+d)$.
13. $(ab-d)(b-c)$. 14. $4.(x-4y)(x-3y)$.

xxiv. (Page 48.)

1. $(x+9)^2$. 2. $(x+13)^2$. 3. $(x+17)^2$. 4. $(y+1)^2$.
5. $(z+100)^2$. 6. $(x^2+7)^2$. 7. $(x+5y)^2$. 8. $(m^2+8n^2)^2$.
9. $(x^3+12)^2$. 10. $(xy+81)^2$.

xxv. (Page 48.)

1. $(x-4)^2$. 2. $(x-14)^2$. 3. $(x-18)^2$. 4. $(y-20)^2$.
5. $(z-50)^2$. 6. $(x^2-11)^2$. 7. $(x-15y)^2$. 8. $(m^2-16n^2)^2$.
9. $(x^3-19)^2$.

xxvi. (Page 50.)

1. $(x+y)(x-y)$. 2. $(x+3)(x-3)$. 3. $(2x+5)(2x-5)$.
4. $(a^2+x^2)(a^2-x^2)$. 5. $(x+1)(x-1)$. 6. $(x^3+1)(x^3-1)$.
7. $(x^4+1)(x^4-1)$. 8. $(m^2+4)(m^2-4)$.
9. $(6y+7z)(6y-7z)$. 10. $(9xy+11ab)(9xy-11ab)$.
11. $(a-b+c)(a-b-c)$. 12. $(x+m-n)(x-m+n)$.
13. $(a+b+c+d)(a+b-c-d)$. 14. $2x \times 2y$.
15. $(x-y+z)(x-y-z)$.
16. $(a-b+m+n)(a-b-m-n)$.
17. $(a-c+b+d)(a-c-b-d)$. 18. $(a+b-c)(a-b+c)$.
19. $(x+y+z)(x+y-z)$. 20. $(a-b+m-n)(a-b-m+n)$.
21. $(ax+by+1)(ax+by-1)$. 22. $2ax \times 2by$.
23. $(1+a-b)(1-a+b)$. 24. $(1+x-y)(1-x+y)$.
25. $(x+y+z)(x-y-z)$. 26. $(a+2b-3c)(a-2b+3c)$.
27. $(a^2+4b)(a^2-4b)$. 28. $(1+7c)(1-7c)$.
29. $(a-b+c+d)(a-b-c-d)$. 30. $(a+b-c-d)(a-b-c+d)$.
31. $3ax(ax+3)(ax-3)$. 32. $(a^2b^3+c^4)(a^2b^3-c^4)$.
33. $12(x-1)(2x+1)$. 34. $(9x+7y)(5x+y)$.
35. 1000×506.

xxvii. (Page 51.)

1. $(a+b)(a^2-ab+b^2)$. 2. $(a-b)(a^2+ab+b^2)$.
3. $(a-2)(a^2+2a+4)$.
4. $(x+7)(x^2-7x+49)$.
5. $(b-5)(b^2+5b+25)$. 6. $(x+4y)(x^2-4xy+16y^2)$.
7. $(a-6)(a^2+6a+36)$. 8. $(2c+3y)(4x^2-6xy+9y^2)$.
9. $(4a-10b)(16a^2+40ab+100b^2)$.
10. $(9x+8y)(81x^2-72xy+64y^2)$.
11. $(x+y)(x^2-xy+y^2)(x-y)(x^2+xy+y^2)$.

12. $(x+1)(x^2-x+1)(x-1)(x^2+x+1)$.
13. $(a+2)(a^2-2a+4)(a-2)(a^2+2a+4)$.
14. $(3+y)(9-3y+y^2)(3-y)(9+3y+y^2)$.

xxviii. (Page 51.)

1. $a+b$. 2. Take b from a and add c to the result.
3. $2x$. 4. $a-5$. 5. $x+1$. 6. $x-2, x-1, x, x+1, x+2$.
7. 0. 8. 0. 9. da. 10. c. 11. $x-y$. 12. $x-y$.
13. $365-6x$. 14. $x-10$. 15. $x+5a$.
16. A has $x+5$ shillings, B has $y-5$ shillings.
17. $x-8$. 18. xy. 19. $12-x-y$. 20. nq. 21. $25-x$.
22. $y-25$. 23. $256m^8$. 24. $4b$. 25. $x-5$. 26. $y+7$.
27. x^2-y^2. 28. $(x+y)(x-y)$. 29. 2. 30. 2.
31. 28. 32. 7. 33. 23. 34. 5. 35. 10.

xxix. (Page 53.)

1. To a add b.
2. From the square of a take the square of b.
3. To four times the square of a add the cube of b.
4. Take four times the sum of the squares of a and b.
5. From the square of a take twice b, and add to the result three times c.
6. To a add the product of m and b, and take c from the result.
7. To a add m. From b take c. Multiply the results together.
8. Take the square root of the cube of x.
9. Take the square root of the sum of the squares of x and y.
10. Add to a twice the excess of 3 above c.
11. Multiply the sum of a and 2 by the excess of 3 above c.

[S.A.]

12. Divide the sum of the squares of a and b by four times the product of a and b.

13. From the square of x subtract the square of y, and take the square root of the result. Then divide this result by the excess of x above y.

14. To the square of x add the square of y, and take the square root of the result. Then divide this result by the square root of the sum of x and y.

XXX. (Page 53.)

1. 2.   2. 0.   3. 17.   4. 31.   5. 20.   6. 33.
7. 105.   8. 27.   9. 14.   10. 120.   11. 210.   12. 1458.
13. 30.   14. 5.   15. 3.   16. 4.   17. 49.   18. 10.
19. 12.   20. 4.   21. 43.   22. 20.   23. 29.   24. 41536.   25. 52.

XXXI. (Page 54.)

1. 0.   2. 0.   3. $2ac$.   4. $2xy$.   5. $a^2 + b^2$.

6. $4x^4 + (6m - 6n)x^3 - (4m^2 + 9mn + 4n^2)x^2$
$\qquad\qquad\qquad\qquad + (6m^2n - 6mn^2)x + 4m^2n^2$.

7. $cr^2 + dr + e$.   8. $-a^4 - b^4 - c^4 + 2a^2b^2 + 2a^2c^2 + 2b^2c^2$. When $c = 0$, this becomes $-a^4 - b^4 + 2a^2b^2$. When $b + c = a$, the product becomes 0. When $a = b = c$, it becomes $3a^4$.   9. 0.   10. 34.

12. (α) $(a + b)x^2 + (c + d)x$.   (β) $(a - b)x^3 - (c + d - 2)x^2$.
(γ) $(4 - a)x^3 - (3 + b)x^2 - (5 + c)x$.   ($\delta$) $a^2 - b^2 + (2a + 2b)x$.
(ϵ) $(m^2 - n^2)x^4 + (2mq - 2nq)x^3 + (2m - 2n)x^2$.

13. $x^3 - (a + b + c)x^2 + (ab + ac + bc)x - abc$.

14. $x^3 + (a + b + c)x^2 + (ab + ac + bc)x + abc$.

15. $(a + b + c)^3 = a^3 + 3a^2b + 3ab^2 + b^3 + c^3 + 3a^2c$
$\qquad\qquad\qquad + 6abc + 3b^2c + 3ac^2 + 3bc^2$.
$(a + b - c)^3 = a^3 + 3a^2b + 3ab^2 + b^3 - c^3 - 3a^2c$
$\qquad\qquad\qquad - 6abc - 3b^2c + 3ac^2 + 3bc^2$.

$(b+c-a)^3 = -a^3 + 3a^2b - 3ab^2 + b^3 + c^3 + 3a^2c$
$\qquad - 6abc + 3b^2c - 3ac^2 + 3bc^2.$
$(c+a-b)^3 = a^3 - 3a^2b + 3ab^2 - b^3 + c^3 + 3a^2c$
$\qquad - 6abc + 3b^2c + 3ac^2 - 3bc^2.$

The sum of the last three subtracted from the first gives $24abc$.

16. $9a^2 + 6ac - 3ab + 4bc - 6b^2.$ 17. $a^{16} - x^{16}.$
18. $2ac - 2bc - 2ad + 2bd.$ The value of the result is $-2bc$.
19. $ab + xy + (b + 1 + 2a)x + (2a - b - 1)y.$
20. 9.   21. $ab + x^2 + (a - b + 1)x - (a + b + 1)y.$
22. 2.   23. $(7m + 4n + 1)x + (1 - 6n - 4m)y.$
25. $4a^2 + 6ac + 2ab + 9bc - 6b^2.$ 26. 3; 128; 3; 118.
27. 9.   28. 44.   29. 20.   30. 35.   31. 18.

xxxii. (Page 60.)

1. 3.   2. 2.   3. 1.   4. 7.   5. 2.   6. 2.   7. 3.
8. 4.   9. 9.   10. *Ans.* 54.
11. 2.   12. 9.   13. 9.   14. -7.   15. 3.   16. 7.
17. 2.   18. 8.   19. 10.   20. 6.   21. 4.   22. 10.
23. 3.   24. 15.   25. 1.   26. 2.   27. 3.   28. 4.
29. 6.   30. -1.

xxxiii. (Page 62.)

1. 70.   2. 43.   3. 23.   4. 7, 21.   5. 36, 26, 18, 12.
6. 12, 8.   7. 50, 30.   8. 10, 14, 18, 22, 26, 30.   9. £68.
10. 12 shillings, 24 shillings.   11. 52.
12. A has £130, B £150, C £130, D £90.
13. 152 men, 76 women, 38 children.   14. £350, £450, £720.
15. 21, 13.   16. £8. 15s.   17. 84, 26.   18. 62, 28.
19. The wife £4000, each son, £1000, each daughter £500.
20. 49 gallons.   21. £14, £24, £38.   22. 31, 17

ANSWERS.

23. £21. 24. 48, 36. 25. 50, 40. 26. 42, 18.
27. 60, 24. 28. 8, 12. 29. 88. 30. 18. 31. 40.
32. 57, 19. 33. 4. 34. 80, 128. 35. 19, 22.
36. 200, 100. 37. 23, 20. 38. 53, 318. 39. 5, 10, 15.

xxxiv. (Page 68.)

1. $a^2 b$. 2. $x^2 y^2 z$. 3. $2x^2 y$. 4. $15 m^2 np$. 5. $18abcd$.
6. $a^2 b^2$. 7. 2. 8. $17pq$. 9. $4x^2 y^2 z^2$. 10. $30 x^2 y^3$.

xxxv. (Page 69.)

1. $a-b$. 2. a^2-b^2. 3. $a-x$. 4. $a+x$. 5. $3x+1$.
6. $1-5a$. 7. $x+y$. 8. $x-y$. 9. $x-1$. 10. $1+a$.

xxxvi. (Page 70.)

1. 3453. 2. 36. 3. 936. 4. 355. 5. 23. 6. 2345.

xxxvii. (Page 74.)

1. $x+4$. 2. $x+10$. 3. $x-7$. 4. $x+12$.
5. $x-3$. 6. $x+3y$. 7. $x-4y$. 8. $x-15y$.
9. $x-y$. 10. $x+y$. 11. $x-y$. 12. $x+y$.
13. $x+y$. 14. $a+b-c$. 15. $4x+y$. 16. $3x-y$.
17. $5x-y$. 18. $x^4+x^3-4x^2+x+1$. 19. x^2-2x+4.
20. x^2+xy+y^2. 21. x^3+x^2-x-1. 22. $3a^2+2ab-b^2$.
23. $3x-y$. 24. $3x-11y$. 25. $3a-b$.
26. $3(a-x)$. 27. $3x-2$. 28. $3x^2+a^2$.
29. x^2+y^2. 30. $x+3$. 31. $(3a+2x)x$.

xxxviii. (Page 76.)

1. $x+2$. 2. $x-1$. 3. $x+1$. 4. $y-1$.
5. x^2-2x+5. 6. $x-2$. 7. y^2-2y+5.

xxxix. (Page 81.)

1. $\dfrac{1}{3a}$.
2. $\dfrac{2x}{9}$.
3. $\dfrac{5b}{12a}$.
4. $\dfrac{2x^2}{5z}$.
5. $\dfrac{a^2b^5c^3}{3}$.
6. $\dfrac{4xy}{3bc}$.
7. $\dfrac{3y}{2az}$.
8. $\dfrac{5b^2c}{4a^2}$.
9. $\dfrac{4}{3c^2y^6}$.
10. $\dfrac{5m}{p}$.
11. $\dfrac{a}{a+b}$.
12. $\dfrac{2mx}{3m^2p-x}$.
13. $\dfrac{1}{3y-5xz}$.
14. $\dfrac{2a+x}{4ax^2-x}$.
15. $\dfrac{y}{bc}$.
16. $\dfrac{a^2}{2x-3y}$.
17. $\dfrac{3ab}{2bc+c}$.
18. $\dfrac{c-2a}{c+2a}$.
19. $\dfrac{3}{5}$.
20. $\dfrac{5}{2x-2y}$.
21. $\dfrac{1}{7ax-7by}$.
22. $\dfrac{2}{9abx-12cdx}$.
23. $\dfrac{xy}{2az}$.
24. $\dfrac{b^2}{2a^2c}$.
25. $\dfrac{1}{2c}$.
26. $\dfrac{2a+2b}{a^2}$.
27. $\dfrac{1}{12}$.
28. $\dfrac{x}{y}$.

xl. (Page 82.)

1. $\dfrac{a+5}{a+3}$.
2. $\dfrac{x-5}{x-3}$.
3. $\dfrac{x+1}{x-7}$.
4. $\dfrac{x-3y}{x+7y}$.
5. x^2-x+1.
6. $\dfrac{x^3+y^3}{x^3-y^3}$.
7. $\dfrac{x-2}{x+4}$.
8. $\dfrac{x-3}{x+1}$.
9. $\dfrac{x^2-5x+6}{3x^2-7x}$.
10. $\dfrac{x^2-5x+6}{3x^2-8x}$.
11. $\dfrac{x^2+xy-y^2}{x^2-xy-y^2}$.
12. $\dfrac{a^2+5a+5}{a^2+a-2}$.
13. $\dfrac{b^2+5b}{b^2+b-5}$.
14. $\dfrac{m^2+4m}{m^2+m-6}$.
15. $\dfrac{a^2-a+1}{a^2+a+1}$.
16. $\dfrac{3ax-7a}{7x^2-3x}$.
17. $\dfrac{14x-6}{9ax-21a}$.
18. $\dfrac{10a-14a^2}{15-9a-6a^2}$.
19. $\dfrac{2ab^2+3ab-5a}{7b^2-5b}$.

20. $\dfrac{a^2-a+1}{a^2-2a+2}$. 21. $\dfrac{3x-1}{x^2-1}$. 22. $\dfrac{a-5}{a-3}$.

23. $\dfrac{x^2-2x+2}{x^2-2}$. 24. 3. 25. $\dfrac{2x^2+3x-5}{7x-5}$.

26. $\dfrac{4x^2+9x+1}{2x^2-3x-2}$. 27. $\dfrac{2x-3a}{4x^2+6ax+9a^2}$. 28. $\dfrac{x-3}{x-2}$.

29. $\dfrac{x-3}{x+1}$. 30. $\dfrac{m-1}{m+1}$. 31. $\dfrac{x^2+5x}{x+3}$.

32. $\dfrac{a-b-c}{a+b-c}$. 33. $\dfrac{5a+2b}{3a+2b}$. 34. $\dfrac{x-5}{2x+3}$.

35. $\dfrac{x^2+4}{x^2+x+1}$. 36. $\dfrac{x^3+x^2-2}{2x^2+2x+1}$. 37. $\dfrac{x^2+x-12}{3x+5}$.

38. $\dfrac{x^2-2x+3}{2x^2+5x-3}$. 39. $\dfrac{x^3-2x^2-2x+1}{4x^2-7x-1}$. 40. $\dfrac{a^2-5a+6}{3a^2-8a}$.

xli. (Page 86.)

1. $\dfrac{7x^2}{12y^2}$. 2. $\dfrac{1}{2}$. 3. $\dfrac{2x^3}{3y^3}$. 4. $\dfrac{by}{9ax}$.

5. ax. 6. $\dfrac{4}{9}$. 7. $\dfrac{3}{8}$. 8. $\dfrac{8a^2c^2}{9d^2}$.

9. $\dfrac{3mnxy}{4pq^2}$. 10. $\dfrac{5km^2}{4pq}$.

xlii. (Page 86.)

1. $\dfrac{a-b}{a^2}$. 2. $\dfrac{4}{3}$. 3. $\dfrac{(x+2)(x-4)}{x(x-2)}$.

4. $\dfrac{(x-1)(x-6)}{x^2}$. 5. $\dfrac{x-6}{x-3}$. 6. $\dfrac{(x-2)(x-5)}{x^2}$.

7. 1. 8. b. 9. $\dfrac{y}{x-y}$. 10. $\dfrac{c-a+b}{c-a-b}$.

11. $\dfrac{x-m+n}{x+m-n}$. 12. 1. 13. $\dfrac{x-y-z}{x+y+z}$.

ANSWERS.

xliii. (Page 87.)

1. $\dfrac{10ac}{3bx}$. 2. $\dfrac{3}{2y}$. 3. $\dfrac{8xy}{b}$. 4. $\dfrac{4}{3bnx}$. 5. $\dfrac{3}{4}$.
6. $\dfrac{5x}{4a}$. 7. $\dfrac{5x}{14}$. 8. $\dfrac{1}{x-2}$. 9. $\dfrac{1}{x-2}$.

xliv. (Page 89.)

1. $12a^3x^2$. 2. $12x^2y^2$. 3. $8a^3b^2$. 4. a^2x^2.
5. $4ax^3$. 6. $a^2b^2c^3$. 7. $a^3x^2y^2$. 8. $102a^2x^4$.
9. $20p^2q^2r$. 10. $72ax^2y^2$.

xlv. (Page 91.)

1. $x^2(a+x)$. 2. x^3-x. 3. $a(a^2-b^2)$.
4. $4x^2-1$. 5. a^3+b^3. 6. x^2-1.
7. $(x^3-1)(x+1)$. 8. $(x^2+1)(x^3+1)$.
9. $(x+1)(x^3-1)$. 10. x^4-1.
11. $x(x^3-1)(x^3+1)$. 12. $x(x+1)(x^3-1)$.
13. $(2a-1)(8a^3+1)$. 14. $2x^2+2xy$.
15. $(a+b)^2(a-b)$. 16. a^2-b^2.
17. $4(1-x^2)$. 18. x^3-1.
19. $(a-b)(a-c)(b-c)$. 20. $(x+1)(x+2)(x+3)$.
21. $(x+y)^2(x-y)^2$. 22. $(a+3)(a^2-1)$.
23. $x^2(x^2-y^2)$. 24. $(x+1)(x+2)(x+3)(x+4)$.
25. $12(x-y)^2(x^3+y^3)$. 26. $120xy(x^2-y^2)$.

xlvi. (Page 93.)

1. $(x+2)(x+3)(x+4)$. 2. $(a-5)(a+4)(a-3)$.
3. $(x+1)(x+2)(x+3)$. 4. $(x+5)(x+6)(x+7)$.
5. $(x-11)(x+2)(x-2)$. 6. $(2x+1)(x+1)(x-2)$.

7. $(x^2+y)(x+y)(x^2+y^2)(x-y)$. 8. $(x-5)(x-3)(x+5)$.
9. $(7x-4)(3x-2)(x^2-3)$. 10. $(x^2+y^2)(x+y)(x-y)$.
11. $(a^2-b^2)(a+2b)(a-2b)$.

xlvii. (Page 94.)

1. $(x-2)(x-1)(x-3)(x-4)$. 2. $(x+4)(x+1)(x+3)$.
3. $(x-4)(x-5)(x-7)$. 4. $(3x-2)(2x+1)(7x-1)$.
5. $(x+1)(x-1)(x+3)(3x-2)(2x+1)$.
6. $(x-3)(x^2+3x+9)(x-12)(x^2-2)$.

xlviii. (Page 95.)

1. $\dfrac{15x}{20}, \dfrac{16x}{20}$. 2. $\dfrac{9x-21}{18}, \dfrac{4x-9}{18}$.

3. $\dfrac{4x-8y}{10x^2}, \dfrac{3x^2-8xy}{10x^2}$. 4. $\dfrac{20a+25b}{10a^2}, \dfrac{6a^2-8ab}{10a^2}$.

5. $\dfrac{48a^2-60ac}{60a^2c}, \dfrac{15a-10c}{60a^2c}$. 6. $\dfrac{ab-b^2}{a^3b^2}, \dfrac{a^4-a^3b}{a^3b^2}$.

7. $\dfrac{3-3x}{1-x^2}, \dfrac{3+3x}{1-x^2}$. 8. $\dfrac{2+2y^2}{1-y^4}, \dfrac{2-2y^2}{1-y^4}$.

9. $\dfrac{5+5x}{1-x^2}, \dfrac{6}{1-x^2}$. 10. $\dfrac{ab+ax}{c(b+x)}, \dfrac{b}{c(b+x)}$.

11. $\dfrac{a-c}{(a-b)(b-c)(a-c)}, \dfrac{b-c}{(a-b)(b-c)(a-c)}$.

12. $\dfrac{c(b-c)}{abc(a-b)(a-c)(b-c)}, \dfrac{b(a-b)}{abc(a-b)(a-c)(b-c)}$.

xlix. (Page 98.)

1. $\dfrac{15x+17}{15}$. 2. $\dfrac{71a-20b-56c}{84}$. 3. $\dfrac{32x+9y}{42}$.

4. $\dfrac{16x^2+55x+4xy-55y}{50x}$. 5. $\dfrac{27x^2-2x^2y-16xy-28y^2}{12x^2}$.

ANSWERS.

6. $\dfrac{180a^2 + 54ab + 331b^2 - 20ab^2}{90b^2}.$

7. $\dfrac{80x^3 + 64x^2 + 84x + 45}{60x^2}.$

8. $\dfrac{35a^2 + 23ab + 21bc - 42c^2}{21ac}.$

9. $\dfrac{4a^2c - 3ac^2 - 3ac + 7c^2}{a^2c^2}.$

10. $\dfrac{11y^2 - 8x^2y^2 - 4xy - 7x^2}{x^3y^3}.$

11. $\dfrac{3a^4 - 7a^3b + 4a^2bc - 5ab^2c + abc^2 - b^2c^2}{a^3b^2c^2}.$

l. (Page 99.)

1. $\dfrac{2x - 1}{(x - 6)(x + 5)}.$
2. $\dfrac{4}{(x - 7)(x - 3)}.$
3. $\dfrac{2}{(1 + x)(1 - x)}.$

4. $\dfrac{4xy}{(x + y)(x - y)}.$
5. $\dfrac{-1}{1 + x}.$
6. $\dfrac{a + bx}{c + dx}.$

7. $\dfrac{2x^2}{(x + y)(x - y)}.$
8. $\dfrac{2x - y}{(x - y)^2}.$
9. $\dfrac{2x + 5a}{(x + a)^2}.$

10. $\dfrac{1}{(a + x)(a - x)}.$

li. (Page 100.)

1. $\dfrac{2}{1 - a}.$
2. $\dfrac{4x}{1 - x^4}.$
3. $\dfrac{2x}{1 - x^4}.$
4. $\dfrac{8b^7}{a^8 - b^8}.$

5. $\dfrac{x + y}{y}.$
6. $\dfrac{3x^3 + 20x^2 - 32x - 235}{(x + 4)(x - 3)(x + 7)}.$

7. $\dfrac{3x^3 - 24x^2 + 60x - 46}{(x - 2)(x - 3)(x - 4)}.$
8. $\dfrac{3x^2 - 2ax - 6a^3}{(x - a)^3}.$

9. $\dfrac{6}{(x - 1)(x + 2)(x + 1)}.$
10. $\dfrac{x}{(x + 1)(x + 2)(x + 3)}.$

11. $\dfrac{3x^2}{x^2 - 1}.$
12. $\dfrac{e - d}{(a + c)(a + d)(a + e)}.$
13. $0.$

14. $2.$
15. $\dfrac{y}{x + y}.$
16. $0.$
17. $\dfrac{x^2 + xy}{x^3 - y^3}.$

18. 0. 19. $\dfrac{b}{a+b}$. 20. 0. 21. 0.

lii. (Page 103.)

1. $\dfrac{y}{x-y}$. 2. $\dfrac{1}{2+x}$. 3. $\dfrac{3x^2}{x^2-1}$. 4. $\dfrac{y+6}{3(1-y^2)}$.

5. 0. 6. $\dfrac{1}{(x+a)(x+b)}$. 7. $\dfrac{a^6-2ab^5+2a^5b+b^6}{a^6-b^6}$.

8. $\dfrac{1}{1-x^4}$. 9. $\dfrac{2}{(x-z)(y-z)}$. 10. $\dfrac{1}{abc}$.

liii. (Page 110.)

1. $\dfrac{2x+11}{(x+4)(x+5)(x+7)}$. 2. $\dfrac{2(x-8)}{(x-6)(x-7)(x-9)}$.

3. $\dfrac{2x-17}{(x-4)(x+11)(x-13)}$. 4. $\dfrac{2}{x+3}$. 5. $\dfrac{m^3+4m^2n+mn^2}{n(m+n)^2}$.

6. 0. 7. $\dfrac{11x^3-x^2+25x-1}{3(1-x^4)}$. 8. 0. 9. $\dfrac{1}{1+x}$.

liv. (Page 107.)

1. 16. 2. 12. 3. 15. 4. 28. 5. 63.
6. 24. 7. 60. 8. 45. 9. 36. 10. 120.
11. 72. 12. 96. 13. 64. 14. 12. 15. 28.
16. 1. 17. 8. 18. 9. 19. 7. 20. 4.
21. 5. 22. 1. 23. 1. 24. $\dfrac{3}{2}$. 25. 100.
26. 24. 27. $\dfrac{2}{3}$. 28. 6. 29. 24. 30. 4.

lv. (Page 108.)

1. 16. 2. 5. 3. $\dfrac{1}{2}$. 4. 1. 5. 8.

ANSWERS. 363

6. $-\dfrac{1}{9}$.   7. 9.   8. 2.   9. 11.   10. 6.   11. 2.
12. 12.   13. 8.   14. 7.   15. 9.   16. 7.   17. 7.
18. 9.   19. 9.   20. 9.   21. 10.

lvi. (Page 109.)

1. $\dfrac{c}{a+b}$.   2. $\dfrac{3c-2a}{5b-c}$.   3. $\dfrac{a^2b-bc+d}{a+f}$.

4. $\dfrac{bc-dm}{a-5}$.   5. $\dfrac{b(a+c)}{1+a}$.   6. $\dfrac{6bd+ab}{3a-12d}$.

7. $\dfrac{3ab-2k-3}{4ac-1}$.   8. 1.   9. $\dfrac{(a+b)^2}{b-a}$.

10. $-\dfrac{a}{2}$.   11. 2.   12. 0.   13. $\dfrac{b}{a-1}$.

14. $\dfrac{3a+1}{2a+b}$.   15. $\dfrac{18a+2b}{4a+3}$.   16. $\dfrac{a-1}{b}$.   17. $\dfrac{p}{q}$.

18. $\dfrac{abd+ac}{ad+d}$.   19. $b-1$.   20. $\dfrac{b}{c}$.   21. $\dfrac{2a^3}{b-1}$.

22. 1.   23. bm.   24. $\dfrac{3a^3bc+2a^3b^4+ab^4}{b^3+3a^3c+3a^2bc+2a^2b^3}$.

25. $\dfrac{bc}{c^2-b}$.   26. $\dfrac{d}{c}$.   27. $\dfrac{ab-1}{bc+d}$.

28. $\dfrac{a(m-3c+3a)}{c-a+m}$.   29. $\dfrac{ac}{b}$.   30. $\dfrac{a^2e(c-d)}{(a^2+b^2)d}$.

lvii. (Page 111.)

1. 2.   2. 15.   3. 1.   4. $\dfrac{6}{13}$.   5. $\dfrac{7}{10}$.

6. $\dfrac{1}{7}$.   7. $\dfrac{3}{2}$.   8. 6.   9. -7.   10. 6.

11. 9.   12. 19.   13. 1.   14. 4.   15. $-\dfrac{35}{6}$.

16. 12.   17. 2.   18. $\dfrac{1}{2}$.   19. $\dfrac{1}{8}$.   20. 3.

lviii. (Page 113.)

1. 20. 2. 3. 3. 40. 4. $\frac{459}{46}$. 5. 60.

6. 10. 7. 5. 8. 20. 9. 3. 10. $-\frac{1}{9}$.

11. 8. 12. 100. 13. 0. 14. ·1. 15. 5.

16. $\frac{5}{6}$. 17. 5.

lix. (Page 114.)

1. 100. 2. 240. 3. 80. 4. 700. 5. 28, 32.

6. $2\frac{6}{7}, 47\frac{1}{7}$. 7. 24, 76. 8. 120. 9. 60.

10. 960. 11. 36. 12. 12, 4. 13. £1897.

14. 540, 36. 15. 3456, 2304. 16. 50. 17. 35, 15.

18. 29340, 1867. 19. 21, 6. 20. $105\frac{1}{3}, 131\frac{2}{3}$.

21. A has £1400, B has £400. 22. 28, 18.

23. $\frac{m(nb-a)}{n-m}, \frac{n(mb-a)}{m-n}$. 24. $\frac{a+b}{2}, \frac{a-b}{2}$. 25. 18.

26. £135, £297, £432. 27. £7200. 28. 47, 23.

29. 7, 32. 30. 112, 96. 31. 78. 32. 75 gallons.

33. 40, 10. 34. 20. 35. 42 years. 36. $1\frac{1}{5}$ days.

37. 20 days. 38. 10 days. 39. 6 hours. 40. $1\frac{1}{29}$ days.

41. $4\frac{6}{11}$ days. 42. $1\frac{5}{7}$ hours. 43. 48'.

44. 2 hours. 45. $\frac{abc}{ab+ac+bc}$ minutes. 46. $48\frac{3'}{4}$.

47. $51\frac{1}{3}, 61\frac{1}{3}, 47\frac{1}{3}$ gallons. 48. $9\frac{1}{7}$ miles from Ely.

ANSWERS.

49. 14 miles. 50. $\dfrac{ac}{b}$, $\dfrac{bd}{a}$. 51. $11\dfrac{13'}{21}$.

52. 42 hours. 53. $30\dfrac{30}{31}$ miles. 54. 50 hours.

55. (1) $38\dfrac{2'}{11}$ past 1. (2) $54\dfrac{6'}{11}$ past 4. (3) $10\dfrac{10'}{11}$ past 8.

56. (1) $27\dfrac{3'}{11}$ past 2. (2) $5\dfrac{5'}{11}$ and also $38\dfrac{2'}{11}$ past 4.

(3) $21\dfrac{9'}{11}$ past 7, and also $54\dfrac{6'}{11}$ past 7.

57. (1) $16\dfrac{4'}{11}$ past 3. (2) $32\dfrac{8'}{11}$ past 6. (3) $49\dfrac{1'}{11}$ past 9.

58. 60.   59. £3.   60. $\dfrac{1}{30}$.   61. $18\tfrac{4}{5}$ days.

62. £600.   63. £275.   64. 60.

65. 90′, 72′, 60′.   66. 126, 63, 56 days.   67. 24.

68. 2, 4, 94.   69. 200.   70. 2^h, $5\dfrac{5'}{11}$.

71. 30000.   72. £200000000.   73. 50.

lx. (Page 127.)

1. $\dfrac{x^2 + ax + 3a}{x}$.   2. $\dfrac{a^2 + 3ax - 2x^2}{x^2}$.

3. $\dfrac{x^2 + y^2}{x(x-y)}$.   4. $\dfrac{2a^3 + 6a^2b + 6ab^2 + 2b^3}{(a-b)(a^2+b^2)}$.

lxi. (Page 128.)

1. $\dfrac{8 - 13x}{70}$.   2. $\dfrac{x + y}{xy}$.   3. $x(1 - x)$.   4. $\dfrac{x+y}{x-y}$.

5. $\dfrac{x^3 + 5x^2 + 1}{2x^2 - x^3 + 1}$.   6. $\dfrac{x^2 - x + 1}{x}$.   7. $\dfrac{a^2 + a + 1}{a}$.

366   ANSWERS.

8. x. 9. $\dfrac{1}{x}$. 10. x. 11. $\dfrac{x^2+y^2}{-2xy}$. 12. x^2.

13. $\dfrac{a(a^2+2ab+2b^2)}{(a+b)^2}$. 14. $m-1$. 15. $\dfrac{1}{c(a-b-c)}$.

lxii. (Page 129.)

1. $\dfrac{1}{2} + \dfrac{3}{2a} + \dfrac{1}{a^2} + \dfrac{5}{2a^3}$.

2. $\dfrac{a}{d} + \dfrac{b}{c} + \dfrac{c}{d} + \dfrac{d}{a}$.

3. $\dfrac{x}{y^2} - \dfrac{3}{y} + \dfrac{3}{x} - \dfrac{y}{x^2}$.

4. $\dfrac{a^3}{12} - \dfrac{a^2}{9} + \dfrac{a}{18} - \dfrac{1}{36}$.

5. $\dfrac{6p}{qrs} + \dfrac{4q}{prs} - \dfrac{12r}{pqs} + \dfrac{24s}{pqr}$.

6. $\dfrac{x^3}{100} - \dfrac{x^2}{40} + \dfrac{3x}{40} - \dfrac{1}{8}$.

lxiii. (Page 131.)

1. $2 - 2a + 2a^2 - 2a^3 + 2a^4 \ldots$

2. $1 - \dfrac{2}{m} + \dfrac{4}{m^2} - \dfrac{8}{m^3} + \dfrac{16}{m^4} \ldots$

3. $1 - \dfrac{2b}{a} + \dfrac{2b^2}{a^2} - \dfrac{2b^3}{a^3} + \dfrac{2b^4}{a^4} \ldots$

4. $1 + \dfrac{2x^2}{a^2} + \dfrac{2x^4}{a^4} + \dfrac{2x^6}{a^6} + \dfrac{2x^8}{a^8} \ldots$

5. $x + \dfrac{x^2}{a} + \dfrac{x^3}{a^2} + \dfrac{x^4}{a^3} + \dfrac{x^5}{a^4} \ldots$

6. $\dfrac{b}{a} - \dfrac{bx}{a^2} + \dfrac{bx^2}{a^3} - \dfrac{bx^3}{a^4} + \dfrac{bx^4}{a^5} \ldots$

7. $1 - 2x + 6x^2 - 16x^3 + 44x^4 \ldots$

8. $1 + 2x + x^2 - x^3 - 2x^4 \ldots$

9. $1 + 3b + 6b^2 + 12b^3 + 24b^4 \ldots$

10. $x^2 - bx + b^2 - \dfrac{2b^3}{x} + \dfrac{2b^4}{x^2} \ldots$

ANSWERS. 367

11. $\dfrac{a^2}{x} + \dfrac{a^2 b}{x^2} + \dfrac{a^2 b^2}{x^3} + \dfrac{a^2 b^3}{x^4} + \dfrac{a^2 b^4}{x^5}$ ......

12. $1 - \dfrac{2x}{a} + \dfrac{3x^2}{a^2} - \dfrac{4x^3}{a^3} + \dfrac{5x^4}{a^4}$ ......

13. $x^3 - 3ax^2 + 2a^2 x + 4a^3$.   14. $m^4 - 10m^2 - 41m - 95$.

lxiv. (Page 132.)

1. $\dfrac{x^3}{9} + \dfrac{x^2}{4} + \dfrac{23x}{120} + \dfrac{1}{20}$.   2. $\dfrac{a^3}{20} - \dfrac{49a^2}{600} + \dfrac{7a}{60} - \dfrac{1}{15}$.

3. $x^4 - \dfrac{1}{x^4}$.   4. $x^4 + 1 + \dfrac{1}{x^4}$.   5. $\dfrac{1}{a^4} - \dfrac{1}{b^4}$.

6. $\dfrac{1}{a^2} + \dfrac{2}{ac} - \dfrac{1}{b^2} + \dfrac{1}{c^2}$.   7. $1 + \dfrac{b^2}{a^2} + \dfrac{b^4}{a^4}$.   8. $1 + \dfrac{x^2}{8} - \dfrac{x^3}{8} - \dfrac{x^5}{64}$.

9. $\dfrac{5}{x^4} + \dfrac{7}{2x^3} - \dfrac{107}{12x^2} + \dfrac{5}{6x} + \dfrac{7}{6}$.   10. $\dfrac{a^4}{b^4} - \dfrac{b^4}{a^4} - \dfrac{4b^2}{a^2} - 4$.

lxv. (Page 134.)

1. $x - \dfrac{1}{x}$.   2. $a + \dfrac{1}{b}$.   3. $m^2 - \dfrac{m}{n} + \dfrac{1}{n^2}$.

4. $c^4 + \dfrac{c^3}{d} + \dfrac{c^2}{d^2} + \dfrac{c}{d^3} + \dfrac{1}{d^4}$.   5. $\dfrac{x}{y} + \dfrac{y}{x}$.

6. $\dfrac{1}{a^2} + \dfrac{1}{ab} + \dfrac{1}{b^2}$.   7. $\dfrac{x^2}{y^2} - 2 + \dfrac{y^2}{x^2}$.   8. $\dfrac{3}{2}x^3 - 5x^2 + \dfrac{1}{4}x + 9$.

9. $\dfrac{a^2}{b^2} - 1 + \dfrac{b^2}{a^2}$.   10. $\dfrac{1}{a^2} - \dfrac{1}{ab} - \dfrac{1}{ac} + \dfrac{1}{b^2} - \dfrac{1}{bc} + \dfrac{1}{c^2}$.

lxvi. (Page 135.)

1. $\cdot 05 x^2 - \cdot 143 x - \cdot 021$.   2. $\cdot 01 x^2 + 1 \cdot 25 x - 21$.

3. $\cdot 12 x^2 + \cdot 13 xy - \cdot 14 y^2$.   4. $\cdot 172 x^2 - \cdot 05 xy - \cdot 312 y^2$.

5. 0.   6. $\cdot 300763$.

lxvii. (Page 135.)

1. $a_1 x \left(1 + \dfrac{a_2}{a_1} x + \dfrac{a_3}{a_1} x^2 + \dfrac{a_4}{a_1} x^3 + \ldots \right).$

2. $xyz \left(\dfrac{1}{z} - \dfrac{1}{y} + \dfrac{1}{x}\right).$ 3. $x^2 \left(1 + \dfrac{y}{x} + \dfrac{y^2}{x^2}\right).$

4. $(a+b)\left\{(a+b)^2 - c(a+b) - d + \dfrac{e}{a+b}\right\}.$

lxix. (Page 138.)

1. 46. 2. $\dfrac{2x^2 + 3x - 5}{7x - 5}$ and $\dfrac{a^2 + 5a - 14}{a + 9}.$ 3. $\dfrac{2ap}{a^2 + p^2}.$

4. $\dfrac{37x^2 - 7y^2 - 19z^2}{24}.$ 5. $\dfrac{11}{9}.$

6. $\dfrac{60x^4 + 42ax^3 - 107a^2x^2 + 10a^3x + 14a^4}{12}.$

8. $\dfrac{x^3 + x^2 y + 2y^3}{x(x^2 - y^2)}.$ 10. $\dfrac{x-8}{x+8}.$ 11. $\dfrac{x^2}{1-x^4}.$ 12. $\dfrac{a}{1-b}.$

13. $l^4 - \dfrac{1}{l^4}.$ 14. $\dfrac{ab + ac + bc + 2a + 2b + 2c + 3}{abc + ab + ac + bc + a + b + c + 1}.$

15. $\dfrac{1}{a} - \dfrac{b}{ax} - \dfrac{b^2}{a^2 c} - \dfrac{b^2}{ax^2}.$ 18. $\dfrac{8a^2 b^2}{a^4 - b^4}.$ 19. $\dfrac{b(a^2 + b^2)}{a(a^2 - b^2)}.$

20. $\dfrac{a^3 + b^3}{(a-b)^2 . (a^2 + b^2)}.$ 21. $\dfrac{1}{2(x+1)^2}.$ 22. $\dfrac{a+b-c}{a-b+c}.$

23. $x.$ 24. $0.$ 25. $1.$ 26. $\dfrac{(x-4)(x+2)^2}{x}.$

27. $\dfrac{a^4 + a^2 + 1}{a^2}.$ 28. $\dfrac{(x-1)^2}{x^3(x^2+1)^2}.$

29. $\dfrac{x^2}{a^2} + \dfrac{a^2}{x^2}.$ 30. $1.$ 31. $3.$

32. $\dfrac{-2 + 5x + 17x^2 - 11x^3 - 21x^4}{(3 - 2x - 7x^2)^4}.$ 33. $\dfrac{xy}{x^2 + y^2}.$ 34. $2.$

35. $\dfrac{2a-b}{a+b}$.   36. 0.   39. $\dfrac{x^2-y^2}{2(x^2+y^2)}$.   40. $\dfrac{x}{a}$.

41. $x^2+3x+3-\dfrac{3}{x}+\dfrac{1}{x^2}$.   43. $\dfrac{(x^2+y^2)^2}{x^4+y^4}$.   44. 1.

46. $\dfrac{p+q}{p-q}$.   47. $\dfrac{1}{(x^2+1)(x^3+1)}$.   48. 1.

49. $2a^2-ax-ay$.   50. $\dfrac{a+b+c}{a+b-c}$.   51. $(a^3-b^3)^2$.

lxx. (Page 145.)

1. $x=10$
   $y=3.$
2. $x=9$
   $y=7.$
3. $x=8$
   $y=5.$
4. $x=6$
   $y=8.$
5. $x=19$
   $y=2.$
6. $x=5$
   $y=3.$
7. $x=16$
   $y=35.$
8. $x=2$
   $y=1.$
9. $x=4$
   $y=3.$

lxxi. (Page 145.)

1. $x=12$
   $y=4.$
2. $x=9$
   $y=2.$
3. $x=49$
   $y=47.$
4. $x=13$
   $y=3.$
5. $x=40$
   $y=3.$
6. $x=7$
   $y=2.$
7. $x=5$
   $y=1.$
8. $x=6$
   $y=4.$
9. $x=7$
   $y=17.$

lxxii. (Page 146.)

1. $x=23$
   $y=10.$
2. $x=8$
   $y=4.$
3. $x=3$
   $y=2.$
4. $x=5$
   $y=9.$
5. $x=2$
   $y=2.$
6. $x=7$
   $y=9.$
7. $x=12$
   $y=9.$
8. $x=2$
   $y=3.$
9. $x=3$
   $y=20.$

[S.A.]

lxxiii. (Page 147.)

1. $x = 7$
   $y = -2$.
2. $x = 9$
   $y = -3$.
3. $x = 12$
   $y = -3$.
4. $x = -2$
   $y = 19$.
5. $x = -5$
   $y = 14$.
6. $x = -3$
   $y = -2$.
7. $x = 7$
   $y = -5$.
8. $x = \dfrac{1}{2}$
   $y = -\dfrac{1}{3}$.
9. $x = -2$
   $y = 1$.

lxxiv. (Page 148.)

1. $x = 6$
   $y = 12$.
2. $x = 20$
   $y = 30$.
3. $x = 42$
   $y = 35$.
4. $x = 10$
   $y = 5$.
5. $x = 9$
   $y = 140$.
6. $x = 4$
   $y = 9$.
7. $x = 5$
   $y = 2$.
8. $x = 40$
   $y = 60$.
9. $x = 12$
   $y = 6$.
10. $x = 19$
    $y = 3$.
11. $x = 6$
    $y = 12$.
12. $x = \dfrac{3201}{708}$
    $y = \dfrac{278}{59}$.
13. $x = 6$
    $y = 5$.
14. $x = 19\dfrac{1}{2}$
    $y = -17$
15. $x = \dfrac{1}{4}$
    $y = \dfrac{1}{5}$.

lxxv. (Page 149.)

1. $x = \dfrac{eq - nf}{mq - np}$
   $y = \dfrac{mf - ep}{mq - np}$.
2. $x = \dfrac{ce + bf}{bd + ae}$
   $y = \dfrac{cd - af}{bd + ae}$.
3. $x = \dfrac{em + bu}{ae + bc}$
   $y = \dfrac{an - cm}{ae + bc}$.
4. $x = \dfrac{de}{c + d}$
   $y = \dfrac{ce}{c + d}$.
5. $x = \dfrac{n'r + nr'}{mn' + m'n}$
   $y = \dfrac{mr' - m'r}{mn' + m'n}$.
6. $x = \dfrac{a + b}{2}$
   $y = \dfrac{a - b}{2}$.
7. $x = \dfrac{c(f - bc)}{af - bd}$
   $y = \dfrac{c(ac - d)}{af - bd}$.
8. $x = \dfrac{1}{ab}$
   $y = \dfrac{1}{cd}$.
9. $x = \dfrac{2b^2 - 6a^2 + d}{3a}$
   $y = \dfrac{3a^2 - b^2 + d}{3b}$.

ANSWERS. 371

10. $x = \dfrac{a}{bc}$

$y = \dfrac{a+2b}{c}.$

11. $x = \dfrac{a^2}{b+c}$

$y = \dfrac{b^2-c^2}{a}.$

12. $x = \dfrac{bm}{b-m}$

$y = \dfrac{bm}{b+m}.$

lxxvi. (Page 151.)

1. $x = \dfrac{1}{2}$

$y = \dfrac{1}{4}.$

2. $x = \dfrac{1}{b-2a}$

$y = \dfrac{2}{3a-b}.$

3. $x = \dfrac{b^2-a^2}{bd-ac}$

$y = \dfrac{b^2-a^2}{bc-ad}.$

4. $x = \dfrac{2a}{m+n}$

$y = \dfrac{2b}{m-n}.$

5. $x = \dfrac{61}{92}$

$y = \dfrac{61}{103}.$

6. $x = \dfrac{1}{3}$

$y = \dfrac{1}{5}.$

7. $x = \dfrac{1}{a}$

$y = \dfrac{1}{b}.$

8. $x = \dfrac{1}{n}$

$y = \dfrac{1}{m}.$

lxxvii. (Page 153.)

1. $x=1$
$y=2$
$z=3.$

2. $x=2$
$y=2$
$z=2.$

3. $x=4$
$y=5$
$z=8.$

4. $x=5$
$y=6$
$z=8.$

5. $x=1$
$y=2$
$z=3.$

6. $x=1$
$y=4$
$z=6.$

7. $x = \dfrac{2}{3}$
$y = -7$
$z = 36\dfrac{1}{3}.$

8. $x=5$
$y=6$
$z=7.$

9. $x=2$
$y=9$
$z=10.$

10. $x=20$
$y=10$
$z=5.$

lxxviii. (Page 155.)

1. 16, 12.
2. 133, 123.
3. 7·25, 6·25.
4. 31, 23.
5. 35, 14.
6. 30, 40, 50.

7. £60, £140, £200.   8. 22s., 26s.   9. £200, £300, £260.
10. 41, 7.   11. 47, 11.   12. 35, 11, 98.   13. £90, £60.
14. 60, 36.   15. 6, 4.   16. 40, 10.   17. 5·03, 1·072
18. 10 barrels.   19. 3s., 1s. 8d.   20. £20, £10.
21. 15s. 10d., 12s. 6d.   22. 4s. 6d., 3s.   23. 35, 65.
24. 26.   25. 28.   26. 45.   27. 24.   28. 45.
29. 84.   30. 75.   31. 36.   32. 12.   33. 333.
34. 584.   35. 759.   36. $\frac{5}{6}$.   37. $\frac{4}{15}$.   38. $\frac{3}{8}$.
39. $\frac{2}{3}$.   40. $\frac{7}{19}$.   41. $\frac{35}{41}$.   42. $\frac{19}{40}$.
43. £1000.   44. £5000, 6 per cent.   45. £4000, 5 per cent.
46. $31\frac{1}{4}$, $18\frac{3}{4}$.   47. 20, 10.   48. 3 miles an hour.
49. 20 miles, 8 miles an hour.   50. 700.   51. 450, 600.
52. 72, 60.   53. 12, 5s.   54. 750, 158, 148.
55. 15 and 2 miles.   56. The second, 320 strokes.   58. 50, 30.
59. 4 yd. and 5 yd.   60. $\frac{5}{6}$, 6, 4 miles an hour respectively.
61. 142857.

lxxix. (Page 164.)

1. $2xy$.   2. $9a^3b^4$.   3. $11m^5n^6r^7$.   4. $8a^2b^5c$.
5. $267a^2bx^3$.   6. $13a^8b^4c^6$.   7. $\frac{3a}{4b}$.   8. $\frac{1}{2ac^2}$.
9. $\frac{5a^2b^3}{11x^4y^6}$.   10. $\frac{16x^6}{17y^2}$.   11. $\frac{25a}{18b}$.

lxxx. (Page 167.)

1. $2a + 3b$.   2. $4k^5 - 3l^3$.   3. $ab + 81$.   4. $y^3 - 19$.
5. $3abc - 17$.   6. $x^2 - 3x + 5$.   7. $3x^2 + 2x + 1$.

ANSWERS.

8. $2r^2 - 3r + 1$. 9. $2n^2 + n - 2$. 10. $1 - 3x + 2x^2$.
11. $x^3 - 2x^2 + 3x$. 12. $2y^2 - 3yz + 4z^2$. 13. $a + 2b + 3c$.
14. $a^3 + a^2b + ab^2 + b^3$. 15. $x^3 - 2x^2 - 2x - 1$.
16. $2x^2 + 2ax + 4b^2$. 17. $3 - 4x + 7x^2 - 10x^3$.
18. $4a^2 - 5ab + 8bx$. 19. $3a^2 - 4ap^3 - 5t$.
20. $2y^2x - 3yx^2 + 2x^3$. 21. $5x^2y - 3xy^2 + 2y^3$.
22. $4x^2 - 3xy + 2y^2$. 23. $3a - 2b + 4c$. 24. $x^2 - 3x + 5$.
25. $5x - 2y + 3z$. 26. $2x^2 - y + y^2$.

lxxxi. (Page 168.)

1. $2a^3 - \dfrac{ab^2}{4}$. 2. $\dfrac{3}{a} - \dfrac{a}{3}$. 3. $a^2 - \dfrac{1}{a^2}$.

4. $\dfrac{a}{b} + \dfrac{b}{a}$. 5. $x^2 - x + \dfrac{1}{2}$. 6. $x^2 + x - \dfrac{1}{2}$.

7. $2a - 3b + \dfrac{b^2}{4}$. 8. $x^2 + 4 + \dfrac{4}{x^2}$. 9. $\dfrac{4}{3}a^3x + 2a^2 - \dfrac{3}{4}$.

10. $\dfrac{1}{x} - \dfrac{2}{y} + \dfrac{3}{z}$. 11. $6m - \dfrac{4}{n} + \dfrac{p}{5}$.

12. $ab - 3cd + \dfrac{ef}{7}$. 13. $\dfrac{2x}{z} - \dfrac{3y}{z} + \dfrac{z}{x}$.

14. $\dfrac{2m}{n} - 4 - \dfrac{3n}{m}$. 15. $\dfrac{a}{3} - \dfrac{b}{4} + \dfrac{c}{5} - \dfrac{d}{2}$.

16. $7x^2 - 2x - \dfrac{3}{2}$. 17. $3x^2 - \dfrac{ax}{2} + bx$.

18. $3x^2 - \dfrac{x}{3} - 3$.

lxxxii. (Page 170.)

1. $2a$. 2. $3x^2y^2$. 3. $-5mn$. 4. $-6a^4b$.
5. $7b^5c^6$. 6. $-10ab^2c^4$. 7. $-12m^7n^8$. 8. $11a^3b^6$.

lxxxiii. (Page 172.)

1. $a-b$.
2. $2a+1$.
3. $a+8b$.
4. $a+b+c$.
5. $x-y+z$.
6. $3x^2-2x+1$.
7. $1-a+a^2$.
8. $x-y+2z$.
9. a^2-4a+2.
10. $2m^2-3m+1$.
11. $x+2y-z$.
12. $2m-3n-r$.
13. $m+1-\dfrac{1}{m}$.

lxxxiv. (Page 173.)

1. $2a-3x$.
2. $1-2a$.
3. $5+4x$.
4. $a-b$.
5. $x+1$.
6. $m-2$.

lxxxv. (Page 175.)

1. ± 8.
2. $\pm ab$.
3. ± 100.
4. ± 7.
5. $\pm \sqrt{(11)}$.
6. $\pm 8a^2c^3$.
7. ± 6.
8. ± 129.
9. ± 52.
10. ± 4.
11. $\pm \sqrt{\left(\dfrac{q-n}{m}\right)}$.
12. $\pm \sqrt{\left(\dfrac{b}{a-1}\right)}$.
13. $\pm \sqrt{6}$.
14. $\pm 2\sqrt{2}$.

lxxxvi. (Page 179.)

1. $6, -12$.
2. $4, -16$.
3. $1, -15$.
4. $2, -48$.
5. $3, -131$.
6. $5, -13$.
7. $9, -27$.
8. $14, -30$.

lxxxvii. (Page 180.)

1. $7, -1$.
2. $5, -1$.
3. $21, -1$.
4. $9, -7$.
5. $8, 4$.
6. $9, 5$.
7. $118, 116$.
8. $10 \pm 2\sqrt{34}$.
9. $12, 10$.
10. $14, 2$.

lxxxviii. (Page 181.)

1. $3, -10$.
2. $12, -1$.
3. $\dfrac{7}{2}, -\dfrac{25}{2}$.
4. $20, -7$.
5. $\dfrac{1}{4}, -\dfrac{5}{4}$.
6. $9, -8$.
7. $45, -82$.
8. $8, -7$.
9. $4, 15$.
10. $290, 1$.

lxxxix. (Page 182.)

1. $\dfrac{7}{3}, -\dfrac{5}{3}.$ 2. $-\dfrac{1}{5}, -\dfrac{3}{5}.$ 3. $3, \dfrac{1}{9}.$

4. $1, -\dfrac{3}{11}.$ 5. $\dfrac{3}{5}, -\dfrac{5}{7}.$ 6. $4, -\dfrac{4}{5}.$

7. $8, \dfrac{2}{3}.$ 8. $7, -\dfrac{45}{7}.$

xc. (Page 182.)

1. $3, -\dfrac{8}{3}.$ 2. $10, -\dfrac{49}{5}.$ 3. $6, -\dfrac{13}{2}.$

4. $8, -\dfrac{19}{2}.$ 5. $5, -\dfrac{16}{5}.$ 6. $4, \dfrac{3}{2}.$

7. $8, -\dfrac{17}{4}.$ 8. $\dfrac{7}{2}, -\dfrac{3}{14}.$

xci. (Page 184.)

1. $-a \pm \sqrt{2} \cdot a.$ 2. $2a \pm \sqrt{11} \cdot a.$ 3. $\dfrac{m}{2}, -\dfrac{7m}{2}.$

4. $3n, -\dfrac{n}{2}.$ 5. $1, -a.$ 6. $b, -a.$ 7. $\dfrac{a^2+ab}{a-b}, \dfrac{a^2-ab}{a+b}.$

8. $\dfrac{d}{c}, -\dfrac{b}{a}.$ 9. $\dfrac{c+\sqrt{(c^2+4ac)}}{2(a+b)}, \dfrac{c-\sqrt{(c^2+4ac)}}{2(a+b)}.$

10. $\dfrac{b^2}{ac}, \dfrac{b^2}{ac}.$ 11. $\dfrac{2a-b}{ac}, -\dfrac{3a+2b}{bc}.$

12. $-\dfrac{ac^2+bd^2}{2a+3d\sqrt{c}}, -\dfrac{ac^2+bd^2}{2a-3d\sqrt{c}}.$

xcii. (Page 185.)

1. $8, -1.$ 2. $6, -1.$ 3. $12, -1.$ 4. $14, -1.$

5. $2, -9.$ 6. $6, \dfrac{9}{4}.$ 7. $5, 4.$ 8. $4, -1.$ 9. $8, -2.$

10. $3, -\frac{7}{3}$.   11. $7, \frac{1}{3}$.   12. $12, -1$.   13. $14, -1$.

14. $\frac{3}{2}, -\frac{5}{6}$.   15. $13, -\frac{13}{3}$.   16. $5, 4$.   17. $36, 12$.

18. $6, 2$.   19. $\frac{25}{18}, -\frac{5}{3}$.   20. $7, -\frac{10}{7}$.   21. $7, -\frac{10}{7}$.

22. $7, -5$.   23. $3, -\frac{1}{2}$.   24. $\frac{1}{2}, -\frac{2}{3}$.   25. $\frac{2}{3}, -\frac{1}{6}$.

26. $15, -14$.   27. $2, -\frac{1}{3}$.   28. $3, -\frac{11}{4}$.   29. $2, \frac{1}{3}$.

30. $2, -\frac{23}{15}$.   31. $3, -\frac{14}{3}$.   32. $4, -\frac{5}{3}$.   33. $3, \frac{21}{11}$.

34. $14, -10$.   35. $2, \frac{58}{13}$.   36. $5, 2$.   37. $-a, -b$.   38. $-a, b$.

39. $a+b, a-b$.   40. $a^2, -a^3$.   41. $\frac{a}{b}, -\frac{2a}{b}$.   42. $\frac{a}{b}, \frac{b}{a}$.

xciii. (Page 187.)

1. $x=30$ or 10
   $y=10$ or 30.
2. $x=9$ or 4
   $y=4$ or 9.
3. $x=25$ or 4
   $y=4$ or 25.
4. $x=22$ or -3.
   $y=3$ or -22.
5. $x=50$ or -5
   $y=5$ or -50.
6. $x=100$ or -1
   $y=1$ or -100.

xciv. (Page 187.)

1. $x=6$ or -2
   $y=2$ or -6.
2. $x=13$ or -3
   $y=3$ or -13.
3. $x=20$ or -6
   $y=6$ or -20.
4. $x=4$
   $y=4$.
5. $x=10$ or 2
   $y=2$ or 10.
6. $x=40$ or 9
   $y=9$ or 40.

xcv. (Page 188.)

1. $x=4$ or 3
   $y=3$ or 4.
2. $x=5$ or 6
   $y=6$ or 5.
3. $x=10$ or 2
   $y=2$ or 10.
4. $x=4$ or -2
   $y=2$ or -4.
5. $x=5$ or -3.
   $y=3$ or -5.
6. $x=7$ or -4
   $y=4$ or -7.

ANSWERS. 377

xcvi. (Page 189.)

1. $x = 5$ or 4
   $y = 4$ or 5.

2. $x = 4$ or 2
   $y = 2$ or 4.

3. $x = \frac{1}{3}$ or $\frac{1}{2}$
   $y = \frac{1}{2}$ or $\frac{1}{3}$.

4. $x = 3$
   $y = 4$.

5. $x = \frac{1}{3}$
   $y = 2$.

6. $x = \frac{1}{5}$
   $y = \frac{1}{2}$.

xcvii. (Page 191.)

1. $x = 4$ or -3
   $y = 3$ or -4.

2. $x = \pm 6$
   $y = \pm 3$.

3. $x = \pm 10$
   $y = \pm 11$.

4. $x = \pm 8$
   $y = \pm 2$.

5. $x = 5$ or 3
   $y = 3$ or 5.

6. $x = 5$ or $-\frac{95}{28}$
   $y = 2$ or $-\frac{33}{7}$.

7. $x = \pm 2$
   $y = \pm 5$.

8. $x = 6$
   $y = 5$.

9. $x = \pm 2$
   $y = \pm 1$.

10. $x = \pm 2$
    $y = \pm 3$.

11. $x = \pm 7$
    $y = \pm 2$.

12. $x = 3$ or $\frac{11}{6}$
    $y = 2$ or $\frac{7}{6}$.

13. $x = 10$ or 12
    $y = 12$ or 10.

14. $x = 4$ or $\frac{85}{8}$
    $y = 9$ or $\frac{19}{8}$.

15. $x = \pm 9$ or ± 12
    $y = \pm 12$ or ± 9.

xcviii. (Page 193.)

1. 72.   2. 224.   3. 18.   4. 50, 15.   5. 85, 76.
6. 29, 13.   7. 30.   8. 107.   9. 75.   10. 20, 6.
11. 18, 1.   12. 17, 15.   13. 12, 4.   14. 1296.   15. $56\frac{1}{4}$.
16. 2601.   17. 6, 4.   18. 12, 5.   19. 12, 7.   20. 1, 2, 3.
21. 7, 8.   22. 15, 16.   23. 10, 11, 12.   24. 12.   25. 16.
26. £2, 5s.   27. 12.   28. 6.   29. 75.   30. 5 and 7 hours.
31. 101 yds. and 100 yds.   32. 63.   33. 63 ft., 45 ft.
34. 16 yds., 2 yds.   35. 37.   36. 100.   37. 1975.

xcix. (Page 199.)

1. $x=3$
   $y=2$.
2. $x=5$
   $y=3$.
3. $x=90, 71, 52$...down to 14
   $y=0, 13, 26$ ...... up to 52.
4. $x=7, 2$
   $y=1, 4$.
5. $x=3, 8, 13$...
   $y=7, 21, 35$...
6. $x=91, 76, 61$...down to 1.
   $y=2, 13, 24$ ...... up to 68
7. $x=0, 7, 14, 21, 28$
   $y=44, 33, 22, 11, 0$.
8. $x=20, 39$...
   $y=3, 7$...
9. $x=40, 49$...
   $y=13, 33$...
10. $x=4, 11$...up to 123
    $y=53, 50$...down to 2.
11. $x=2$
    $y=0$.
12. $x=92, 83$....2
    $y=1, 8$...71.
13. $\dfrac{4}{7}$ and $\dfrac{3}{9}$.
14. $\dfrac{8}{11}$ and $\dfrac{2}{13}$.
15. 3 ways, viz. 12, 7, 2; 2, 6, 10.
16. 7.
17. 12, 57, 102...
18. 3.
19. 2.
21. 19 oxen, 1 sheep and 80 hens. There is but one other solution, that is, in the case where he bought no oxen, and no hens, and 100 sheep.
22. A gives B 11 sixpences, and B gives A 2 fourpenny pieces.
23. 2, 106, 27.
24. 3.
25. A gives 6 sovereigns and receives 28 dollars.
26. 22, 3; 16, 9; 10, 15; 4, 21.
27. 5.
28. 56, 44.
29. 82, 18; 47, 53; 12, 88.
30. 301.

c. (Page 205.)

(1) 1. $x^{\frac{5}{2}} + x^{\frac{2}{3}} + x^{\frac{7}{2}}$.
    2. $x^{\frac{2}{3}}y + x^2 y^{\frac{2}{5}} + x^{\frac{2}{7}} y^{\frac{2}{7}}$.
    3. $a^{\frac{4}{3}} + a^{\frac{5}{3}} + a^{\frac{5}{2}}$.
    4. $x^{\frac{1}{3}} y z^{\frac{2}{3}} + a^{\frac{1}{2}} y^{\frac{3}{4}} z + a^{\frac{1}{5}} y z^{\frac{2}{5}}$.

(2) 1. $x^{-1} + a x^{-2} + b^2 x^{-3} + 3 x^{-4}$.
    2. $x^2 y^{-2} + 3 x y^{-3} + 4 y^{-4}$.
    3. $\dfrac{x^3 y^{-2} z^{-2}}{4} + \dfrac{5 x^2 y^{-1} z^{-3}}{7} + x y^{-1} z^{-1}$.
    4. $\dfrac{x y z^{-2}}{3} + \dfrac{x^{-2} y^{-3}}{5} + x^{-3} y^{-4} z$.

(3) 1. $\dfrac{1}{a} + \dfrac{1}{a x} + \dfrac{1}{b^2 x^{-3}} + \dfrac{1}{3 x^{-4}}$.
    2. $\dfrac{1}{x^2 y^{-2}} + \dfrac{1}{3 x y^{-3}} + \dfrac{1}{5 y^{-5}}$.

3. $\dfrac{4}{a^{-2}b^{-2}c^3} + \dfrac{3}{a^{-1}b^{\frac{1}{2}}c^{\frac{1}{2}}} + \dfrac{1}{x^{-\frac{1}{3}}y}$.

4. $\dfrac{1}{3x^{-\frac{1}{4}}y^{-\frac{1}{4}}z} + \dfrac{1}{a^{-1}b^{-\frac{2}{3}}c^2} + \dfrac{1}{a^{-2}b^{-1}c}$.

(4) 1. $2\sqrt[3]{c^2} + 3\sqrt[3]{(xy^2)} + \dfrac{1}{xy^2}$.   2. $\dfrac{1}{\sqrt[3]{x}} + \dfrac{1}{\sqrt[3]{y^2}} + \dfrac{1}{z^3}$.

3. $\dfrac{\sqrt[3]{y^2}}{\sqrt[3]{x}} + \dfrac{3\sqrt[4]{y^3}}{x^2} + \dfrac{\sqrt[3]{y}}{3\sqrt[3]{x^2}}$.   4. $\dfrac{1}{x^2\sqrt[3]{y}} + \dfrac{y}{\sqrt[3]{x}} + \dfrac{\sqrt[3]{y}}{\sqrt[3]{x^2}}$.

ci. (Page 206.)

1. $x^{4p} + x^{2p}y^{2p} + y^{4p}$.   2. $a^{4m} - 81y^{4n}$.

3. $x^{8d} + 4a^2x^{4d} + 16a^4$.   4. $a^{2m} + 2a^m c^r - b^{2n} + c^{2r}$.

5. $2a^{2n} + 2a^m b^n - 4a^m c^r - a^m b - b^{n+1} + 2bc^r + a^m c^2 + b^n c^2 - 2c^{r+2}$.

6. $x^{mn} + x^{mn-n} \cdot y^{mn-n} - x^n y^m - y^{mn-n+m}$.   7. $x^{4n} + x^{2n}y^{2n} + y^{4n}$.

8. $a^{2p^2} - a^{p^2-p}b^{p^2} + a^{p^2-p}c^p + a^{p^2+p} \cdot b^{1-p^2} - b + b^{1-p^2}c^p + a^{p^2+p}c^{1-p} - b^{p^2}c^{1-p} + c$.

9. $x^{4p} + 2x^{3p} + 3x^{2p} + 2x^p + 1$.   10. $x^{4p} - 2x^{3p} + 3x^{2p} - 2x^p + 1$.

cii. (Page 207.)

1. $x^{3m} + x^{2m}y^m + x^m y^{2m} + y^{3m}$.

2. $x^{4n} - x^{3n}y^n + x^{2n}y^{2n} - x^n y^{3n} + y^{4n}$.

3. $x^{5r} + x^{4r}y^r + x^{3r}y^{2r} + x^{2r}y^{3r} + x^r y^{4r} + y^{5r}$.

4. $a^{12p} - a^{9p}b^{2q} + a^{6p}b^{4q} - a^{3p}b^{6q} + b^{8q}$.

5. $x^{4d} + 3x^{3d} + 9x^{2d} + 27x^d + 81$.

6. $a^{2m} - 2a^m x^n + 4x^{2n}$.   7. $2 - x^p + 3x^{2p}$.

8. $4b^m c^m - 5b^{2m}$.

9. $a^{3m} + 3a^{2m} + 3a^m + 1$.   10. $a^m + b^n + c^r$.

ciii. (Page 208.)

1. $x - 3x^{\frac{2}{3}} + 3x^{\frac{1}{3}} - 1$.
2. $y - 1$.
3. $a^2 - x^2$.
4. $a + b + c - 3a^{\frac{1}{3}}b^{\frac{1}{3}}c^{\frac{1}{3}}$.
5. $10x - 11x^{\frac{3}{4}}y^{\frac{1}{4}} + 5x^{\frac{1}{4}}y^{\frac{3}{4}} - 21y$.
6. $m - n$.
7. $m^{\frac{4}{3}} + 4d^{\frac{2}{3}}m^{\frac{2}{3}} + 16d$.
8. $16a + 8a^{\frac{6}{7}}b^{\frac{1}{7}} + 10a^{\frac{5}{7}}b^{\frac{2}{7}} + 18a^{\frac{4}{7}}b^{\frac{3}{7}} - 24a^{\frac{3}{7}}b^{\frac{4}{7}} - 12a^{\frac{2}{7}}b^{\frac{5}{7}} - 15a^{\frac{1}{7}}b^{\frac{6}{7}} - 27b$.
9. $x^{\frac{2}{3}} + 2a^{\frac{1}{3}}x^{\frac{1}{3}} + a^{\frac{2}{3}}$.
10. $x^{\frac{2}{3}} - 2a^{\frac{1}{3}}x^{\frac{1}{3}} + a^{\frac{2}{3}}$.
11. $x^{\frac{4}{5}} + 2x^{\frac{2}{5}}y^{\frac{2}{5}} + y^{\frac{4}{5}}$.
12. $a^2 + 2ab^{\frac{1}{4}} + b^{\frac{1}{2}}$.
13. $x - 4x^{\frac{3}{4}} + 10x^{\frac{1}{2}} - 12x^{\frac{1}{4}} + 9$.
14. $4x^{\frac{4}{7}} + 12x^{\frac{3}{7}} + 25x^{\frac{2}{7}} + 24x^{\frac{1}{7}} + 16$.
15. $x^{\frac{2}{3}} - 2x^{\frac{1}{3}}y^{\frac{1}{3}} + 2x^{\frac{1}{3}}z^{\frac{1}{3}} + y^{\frac{2}{3}} - 2y^{\frac{1}{3}}z^{\frac{1}{3}} + z^{\frac{2}{3}}$.
16. $x^{\frac{1}{2}} + 4x^{\frac{1}{4}}y^{\frac{1}{4}} - 2x^{\frac{1}{4}}z^{\frac{1}{4}} + 4y^{\frac{1}{2}} - 4y^{\frac{1}{4}}z^{\frac{1}{4}} + z^{\frac{1}{2}}$.

civ. (Page 209.)

1. $x^{\frac{1}{2}} + y^{\frac{1}{2}}$.
2. $a^{\frac{1}{2}} - b^{\frac{1}{2}}$.
3. $x^{\frac{2}{3}} + x^{\frac{1}{3}}y^{\frac{1}{3}} + y^{\frac{2}{3}}$.
4. $a^{\frac{2}{3}} - a^{\frac{1}{3}}b^{\frac{1}{3}} + b^{\frac{2}{3}}$.
5. $x^{\frac{4}{5}} - x^{\frac{3}{5}}y^{\frac{1}{5}} + x^{\frac{2}{5}}y^{\frac{2}{5}} - x^{\frac{1}{5}}y^{\frac{3}{5}} + y^{\frac{4}{5}}$.
6. $m^{\frac{5}{6}} + m^{\frac{3}{3}}n^{\frac{1}{6}} + m^{\frac{2}{3}}n^{\frac{1}{3}} + m^{\frac{1}{3}}n^{\frac{2}{3}} + m^{\frac{1}{6}}n^{\frac{2}{3}} + n^{\frac{5}{6}}$.
7. $x^{\frac{3}{4}} + 3x^{\frac{1}{2}}y^{\frac{1}{4}} + 9x^{\frac{1}{4}}y^{\frac{1}{2}} + 27y^{\frac{3}{4}}$.
8. $27a^{\frac{3}{4}} + 18a^{\frac{1}{2}}b^{\frac{1}{4}} + 12a^{\frac{1}{4}}b^{\frac{1}{2}} + 8b^{\frac{3}{4}}$.
9. $a^{\frac{1}{2}} - x^{\frac{1}{2}}$.
10. $m^{\frac{4}{5}} + 3m^{\frac{3}{5}} + 9m^{\frac{2}{5}} + 27m^{\frac{1}{5}} + 81$.
11. $x^{\frac{1}{2}} + 10$.
12. $x^{\frac{1}{3}} + 4$.
13. $-b + 2b^{\frac{2}{3}} - b^{\frac{1}{3}}$.
14. $x^{\frac{2}{3}} - x^{\frac{1}{3}}y^{\frac{1}{3}} - x^{\frac{1}{3}}z^{\frac{1}{3}} + y^{\frac{2}{3}} + z^{\frac{2}{3}} - y^{\frac{1}{3}}z^{\frac{1}{3}}$.
15. $x^{\frac{2}{3}} - 9x^{\frac{1}{3}} - 10$.
16. $m^{\frac{1}{2}} + m^{\frac{1}{4}}n^{\frac{1}{4}} + n^{\frac{1}{2}}$.
17. $p^{\frac{1}{2}} - 2p^{\frac{1}{4}} + 1$.
18. $x^{\frac{1}{2}} - y^{\frac{1}{2}} - z^{\frac{1}{2}}$.
19. $x^{\frac{1}{3}} + y^{\frac{1}{3}}$.

cv. (Page 210.)

1. $a^{-2} - b^{-2}$.
2. $x^{-6} - b^{-4}$.
3. $x^4 - x^{-4}$.
4. $x^4 + 1 + x^{-4}$.
5. $a^{-4} - b^{-4}$.
6. $a^{-2} + 2a^{-1}c^{-1} - b^{-2} + c^{-2}$.
7. $1 + a^2b^{-2} + a^4b^{-4}$.
8. $a^4b^{-4} - a^{-4}b^4 - 4a^{-2}b^2 - 4$.
9. $4x^{-5} - x^{-4} + 3x^{-3} + 2x^{-2} + x^{-1} + 1$.
10. $5x^{-4} + \dfrac{7x^{-3}}{2} - \dfrac{107x^{-2}}{12} + \dfrac{5x^{-1}}{6} + \dfrac{7}{6}$.

cvi. (Page 211.)

1. $x - x^{-1}$.
2. $a + b^{-1}$.
3. $m^2 - mn^{-1} + n^{-2}$.
4. $c^4 + c^3d^{-1} + c^2d^{-2} + cd^{-3} + d^{-4}$.
5. $xy^{-1} + x^{-1}y$.
6. $a^{-2} + a^{-1}b^{-1} + b^{-2}$.
7. $x^2y^{-2} - 2 + x^{-2}y^2$.
8. $\dfrac{3}{2}x^{-3} - 5x^{-2} + \dfrac{1}{4}x^{-1} + 9$.
9. $a^2b^{-2} - 1 + a^{-2}b^2$.
10. $a^{-2} - a^{-1}b^{-1} - a^{-1}c^{-1} + b^{-2} - b^{-1}c^{-1} + c^{-2}$.

cvii. (Page 211.)

1. $x^{\frac{2}{3}} - 2x^{\frac{1}{2}}y^{\frac{1}{2}} + 2y$.
2. $x^{\frac{15ab+12}{3a-2}}$.
3. $x^{\frac{10b+18a}{3a-2}}$.
4. $-\dfrac{2a}{(x^4 - a^4)^{\frac{1}{2}}}$.
5. $7x^{-4} + \dfrac{22}{3}x^{-3} - \dfrac{421}{42}x^{-2} - \dfrac{10}{7}x^{-1} + \dfrac{1}{7}$.
6. x^a.
7. $x^n - y^n$.
8. $a^2 + 2a^{\frac{3}{2}}b^{\frac{3}{5}} - 2a^{\frac{1}{2}}b^{\frac{9}{5}} - b^{\frac{12}{5}}$.
9. $a^{\frac{2}{3}} + a^{\frac{1}{3}}b^{\frac{1}{3}} + b^{\frac{2}{3}}$.
11. $m = n^{\frac{1}{n-1}}$.
12. $x^{2a+2b+2c}$.
13. x^{pq}.
14. $16a^{2x}$.
15. a^{2mnp}.
16. $2a^{2m} + 2a \, b^n - 4a \, c^n - 3a \, b - 3b^{p+1} + 6bc^n$.
17. c.

19. $x^{\frac{4}{3}} + x^{\frac{2}{3}} + 1$.

20. $a^{m+n} + 2a^{m+n-3} \cdot bcx^3 - a^{m+n-2} b^2 x^2 - a^{m+n-4} c^2 x^4$.

21. $x^{p(q-1)} - y^{q(p-1)}$.   22. a^{m-1}.   23. $x^{4r} - y^{4p}$.

24. $5, \dfrac{1}{144}$.   25. $x^{mn} - x^n y^{(n-1)m} - x^{(m-1)n} y^m + y^{mn}$.

26. $x + 3x^{\frac{3}{4}} - 2x^{\frac{1}{2}} - 7x^{\frac{1}{4}} + 2x^{-\frac{1}{4}}$.

cviii. (Page 215.)

1. $\sqrt[6]{x^3}, \sqrt[6]{y^2}$.   2. $\sqrt[10]{(1024)}, \sqrt[10]{8}$.

3. $\sqrt[6]{(5832)}, \sqrt[6]{(2500)}$.   4. $\sqrt[mn]{2^n}, \sqrt[mn]{2^m}$.   5. $\sqrt[mn]{a^n}, \sqrt[mn]{b^m}$.

6. $\sqrt[6]{(a^2 + 2ab + b^2)}, \sqrt[6]{(a^3 - 3a^2 b + 3ab^2 - b^3)}$.

cix. (Page 217.)

1. $2\sqrt{6}$.   2. $5\sqrt{2}$.   3. $2a\sqrt{a}$.   4. $5a^2 d \sqrt{(5d)}$.

5. $4z\sqrt{(2yz)}$.   6. $10\sqrt{(10a)}$.   7. $12c\sqrt{5}$.

8. $42\sqrt{(11x)}$.   9. $6x\sqrt{\dfrac{5x}{3}}$.   10. $a^2 \sqrt{\dfrac{a}{b}}$.

11. $(a+x)\cdot\sqrt{a}$.   12. $(x-y)\sqrt{x}$.   13. $5(a-b)\cdot\sqrt{2}$.

14. $(3c^2 - y)\cdot\sqrt{(7y)}$.   15. $3a^2 \sqrt[3]{(2b^2)}$.

16. $2xy^2 \cdot \sqrt[3]{(20xy)}$.   17. $3m^3 n^3 \sqrt[3]{(4n)}$.

18. $7a^5 b^5 \sqrt[3]{(4b)}$.   19. $(x+y)\cdot\sqrt[3]{x}$.   20. $(a-b)\cdot\sqrt[3]{a}$.

cx. (Page 217.)

1. $\sqrt{(48)}$.   2. $\sqrt{(63)}$.   3. $\sqrt[3]{(1125)}$.   4. $\sqrt[3]{(96)}$.

5. $\sqrt[3]{\dfrac{81}{7}}$.   6. $\sqrt{(9a)}$.   7. $\sqrt{(48a^2 x)}$.   8. $\sqrt{(3a^3 x)}$.

9. $\sqrt{(m^2 - n^2)}$.   10. $\left(\dfrac{a+b}{a-b}\right)^{\frac{3}{2}}$.   11. $\left(\dfrac{x}{x+y}\right)^{\frac{1}{2}}$.

ANSWERS. 383

cxi. (Page 218.)

The numbers are here arranged in order, the highest on the *left* hand.

1. $\sqrt{3}, \sqrt[3]{4}$. 2. $\sqrt{10}, \sqrt[3]{15}$. 3. $3\sqrt{2}, 2\sqrt{3}$.
4. $\sqrt[3]{\left(\frac{14}{15}\right)}, \sqrt{\frac{3}{5}}$. 5. $3\sqrt{7}, 4\sqrt{3}$.
6. $2\sqrt{87}, 3\sqrt{33}$. 7. $3\sqrt[3]{7}, 4\sqrt{2}, 2\sqrt[3]{22}$.
8. $5\sqrt[3]{18}, 3\sqrt{19}, 3\sqrt[3]{82}$. 9. $5\sqrt[3]{2}, 2\sqrt[3]{14}, 3\sqrt[3]{3}$.
10. $\frac{1}{2}\sqrt{2}, \frac{1}{3}\sqrt{3}, \frac{1}{4}\sqrt{4}$.

cxii. (Page 219.)

1. $29\sqrt{3}$. 2. $30\sqrt{10} + 164\sqrt{2}$. 3. $(a^2 + b^2 + c^2)\sqrt{c}$.
4. $13\sqrt[3]{2}$. 5. $33\sqrt[3]{2}$. 6. $\sqrt{6}$. 7. $5\sqrt{3}$.
8. $48\sqrt{2}$. 9. $4\sqrt[3]{2}$. 10. 0. 11. $4\sqrt{3}$.
12. $2\sqrt{(70)}$. 13. 100. 14. $3ab$. 15. $2ab\sqrt[3]{(12b)}$.
16. 2. 17. $\frac{3}{5}$. 18. $\sqrt[3]{\frac{a}{b}}$. 19. $\sqrt{\frac{a}{b}}$.
20. $\sqrt{\frac{x}{1+xy}}$.

cxiii. (Page 220.)

1. $\sqrt{(xy)}$. 2. $\sqrt{(xy-y^2)}$. 3. $x+y$. 4. $\sqrt{(x^2-y^2)}$.
5. $18x$. 6. $56(x+1)$. 7. $90\sqrt{(c^2-x)}$. 8. $2c\sqrt{3}$.
9. $-x$. 10. $1-x$. 11. $-12x$. 12. $6a$.
13. $-\sqrt{(x^2-7x)}$. 14. $6\sqrt{(x^2+7x)}$. 15. $8(a^2-1)$.
16. $-6a^2 + 12a - 18$.

cxiv. (Page 221.)

1. $x + 9\sqrt{x} + 14$. 2. $x - 2\sqrt{x} - 15$. 3. a.
4. $a - 53$. 5. $3x + 5\sqrt{x} - 28$. 6. $6x - 54$. 7. 6.
8. $\sqrt{(9x^2+3x)} + \sqrt{(6x^2-3x)} - \sqrt{(6x^2-x-1)} - 2x + 1$.

9. $\sqrt{(ax)} + \sqrt{(ax - x^2)} - \sqrt{(a^2 - ax)} - a + x$.
10. $3 + x + \sqrt{(3x + x^2)}$.
11. $x - y + z + 2\sqrt{xz}$.
12. $2x + 2\sqrt{(ax)}$.
13. $432 + 42\sqrt{(x^2 - 9)} + x^2$.
14. $2x + 11 + 2\sqrt{(x^2 + 11x + 24)}$.
15. $2x - 4 + 2\sqrt{(x^2 - 4x)}$.
16. $2x - 6 + 2\sqrt{(x^2 - 6x)}$.
17. $4x + 9 - 12\sqrt{x}$.
18. $2x - 2\sqrt{(x^2 - y^2)}$.
19. $x^2 + 2x - 1 - 2\sqrt{(x^3 - x)}$.
20. $x^2 + 1 + 2\sqrt{(x^3 - x)}$.

CXV. (Page 222.)

1. $(\sqrt{c} + \sqrt{d})(\sqrt{c} - \sqrt{d})$.
2. $(c + \sqrt{d})(c - \sqrt{d})$.
3. $(\sqrt{c} + d)(\sqrt{c} - d)$.
4. $(1 + \sqrt{y})(1 - \sqrt{y})$.
5. $(1 + \sqrt{3}.x)(1 - \sqrt{3}.x)$.
6. $(\sqrt{5}.m + 1)(\sqrt{5}.m - 1)$.
7. $\{2a + \sqrt{(3x)}\}\{2a - \sqrt{(3x)}\}$.
8. $\{3 + 2\sqrt{(2n)}\}\{3 - 2\sqrt{(2n)}\}$.
9. $\{\sqrt{(11)}.n + 4\}\{\sqrt{(11)}.n - 4\}$.
10. $(p + 2\sqrt{r})(p - 2\sqrt{r})$.
11. $(\sqrt{p} + \sqrt{3}.q)(\sqrt{p} - \sqrt{3}.q)$.
12. $\{a^m + b^{\frac{n}{2}}\}\{a^m - b^{\frac{n}{2}}\}$.
13. $\dfrac{a + \sqrt{b}}{a^2 - b}$.
14. $\dfrac{a + \sqrt{(ab)}}{a - b}$.
15. $24 + 17\sqrt{2}$.
16. $2 + \sqrt{2}$.
17. $3 + 2\sqrt{3}$.
18. $3 - 2\sqrt{2}$.
19. $\dfrac{a + x + 2\sqrt{(ax)}}{a - x}$.
20. $\dfrac{1 + x + 2\sqrt{x}}{1 - x}$.
21. $\dfrac{a + \sqrt{(a^2 - x^2)}}{x}$.
22. $m^2 - \sqrt{(m^4 - 1)}$.
23. $2a^2 - 1 + 2a\sqrt{(a^2 - 1)}$.
24. $\dfrac{2a^2 - x^2 + 2a\sqrt{(a^2 - x^2)}}{x^2}$.

CXVI. (Page 224.)

1. 19.
2. 11.
3. $8 - 26\sqrt{(-1)}$.
4. $5 + 4\sqrt{3}$.
5. $2b + 2\sqrt{(ab)} - 12a$.
6. $a^2 + a$.
7. $b^3 - a^3$.
8. $a^2 + \beta^2$.
9. c^2.
10. $e^{2p\sqrt{(-1)}} - e^{-2p\sqrt{(-1)}}$.

1. $\dfrac{x+y}{3\sqrt{(xy)}}.$ 3. $\dfrac{x+y}{2\sqrt{(xy)}}.$ 5. $x^2 - \sqrt{2}.ax + a^2.$

6. $m^2 + \sqrt{2}.mn + n^2.$ 7. $2x\sqrt{x}.$ 8. $\dfrac{2a\sqrt{b} - 2b\sqrt{a}}{a-b}.$

9. $\dfrac{a^2c}{b} + cd - 2ac\sqrt{\dfrac{d}{b}}.$ 10. $a^2\sqrt{2} - 2 + \dfrac{1}{a^2\sqrt{2}}.$

11. $\dfrac{2x^2}{a^2}.$ 12. $\sqrt{(1-x)}.$ 13. $\dfrac{x-1}{\sqrt{x}}.$

14. $\dfrac{x}{2} - 2\sqrt{\left(\dfrac{x^2}{16} - 9\right)}.$ 15. $2x - 2\sqrt{(x^2 - a^2)}.$

16. $a^2b^6c.$ 17. $-1 + 5a^2(2-a^2) + a(10a^2 \; a^4 - 5)\sqrt{(-1)}.$

18. $8 + 7\sqrt[3]{3}.$ 19. $4\sqrt{(3cx)}.$ 20. $x\sqrt[3]{(3y^2)}.$

21. $\dfrac{1}{n}\sqrt[3]{(-4n^2)}.$ 22. $(9n-10).\sqrt{7}.$ 23. $0.$

cxviii. (Page 228.)

1. $\sqrt{7} + \sqrt{3}.$ 2. $\sqrt{11} + \sqrt{5}.$ 3. $\sqrt{7} - \sqrt{2}.$ 4. $7 - 3\sqrt{5}.$

5. $\sqrt{10} - \sqrt{3}.$ 6. $2\sqrt{5} - 3\sqrt{2}.$ 7. $2\sqrt{3} - \sqrt{2}.$ 8. $3\sqrt{11} - 2.$

9. $3\sqrt{7} - 2\sqrt{3}.$ 10. $3\sqrt{7} - 2\sqrt{6}.$ 11. $\dfrac{1}{2}(\sqrt{10} - 2).$ 12. $3\sqrt{5} - 2\sqrt{3}.$

cxix. (Page 229.)

1. 49. 2. 81. 3. 25. 4. 8. 5. 27. 6. 256.

7. 27. 8. 56. 9. 79. 10. 153. 11. 6. 12. 36.

13. 12. 14. $\dfrac{(a-b)^2}{c^2}.$ 15. 5. 16. 6.

17. 3. 18. 10. 19. $\dfrac{b-a^2}{3a}.$ 20. $\dfrac{n-m^2}{13m}.$

[s.a.]

cxx. (Page 231.)

1. 9. 2. 25. 3. 49. 4. 121. 5. $1\frac{4}{9}$. 6. 8, 0.

7. 0, −8. 8. $\left(\frac{a+1}{2}\right)^2$. 9. $\left(\frac{m+4}{4}\right)^2$. 10. 5.

cxxi. (Page 231.)

1. 25. 2. 25. 3. 9. 4. 64. 5. $\frac{36}{5}$.

6. $\frac{12a}{5}$. 7. a. 8. $\frac{1}{4}$ or 0. 9. 64. 10. 100.

cxxii. (Page 232.)

1. 16, 1. 2. 81, 25. 3. 3, $2\frac{5}{9}$. 4. 10, −13.

5. 5, $\frac{5}{9}$. 6. −4, −32. 7. 9, $-3\frac{3}{5}$. 8. 28, $-\frac{12252}{529}$.

9. 49. 10. 729. 11. 4, −21. 12. 1 or $\frac{1}{21}$. 13. ±24.

14. 5 or 221. 15. 5 or $\frac{145}{121}$. 16. 5 or 0. 17. $\frac{25}{36}$. 18. 25.

19. ±9√2. 20. ±√65 or ±√5. 21. $2a$.

22. $-2a$. 23. $\frac{1}{2}$ or $-4\frac{1}{6}$. 24. $\frac{1}{4}$. 25. $\frac{1}{12}$.

26. $\frac{1276}{81}$. 27. $\frac{36}{5}$. 28. ±5 or ±3√2. 29. ±14.

30. 6 or $-\frac{10}{9}$. 31. 1. 32. $\frac{5}{4}$. 33. 2 or 0. 34. 0 or $\frac{9a}{16}$.

cxxiii. (Page 235.)

1. 2, 5. 2. 3, −7. 3. −9, −2. 4. $5a, 6b$.

5. $-\frac{7}{2}, \frac{5}{3}$. 6. $\frac{227}{19}, -\frac{83}{14}$. 7. $\frac{4m}{5}, \frac{11n}{6}$.

ANSWERS. 387

8. $-2a, -3a$ and $3a, 4a$.   9. $\pm 2, a$.   10. $0, 5$.

11. $\dfrac{2a-b}{ac}, \dfrac{b-3a}{bc}$.   12. $\dfrac{d}{c}, \dfrac{e}{c}$.

cxxv. (Page 239.)

1. $x^2 - 11x + 30 = 0$.   2. $x^2 + x - 20 = 0$.   3. $x^2 + 9x + 14 = 0$.
4. $6x^2 - 7x + 2 = 0$.   5. $9x^2 - 58x - 35 = 0$.   6. $x^2 - 3 = 0$.
7. $x^2 - 2mx + m^2 - n^2 = 0$.   8. $x^2 - \dfrac{\alpha + \beta}{\alpha\beta}x + \dfrac{1}{\alpha\beta} = 0$.
9. $x^2 + \dfrac{\alpha^2 - \beta^2}{\alpha\beta}x - 1 = 0$.

cxxvi. (Page 240.)

1. $(x-2)(x-3)(x-6)$.   2. $(x-1)(x-2)(x-4)$.
3. $(x-10)(x+1)(x+4)$.   4. $4(x+1)\left(x+\dfrac{1-\sqrt{5}}{4}\right)\left(x+\dfrac{1+\sqrt{5}}{4}\right)$.
5. $(x+2)(x+1)(6x-7)$.
6. $(x+y+z)(x^2+y^2+z^2-xy-xz-yz)$.
7. $(a-b-c)(a^2+b^2+c^2+ab+ac-bc)$.
8. $(x-1)(x+3)(3x-7)$.   9. $(x-1)(x-4)(2x+5)$.
10. $(x+1)(3x+7)(5x-3)$.

cxxvii. (Page 242.)

1. $\sqrt{13}$ or $\sqrt{-1}$.   2. $\sqrt[3]{-2}$ or $\sqrt[3]{-12}$.   3. $\sqrt[4]{-1}$ or $\sqrt[4]{-21}$.
4. 1 or $\sqrt[m]{-4}$.   5. $\sqrt[2n]{\dfrac{5}{2}}$ or $\sqrt[2n]{-\dfrac{5}{6}}$.   6. 25 or $\dfrac{1}{4}$.
7. $\dfrac{6}{-9 \pm \sqrt{97}}$.   8. $\left(\dfrac{1}{5}\right)^{\frac{1}{n}}$ or $\left(-\dfrac{1}{4}\right)^{\frac{1}{n}}$.   9. 1 or $1 \pm 2\sqrt{15}$.
10. 3 or $-\dfrac{1}{2}$ or $\dfrac{5 \pm \sqrt{1329}}{4}$.

11. $a+2$, or $-\dfrac{a+6}{3}$, or $\dfrac{a\pm 2\sqrt{(a^2-3a)}}{3}$.

12. 0, or a, or $\dfrac{a\pm\sqrt{(a^2-16a+16)}}{2}$.

cxxviii. (Page 245.)

1. $6:7,\ 7:9,\ 2:3$. 2. The second is the greater.
3. The second is the greater.
4. $\dfrac{ad-bc}{c-d}$. 5. $10:9$ or $9:10$.

cxxix. (Page 246.)

1. $2:3$. 2. $b:a$. 3. $b+d:a-c$. 4. $\pm\sqrt{6}-1:1$.
5. $13:1$, or, $-1:1$. 6. $\pm\sqrt{(m^2+4n^2)}-m:2$. 7. $6,8$.
8. $12,14$. 9. $35,65$. 10. $13,11$. 11. $4:1$. 12. $1:5$.

cxxx. (Page 247.)

1. $\dfrac{8}{15}$. 2. $\dfrac{8}{9}$. 3. $\dfrac{x-y}{x+y}$. 4. $\dfrac{a-b+c}{a-b-c}$.
5. $\dfrac{m^2-mn+n^2}{m^2+mn+n^2}$. 6. $\dfrac{(x+2)y}{(y-4)x}$.

cxxxii. (Page 255.)

6. $x=4$ or 0. 8. 440 yds. and 352 yds. per minute.
11. $x=30,\ y=20$. 13. $\dfrac{b^2}{d}$. 15. $\dfrac{9}{41}$.
16. 50, 75 and 80 yards. 17. 120, 160, 200 yards.
19. $1\tfrac{1}{3}$ miles per hour. 20. $1:7$.
21. 160 quarters, £2. 22. £80. 23. £60.
24. £20. 25. $90:79$. 26. 45 miles and 30 miles.

ANSWERS. 389

cxxxiii. (Page 262.)

4. $16\frac{4}{5}$. 5. 5. 6. 12. 7. $3\frac{3}{14}$. 8. $\frac{2}{5}$.

9. $A \propto C^{\frac{2}{3}}$. 10. 5. 11. $A = \frac{2}{3}B$. 12. $64c^2 = 9y^3$.

13. $x^2 = \frac{108}{y^3}$. 14. $4x^3 = 27y^2$. 18. $y = 3 + 2x + x^2$. 19. 18 ft.

cxxxiv. (Page 266.)

1. 50. 2. 200. 3. $10\frac{3}{4}$. 4. $-32\frac{1}{2}$.

5. $-2\frac{5}{6}$. 6. 40. 7. 117. 8. 0.

9. $x^2 + y^2 - 2(n-2)xy$. 10. $\dfrac{3an - 2bn - 2a + b}{a+b}$.

cxxxv. (Page 268.)

1. 5050. 2. 2550. 3. 820. 4. 30.

5. 24. 6. $-31\frac{1}{6}$. 7. $\dfrac{n \cdot (n+1)}{2}$. 8. $\dfrac{3n^2 - n}{2}$.

$\dfrac{7n^2 - 5n}{2}$. 10. $\dfrac{n-1}{2}$.

cxxxvi. (Page 269.)

1. -6. 2. $-\dfrac{x}{25}$. 3. $\dfrac{1}{8}$. 4. $-\dfrac{7}{8}$.

5. -2. 6. $-1\dfrac{2}{3}$.

cxxxvii. (Page 269.)

1. (1) -46. (2) $3b - 2$. (3) $\dfrac{2}{5}$. (4) 4·4.

2. 155. 3. 112. 4. 888. 5. 100.

6. $6433\frac{1}{3}$.   7. £135. 4s.

8. (1) 355, 7175.   (2) $-156a^2$, $-3110a$.

   (3) $161+81x$, $3321+1681x$.   (4) $119\frac{1}{2}$, $2357\frac{1}{2}$.

   (5) $8\frac{1}{4}$, $174\frac{1}{4}$.

9. (1) 126, 63252.   (2) 25, 2250.

   (3) $45, -1570·5x$.   (4) $99, -1163\frac{1}{4}$.

   (5) 71, $4899(1-m)$.   (6) $65, 65x+8190$.

cxxxviii. (Page 271.)

1. 6, 9, 12, 15.   2. $1\frac{1}{3}$, $\frac{2}{3}$, 0, $-\frac{2}{3}$, $-1\frac{1}{3}$.

3. $2\frac{5}{12}$, $1\frac{5}{6}$, $1\frac{1}{4}$.   4. $\frac{7}{15}$, $\frac{13}{30}$, $\frac{2}{5}$, $\frac{11}{30}$.

cxxxix. (Page 272.)

1. $\frac{3m+n}{4}$, $\frac{m+n}{2}$, $\frac{m+3n}{4}$.

2. $\frac{5m+3}{5}$, $\frac{5m+1}{5}$, $\frac{5m-1}{5}$, $\frac{5m-3}{5}$.

3. $\frac{5n^2+1}{5}$, $\frac{5n^2+2}{5}$, $\frac{5n^2+3}{5}$, $\frac{5n^2+4}{5}$.

4. $\frac{2x^2+y^2}{2}$, x^2, $\frac{2x^2-y^2}{2}$.

cxl. (Page 275.)

1. 64.   2. 78732.   3. 327680.   4. $\frac{1}{2048}$.

5. 13122.   6. 16384.   7. $-\frac{1}{96}$.

cxli. (Page 276.)

1. 65534.　　　2. 364.　　　3. $\dfrac{a(x^{26}-1)}{x^2-1}$.

4. $\dfrac{a(x^9-1)}{x^8(x-1)}$.　　5. $\dfrac{(a-x)\{1-(a+x)^7\}}{(a+x)^5 \cdot (1-a-x)}$.　　6. 3^n-1.

7. $7(2^n-1)$.　　8. -425.　　9. $-\dfrac{43}{96}$.

cxlii. (Page 278.)

1. 2.　　2. $\dfrac{4}{3}$.　　3. $\dfrac{27}{8}$.　　4. $\dfrac{4}{3}$.　　5. $1\dfrac{1}{8}$.

6. 3.　　7. $8\dfrac{8}{11}$.　　8. $2\dfrac{1}{4}$.　　9. $85\dfrac{1}{3}$.　　10. $\dfrac{16x^5}{8x^2+1}$.

11. $\dfrac{a^2}{a-b}$.　　12. $\dfrac{1}{9}$.　　13. $\dfrac{x^2}{x+y}$.　　14. $\dfrac{86}{99}$.

15. $\dfrac{49}{90}$.　　　　16. $\dfrac{46}{55}$.

cxliii. (Page 279.)

1. 9, 27, 81.　　2. 4, 16, 64, 256.　　3. 2, 4, 8.

4. $\dfrac{3}{4}, \dfrac{9}{8}, \dfrac{27}{16}, \dfrac{81}{32}$.

cxliv. (Page 279.)

1. (1) 558.　　(2) 800.　　(3) $\dfrac{18}{5}$.　　(4) $\dfrac{16}{9}$.

(5) $-\dfrac{169}{2}$.　　(6) $\dfrac{133}{486}$.　　(7) $-\dfrac{1189}{2}$.　　(8) $13\dfrac{5}{7}$.

(9) 1.　　(10) -84.　　(11) $-\dfrac{9999\sqrt{3}}{(\sqrt{10}+1)\cdot\sqrt{5}}$.

(12) $-\dfrac{3157}{80}$.

5. 42　　6. $ac=b^2$.　　7. ± 1.　　8. $n+\dfrac{1}{4n}$.

ANSWERS.

9. 4.   10. 10.   13. 4.   14. 642.

16. 49, 1.   17. $3\frac{1}{2}$, 6, $8\frac{1}{2}$.   18. 60.

19. $\frac{4}{5}, \frac{3}{5}, \frac{2}{5}, \frac{1}{5}, 0, -\frac{1}{5}, -\frac{2}{5}, -\frac{3}{5}, -\frac{4}{5}$.

22. 3, 7, 11, 15, 19.   23. 5, 15, 45, 135, 405.

25. 139.   26. 10 per cent.

cxlv. (Page 285.)

1. 8, 12.   2. $\frac{15}{7}, \frac{30}{13}, \frac{5}{2}, \frac{30}{11}$.   3. $\frac{12}{29}, \frac{6}{11}, \frac{4}{5}$.

4. $\frac{1}{6}, \frac{1}{9}, \frac{1}{12}, \frac{1}{15}$.   5. $-2, \infty, 2, 1, \frac{2}{3}$.

6. $\frac{3}{4}, \frac{3}{2}, \infty, -\frac{3}{2}, -\frac{3}{4}$.   7. $\frac{6}{5}, \frac{3}{4}, \frac{6}{11}, \frac{3}{7}, \frac{6}{17}, \frac{3}{10}$.

8. $\dfrac{6xy(n+1)}{3ny+2x}, \dfrac{6xy(n+1)}{3ny+4x-3y}, \ldots\ldots, \dfrac{6xy(n+1)}{2nx+3y}$.

9. $-\frac{1}{4}, -\frac{1}{2}, \infty, \frac{1}{2}, \frac{1}{4}, \frac{1}{6}$, or $\frac{5}{31}, \frac{5}{24}, \frac{5}{17}, \frac{1}{2}, \frac{5}{3}, -\frac{5}{4}$.

10. 104, 234.   13. 2, 3, 6.

cxlvi. (Page 290.)

1. 132.   2. 3360.   3. 116280.   4. 6720.

5. $\dfrac{\lfloor 11}{8}$.   6. 40320.   7. 3628800.   8. 125.   9. 2520.

10. 6.   11. 4.   12. 120.   13. 1260.

14. 2520, 6720, 5040, 1663200, 34650.

cxlvii. (Page 295.)

1. 3921225.   2. 6.   3. 126.   4. 116280.

5. 12.   6. 12.   7. 816000.   8. 3353011200.

9. 7.   10. 63.   11. 52.   12. 123200.   13. 376992 : 52360

cxlviii. (Page 300.)

1. $a^4 + 4a^3x + 6a^2x^2 + 4ax^3 + x^4$.
2. $b^6 + 6b^5c + 15b^4c^2 + 20b^3c^3 + 15b^2c^4 + 6bc^5 + c^6$.
3. $a^7 + 7a^6b + 21a^5b^2 + 35a^4b^3 + 35a^3b^4 + 21a^2b^5 + 7ab^6 + b^7$.
4. $x^8 + 8x^7y + 28x^6y^2 + 56x^5y^3 + 70x^4y^4 + 56x^3y^5 + 28x^2y^6 + 8xy^7 + y^8$.
5. $625 + 2000a + 2400a^2 + 1280a^3 + 256a^4$.
6. $a^{10} + 5a^8bc + 10a^6b^2c^2 + 10a^4b^3c^3 + 5a^2b^4c^4 + b^5c^5$.

cxlix. (Page 301.)

1. $a^6 - 6a^5x + 15a^4x^2 - 20a^3x^3 + 15a^2x^4 - 6ax^5 + x^6$.
2. $b^7 - 7b^6c + 21b^5c^2 - 35b^4c^3 + 35b^3c^4 - 21b^2c^5 + 7bc^6 - c^7$.
3. $32x^5 - 240x^4y + 720x^3y^2 - 1080x^2y^3 + 810xy^4 - 243y^5$.
4. $1 - 10x + 40x^2 - 80x^3 + 80x^4 - 32x^5$.
5. $1 - 10x + 45x^2 - 120x^3 + 210x^4 - 252x^5 + 210x^6 - 120x^7 + 45x^8 - 10x^9 + x^{10}$.
6. $a^{24} - 8a^{21}b^2 + 28a^{18}b^4 - 56a^{15}b^6 + 70a^{12}b^8 - 56a^9b^{10} + 28a^6b^{12} - 8a^3b^{14} + b^{16}$.

cl. (Page 302.)

1. $a^3 + 6a^2b - 3a^2c + 12ab^2 - 12abc + 3ac^2 + 8b^3 - 12b^2c + 6bc^2 - c^3$.
2. $1 - 6x + 21x^2 - 44x^3 + 63x^4 - 54x^5 + 27x^6$.
3. $x^9 - 3x^8 + 6x^7 - 7x^6 + 6x^5 - 3x^4 + x^3$.
4. $27x + 54x^{\frac{5}{6}} + 63x^{\frac{2}{3}} + 44x^{\frac{1}{2}} + 21x^{\frac{1}{3}} + 6x^{\frac{1}{6}} + 1$.
5. $x^3 + 3x^2 - 5 + \dfrac{3}{x^2} - \dfrac{1}{x^3}$.
6. $a^{\frac{3}{4}} + b^{\frac{3}{4}} - c^{\frac{3}{4}} + 3a^{\frac{1}{2}}b^{\frac{1}{4}} + 3a^{\frac{1}{4}}b^{\frac{1}{2}} - 3a^{\frac{1}{2}}c^{\frac{1}{4}} - 3b^{\frac{1}{2}}c^{\frac{1}{4}} + 3a^{\frac{1}{4}}c^{\frac{1}{2}} + 3b^{\frac{1}{4}}c^{\frac{1}{2}} - 6a^{\frac{1}{4}}b^{\frac{1}{4}}c^{\frac{1}{4}}$.

cli. (Page 303.)

1. $330x^7$.   2. $495a^{10}b^8$.   3. $-161700a^{97}b^3$.
4. $192192a^6b^6c^8d^8$.   5. $12870a^8b^8$.
6. $70a^{\frac{1}{2}}b^{\frac{1}{2}}$.   7. $-92378a^{10}b^9$ and $92378a^9b^{10}$.
8. $1716a^7x^6$ and $1716a^6x^7$.

clii. (Page 311.)

1. $1 + \frac{1}{2}x - \frac{1}{8}x^2 + \frac{1}{16}x^3 - \frac{5}{128}x^4$.

2. $1 + \frac{2a}{3} - \frac{a^2}{9} + \frac{4a^3}{81}$.

3. $a^{\frac{1}{3}} + \frac{x}{3a^{\frac{2}{3}}} - \frac{x^2}{9a^{\frac{5}{3}}} + \frac{5x^3}{81a^{\frac{8}{3}}} - \frac{10x^4}{243a^{\frac{11}{3}}}$.

4. $1 + x - \frac{1}{2}x^2 + \frac{1}{2}x^3 - \frac{5}{8}x^4$.

5. $a^{\frac{3}{4}} + a^{-\frac{1}{4}}x - \frac{1}{6}a^{-\frac{5}{4}}x^2 + \frac{5}{54}a^{-\frac{9}{4}} \cdot x^3$.

6. $a^{\frac{1}{5}} + \frac{4}{5} \cdot a^{-\frac{1}{20}}x^{\frac{1}{4}} - \frac{2}{25} \cdot a^{-\frac{3}{10}}x^{\frac{1}{2}} + \frac{4}{125} \cdot a^{-\frac{11}{20}}x^{\frac{3}{4}}$.

7. $1 - \frac{x^2}{2} - \frac{x^4}{8} - \frac{x^6}{16} - \frac{5x^8}{128}$.

8. $1 - \frac{7}{3}a^2 + \frac{14}{9}a^4 - \frac{14}{81}a^6$.

9. $1 - \frac{9x}{4} - \frac{27x^2}{32} - \frac{135}{128} \cdot x^3$.

10. $x^3 - xy + \frac{y^2}{6x} + \frac{y^3}{54c^3}$.

cliii. (Page 312.)

1. $1 - 2a + 3a^2 - 4a^3 + 5a^4$.
2. $1 + 3x + 9x^2 + 27x^3 + 81x^4$.
3. $1 + x + \dfrac{5}{8}x^2 + \dfrac{5}{16}x^3$.
4. $1 + x + \dfrac{3x^2}{4} + \dfrac{x^3}{2} + \dfrac{5x^4}{16}$.
5. $a^{-10} + 10a^{-12}x + 60a^{-14}x^2 + 280a^{-16}x^3 + 1120a^{-18}x^4$.
6. $\dfrac{1}{a^2} + \dfrac{6x^{\frac{1}{3}}}{a^{\frac{7}{3}}} + \dfrac{21x^{\frac{2}{3}}}{a^{\frac{8}{3}}} + \dfrac{56x}{a^3}$.

cliv. (Page 313.)

1. $1 - \dfrac{x^2}{2} + \dfrac{3x^4}{8} - \dfrac{5x^6}{16} + \dfrac{35x^8}{128}$.
2. $1 + \dfrac{3x^2}{2} + \dfrac{15x^4}{8} + \dfrac{35x^6}{16} + \dfrac{315x^8}{128}$.
3. $x^{-2} - \dfrac{2}{5}x^{-7}z^5 + \dfrac{7}{25}x^{-12}z^{10} - \dfrac{28}{125}x^{-17}z^{15}$.
4. $1 - x + \dfrac{3x^2}{2} - \dfrac{5x^3}{2} + \dfrac{35x^4}{8}$.
5. $\dfrac{1}{a} - \dfrac{x^2}{2a^3} + \dfrac{3x^4}{8a^5} - \dfrac{5x^6}{16a^7}$.
6. $\dfrac{1}{a} - \dfrac{x^3}{3a^4} + \dfrac{2x^6}{9a^7} - \dfrac{14x^9}{81a^{10}}$.

clv. (Page 314.)

1. $\dfrac{7 \cdot 6 \ldots (9-r)}{1 \cdot 2 \ldots (r-1)} \cdot x^{r-1}$.
2. $(-1)^{r-1} \cdot \dfrac{12 \cdot 11 \ldots (14-r)}{1 \cdot 2 \ldots (r-1)} \cdot x^{r-1}$.
3. $(-1)^{r-1} \cdot \dfrac{8 \cdot 7 \ldots (10-r)}{1 \cdot 2 \ldots (r-1)} \cdot a^{9-r} \cdot x^{r-1}$.

4. $\dfrac{9.8\ldots(11-r)}{1.2\ldots(r-1)} \cdot (5x)^{10-r} \cdot (2y)^{r-1}$.   5. $(-1)^{r-1} \cdot r \cdot x^{r-1}$.

6. $\dfrac{r \cdot (r+1) \cdot (r+2)}{6} \cdot (3x)^{r-1}$.   7. $\dfrac{1.3.5\ldots(2r-3)}{1.2.3\ldots(r-1)} \cdot \left(\dfrac{x}{2}\right)^{r-1}$.

8. $\dfrac{1.2.5\ldots(3r-7)}{1.2.3\ldots(r-1)} \cdot \left(-\dfrac{x}{3a}\right)^{r-1} \cdot a^{\frac{1}{3}}$.

9. $\dfrac{7.9.11\ldots(2r+3)}{1.2.3\ldots(r-1)} \cdot x^{r-1}$.

10. $\dfrac{a^{-\frac{3}{2}}}{4^{r-1}} \cdot \dfrac{3.7.11\ldots(4r-5)}{1.2.3\ldots(r-1)} \cdot \left(\dfrac{x}{a}\right)^{2(r-1)}$.

11. $\dfrac{(r+1)(r+2)}{2} \cdot x^r$.   12. $\dfrac{1.3.5\ldots(2r-1)}{1.2.3\ldots r} \cdot (2x)^r$.

13. $\dfrac{1.3.5\ldots(2r-1)}{1.2.3\ldots r} \cdot (2x)^r$.   15. $\dfrac{5}{16} \cdot \dfrac{1}{a^{\frac{7}{2}} \cdot a^3}$.

16. $\dfrac{3}{128} \cdot a^{-5} b^3$.   17. $-\dfrac{429}{128} \cdot \dfrac{x^{16}}{a^{15}}$.

18. $-\dfrac{m \cdot (m+1)\ldots(m+8)}{1.2\ldots 9} \cdot a^{-(m+9)} \cdot b^9$.

19. $\dfrac{(1-5m)(1-4m)\ldots(1-m)}{1.2\ldots 6m^6} \cdot a^{\frac{1}{m}-6}$.

clvi. (Page 315.)

1. $3\cdot 14137\ldots$ 2. $1\cdot 95204\ldots$
3. $3\cdot 04084\ldots$ 4. $1\cdot 98734\ldots$

clvii. (Page 319.)

1. 1045032.   2. 10070344.   3. 80451.
4. 31134.   5. 51117344.   6. 14332216.
7. 31450 and remainder 2.   8. 522256 and remainder 1.
9. 4112.   10. 2437.

ANSWERS. 397

clviii. (Page 321.)

1. 5221. 2. 12232. 3. 2139e. 4. 104300.
5. 1110111001111. 6. ttteе. 7. 6500445.
8. 211021. 9. 6t12. 10. 814. 11. 61415.
12. 123130. 13. 16430335. 14. 27t.

clix. (Page 327.)

1. ·41. 2. ·16̇2355043̇. 3. 25·1̇.
4. 12232·20052̇. 5. Senary. 6. Octonary.

clx. (Page 336.)

1. $\bar{1}$·2187180. 2. $\bar{7}$·7074922. 3. 2·4036784.
4. 4·740378. 5. 2·924059. 6. $\bar{3}$·724833.
7. $\bar{5}$·3790163. 8. $\overline{40}$·578098. 9. $\overline{62}$·9905319.
10. 2·1241803. 11. $\bar{3}$·738827. 12. 1·61514132.

clxi. (Page 339.)

1. 2·1072100 ; 2·0969100 ; 3·3979400.
2. 1·6989700 ; $\bar{3}$·6989700 ; 2·2922560.
3. ·7781513 ; 1·4313639 ; 1·7323939 ; 2·7604226.
4. 1·7781513 ; 2·4771213 ; ·0211893 ; $\bar{5}$·6354839.
5. $\bar{4}$·8750613 ; 1·4983106.
6. ·3010300 ; $\bar{2}$·8061800 ; ·2916000.
7. ·6989700 ; $\bar{1}$·0969100 ; 3·3910733.
8. $-2, 0, 2 : 1, 0, -1$.
9. (1) 3. (2) 2. 10. $x = \dfrac{9}{2}, y = \dfrac{3}{2}$.

11. (a) ·3010300; 1·3979400; 1·9201233; $\bar{1}$·9970588. (b) 103.

12. (a) ·6989700; ·6020600; 1·7118072; $\bar{1}$·9880618.

    (b) 8.

13. 3·8821260; $\bar{1}$·4093694; $\bar{3}$·7455326.

14. (1) $x = \dfrac{1}{6}$. (2) $x = 2$. (3) $x = \dfrac{\log m}{\log a + \log b}$.

    (4) $x = \dfrac{\log c}{m \log a + 2 \log b}$.

    (5) $x = \dfrac{4 \log b + \log c}{2 \log c + \log b - 3 \log a}$.

    (6) $x = \dfrac{\log c}{\log a + m \log b + 3 \log c}$.

clxii. (Page 343.)

1. 17·6 ...... years.
2. 23·4 ...... years.
3. 7·2725 years nearly.
4. 22·5 years nearly.
6. 12 years nearly.
7. 11·724 ...... years.

APPENDIX.

The following papers are from those set at the Matriculation Examinations of Toronto, Victoria, and McGill Universities, and at the Examinations for Second Class Provincial Certificates for Ontario.

UNIVERSITY OF TORONTO.

Junior Matric., 1872   *Pass.*

1. Multiply $\frac{1}{3} x^2 - \frac{1}{4} xy + y^2$ by $\frac{1}{3} x^2 + \frac{1}{4} xy - y^2$.

   Divide $a^4 - 81b^4$ by $a \pm 3b$ and $(x+a)^3 - (y-b)^3$ by $x + a - y + b$.

2. What quantity subtracted from $x^2 + px + q$ will make the remainder exactly divisible by $x - a$?

   Shew that
   $(a + b + c)^3 - (a + b + c)(a^2 + b^2 + c^2 - ab - bc - ca) - 3abc = 3(a+b)(b+c)(c+a)$.

3. Solve the following equations:

   (a) $\frac{1}{3}(2x - 3) + \frac{1}{4}(6x - 7) = \frac{1}{6}(x - \frac{1}{2})$.

   (b) $\dfrac{4x - 7}{\frac{1}{2}x - 1} + \dfrac{3x - 5}{\frac{1}{4}x - 2} = 20$.

   (c) $\dfrac{1}{x - 3} - \dfrac{1}{x - 4} = \dfrac{1}{x - 5} - \dfrac{1}{x - 6}$.

   (d) $x + \dfrac{y + \frac{2}{3}}{2} = 1, \; \dfrac{y}{3} + \dfrac{x + 2}{5} = \dfrac{11}{18}$.

4. In a certain constituency are 1,300 voters, and two candidates, A and B. A is elected by a

certain majority. But the election having been declared void, in the second contest (A and B being again the candidates), B is elected by a majority of 10 more than A's majority in the first election; find the number of votes polled for each in the second election; having given that, the number of votes polled for B in the first case : number polled in the second case : : 43 : 44.

Junior Matric., 1872.   *Pass and Honor.*

1. Multiply $x + y + z^{\frac{1}{2}} - 2y^{\frac{1}{2}} z^{\frac{1}{2}} + 2z^{\frac{1}{2}} x^{\frac{1}{2}} - 2x^{\frac{1}{2}} y^{\frac{1}{2}}$ by $x + y + z^{\frac{1}{2}} + 2y^{\frac{1}{2}} z^{\frac{1}{2}} - 2z^{\frac{1}{2}} x^{\frac{1}{2}} - 2x^{\frac{1}{2}} y^{\frac{1}{2}}$, and divide $a^3 + 8b^3 + 27 c^3 - 18abc$ by $a^2 + 4b^2 + 9 c^2 - 2ab - 3ac - 6bc$.

2. Investigate a rule for finding the H. C. D. of two algebraical expressions.

If $x + c$ be the H. C. D. of $x^2 + px + q$, and $x^2 + p'x + q'$, show that
$$(q - q')^2 - p(q - q')(p - p') + q(p - p')^2 = 0.$$

3. Shew how to find the square root of a binomial, one of whose terms is rational and the other a quadratic surd. What is the condition that the result may be more simple than the indicated square root of the given binomial? Does the reasoning apply if one of the terms is imaginary? Show that $\sqrt{-4m^2} = \sqrt{m} + \sqrt{-m}$.

4. Shew how to solve the quadratic equation $ax^2 + bx + c = 0$, and discuss the results of giving different values to the coefficients.

If the roots of the above equation be as p to q show that $\dfrac{b^2}{ac} = \dfrac{(p+q)^2}{}$.

5. Solve the equations

(a) $\dfrac{x}{2} + \sqrt{x^2 + 3x - 3} = 14\tfrac{1}{3} - \dfrac{2x^2 + 3x}{6}$.

(b) $x^2 - 3xy + 2y^2 + 1 = 0$
$xy + y^2 - 10 = 0$.

(c) $\dfrac{x^2 + 6x + 2}{x^2 + 6x + 4} - \dfrac{x^2 + 6x + 6}{x^2 + 6x + 8} = \dfrac{x^2 + 6x + 4}{x^2 + 6x + 6} - \dfrac{x^2 + 6x + 8}{x^2 + 6x + 10}$.

(d) $6x^4 - 5x^3 - 38x^2 - 5x + 6 = 0$

6. Shew how to find the sum of n terms of a geometric series. What is meant by the sum of an infinite series? When can such a series be said to have a sum?

Sum to infinity the series $1 + 2r + 3r^2 + \&c.$, and find the series of which the sum of n terms is $a^p \dfrac{a^{nr} - 1}{a - 1}$.

7. Find the condition that the equations
$ax + by - cz = 0$.
$a_1 x + b_1 y - c_1 z = 0$.
$a_2 x + b_2 y - c_2 z = 0$.
may be satisfied by the same values of x, y, z.

8. A number of persons were engaged to do a piece of work which would have occupied them m hours if they had commenced at the same time; instead of doing so, they commenced at equal intervals, and then continued to work till the whole was finished, the payments being proportional to the work done by each; the first comer received r times as much as the last: find the time occupied.

Junior Matric., 1872. Honor.

1. There are three towns, A, B, and C; the road from B to A forming a right angle with that from B to C. A person travels a certain distance from B towards A, and then crosses by the nearest way to the road leading from C to A, and finds himself three miles from A and seven from C. Arriving at A, he finds he has gone farther by one-fourth of the distance from B to C than he would have done had he not left the direct road. Required the distance of B from A and C.

2. If $\dfrac{ay+bx}{c} = \dfrac{cx+az}{b} = \dfrac{bz+cy}{a}$, then will

$$\dfrac{x}{a} \cdot \dfrac{y}{b} \cdot \dfrac{z}{c}$$
$$\overline{b^2+c^2-a^2} = \overline{c^2+a^2-b^2} = \overline{a^2+b^2-c^2}.$$

3. Solve the equations $x^2 - yz = a^2$, $y^2 - zx = b^2$, $z^2 - xy = c^2$.

4. If a, b, and c be positive quantities, shew that
$$a^2(b+c) + b^2(c+a) + c^2(a+b) > 6abc.$$

5. Find the values of x and y from the equations
$$2y + \dfrac{5y+3}{x} = 1,$$
$$x^2 + 5x + y(y-1) = 24.$$

6. A steamer made the trip from St. John to Boston via Yarmouth in 33 hours; on her return she made two miles an hour less between Boston and Yarmouth, but resumed her former speed between the latter place and St. John, thereby making the entire return passage in $\frac{32}{35}$ of the time she would have required had her diminished speed lasted throughout; had she made her usual time between Boston and Yarmouth, and two miles an hour less between Yarmouth and

St. John, her return trip would have been made in $\frac{17}{17}$ of the time she would have taken had the whole of her return trip been made at the diminished rate. Find the distance between St. John and Yarmouth, and between the latter place and Boston.

Junior Matric., Honor.
Senior Matric., Pass. } 1874.

1. Solve the following equations:

   (a) $\begin{cases} x^3 - 2xy + 2y^2 = xy. \\ x^2 + xy + y^2 = 63. \end{cases}$

   (b) $\begin{cases} 4x - 3xy = 171. \\ 3y - 4xy = 150. \end{cases}$

   (c) $\begin{cases} \dfrac{1}{x^2} + \dfrac{1}{xy} + \dfrac{1}{y^2} = 19. \\ \dfrac{1}{x^4} + \dfrac{1}{x^2y^2} + \dfrac{1}{y^4} = 133. \end{cases}$

And find one solution of the equations:

   (d) $\begin{cases} y^3 - x^4 = 68. \\ x^2 + \sqrt{x} = y. \end{cases}$

2. Find a number whose cube exceeds six times the next greater number by three.

3. Explain the meaning of the terms Highest common measure and Lowest common multiple as applied to algebraical quantities, and prove the rule for finding the Highest common measure of two quantities.

4. Reduce to their lowest terms the following fractions:

   (a) $\begin{cases} 99x^4 + 117x^3 - 257x^2 - 325x - 50 \\ 3x^3 + 4x^2 - 9x - 10. \end{cases}$

   (b) $\begin{cases} x^4 + 10x^3 + 35x^2 + 50x + 24 \\ x^4 + 18x^3 + 119x^2 + 312x + 360 \end{cases}$

5. Find the sum of n terms of the series $-\frac{1}{2}, \frac{1}{3}, -\frac{1}{4}, \&c$, and the xth term of the series

$$\frac{x+1}{x-1}, \frac{2}{x-1}, \frac{3-x}{x-1}, \&c.$$

6. Find the relations between the roots and coefficients of the equation $ax^2 + px + q = 0$.

Solve the equation

$$x^4 + 6x^3 + 10x^2 + 3x = 110.$$

7. A cask contains 15 gallons of a mixture of wine and water, which is poured into a second cask containing wine and water in the proportion of two of the former to one of the latter, and in the resulting mixture the wine and water are found to be equal. Had the quantity in the second cask originally been only one-half of what it was, the resulting mixture would have been in the proportion of seven of wine to eight of water. Find the quantity in the second cask.

8. What rate per cent. per annum, payable half-yearly, is equivalent to ten per cent. per annum, payable yearly.

9. A is engaged to do a piece of work and is to receive $3 for every day he works, but is to forfeit one dollar for the first day he is absent, two for the second, three for the third, and so on. Sixteen days elapse before he finishes the work and he receives $26. Find the number of days he is absent.

Change the enunciation of this problem so as to apply to the negative solution.

Junior Matric., 1876. *Pass.*

1. Explain the use of negative and fractional indices in Algebra.

Multiply $\frac{1}{\sqrt{a}}$ by $\sqrt[6]{a^7}$, and the product by $\sqrt[12]{a}$.

Simplify $\dfrac{a^m b^n c d^2}{a^n b^2 c^2 d}$, writing the factors all in one line.

2. Multiply together $a^2 + ax + x^2$, $a + x$, $a^2 - ax + x^2$, $a - x$, and divide the product by $a^3 - x^3$.

3. Divide 1 by $1 - 2x + x^2$ to six terms, and give the remainder. Also divide $27x^4 - 6x^2 + \tfrac{1}{3}$ by $3x^2 + 2x + \tfrac{1}{3}$.

4. Multiply $a^{+\frac{m+n}{}} + b^{\frac{m-n}{}}$ by $a^{-\frac{m}{}} + b^{+\frac{n}{}}$

5. Solve the equations:

(1). $\dfrac{3x+4}{5} - \dfrac{7x-3}{2} = \dfrac{x-16}{4}$.

(2). $\begin{cases} x(y+z) = 24, \\ y(z+x) = 45, \\ z(x+y) = 49. \end{cases}$

Junior Matric., 1876. *Honor.*

1. An oarsman finds that during the first half of the time of rowing over any course he rows at the rate of five miles an hour, and during the second half, at the rate of four and a half miles. His course is up and down a stream which flows at the rate of three miles an hour, and he finds that by going down the stream first, and up afterwards, it takes him one hour longer to go over the course than by going first up and then down. Find the length of the course.

2. Shew that if a^2, b^2, c^2 be in *A.P.*, then will $b + c$, $c + a$, $a + b$ be in *H.P.*

Also, if a, b, c be in *A.P.*, then will

$$a + \dfrac{bc}{b+c},\ b + \dfrac{ca}{c+a},\ c + \dfrac{ab}{a+b}$$

be in *H.P.*

3. If $s = a+b+c$, then
$$\sqrt{(as+bc)(bs+ac)(cs+ab)} = (s-a)(s-b)(s-c)$$

4. If $a_1 + a_2 + \ldots + a_n = \dfrac{ns}{2}$, then
$$(s-a_1)^2 + \ldots + (s-a_n)^2 = a_1^2 + a_2^2 + \ldots + a_n^2.$$

5. If the fraction $\dfrac{1}{2n+1}$, when reduced to a repetend, contains $2n$ figures, shew how to infer the last n digits after obtaining the first n.

Find the value of $\frac{1}{17}$ by dividing to 8 digits.

6. Solve the equations
$$\left.\begin{array}{l} x - y + z = 3, \\ xy + xz = 2 + yz, \\ x^2 + y^2 + z^2 = 29. \end{array}\right\}$$

Junior Matric., 1876.   Honor.

1. Shew that the method of finding the square root of a number is analagous to that of finding the square root of an algebraic quantity.

Fencing of given length is placed in the form of a rectangle, so as to include the greatest possible area, which is found to be 10 acres. The shape of the field is then altered, but still remains a rectangle, and it is found that with 162 yards more fencing, the same area as before may be enclosed. Find the sides of the latter rectangle.

2. Prove the rule for finding the Lowest Common Multiple of two compound algebraic quantities.

Find the L.C.M. of $a^3 - b^3 + c^3 + 3abc$ and $a^2(b+c) - b^2(c+a) + c^2(a+b) + abc$.

3. If α, β be the roots of the equation $x^2 + px + q = 0$, shew that the equation may be thrown into the form $(x-\alpha)(x-\beta) = 0$.

$3 + \sqrt{2}$ is a root of the equation $x^4 - 5x^3 + 2x^2 + x + 7 = 0$: find the other roots.

4. (1) Shew how to extract the square root of a binomial, one of whose terms is rational and the other a quadratic surd.

(2) Find a factor which will rationalize $x^{\frac{1}{3}} - y^{\frac{1}{4}}$.

5. a, b are the first two terms of an $H.P.$, what is the nth term?

If a, l, c be in $H.P.$, shew that
$$b^2(a-c)^2 = 2c^2(b-a)^2 + 2a^2(c-b)^2.$$

6. A and B are to race from M to N and back. A moves at the rate of 10 miles an hour, and gets a start of 20 minutes. On A's returning from N, he meets B moving towards it, and one mile from it; but A is overtaken by B when one mile from M. Find the distance from M to N.

7. Solve the equations

(1). $x^3 + 8 = 2x^2 + 11x + 14$

(2). $\begin{cases} \dfrac{x}{y} = \dfrac{51}{4} - xy, \\ \dfrac{y}{x} = \dfrac{17}{12} - \dfrac{1}{xy}. \end{cases}$

Second Class Certificates, 1873.

1. Multiply $\dfrac{a}{b} + \dfrac{b}{a} + 1$ by $\dfrac{a}{b} + \dfrac{b}{a} - 1$.

2. Shew that $\dfrac{a^2 - 3ab + 2b^2}{a - 2b} - \dfrac{a^2 - 7ab + 12b^2}{a - 3b}$

can be reduced to the form $3b$.

3. Reduce to its lowest terms the fraction,
$$\frac{x^4 + \frac{5x^2}{12} + \frac{1}{9}}{x^4 - x^3 + \frac{x^2}{4} - \frac{1}{9}}.$$

4. (a) Prove that $x^m - y^m$ is divisible by $x - y$ without remainder, when m is any positive integer.

(b) Is there a remainder when $x^{100} - 100$ is divided by $x - 1$? If so, write it down.

5. Given $ax + by = 1$,

and $\dfrac{x}{a} + \dfrac{y}{b} = \dfrac{1}{ab}$.

Find the difference between x and y.

6. Given $3 - \dfrac{7\{3x - 2(m - \frac{1}{2})\}}{8(x-1)} - \dfrac{\frac{8}{6}(x-4)}{3(x+1)} = 0$.

Find x in terms of m.

7. Given $\dfrac{x}{y} = \dfrac{2}{3}$. Find the value of $\dfrac{7x + 16}{7y + 24}$.

8. Given $\dfrac{2}{x-y} - \dfrac{5}{x+y} = 1$,

and $\dfrac{5}{x-y} - \dfrac{10}{x+y} = 3$. Find x and y.

9. There is a number of two digits. By inverting the digits we obtain a number which is less by 8 than three times the original number; but if we increase each of the digits of the original number by unity, and invert the digits thus augmented, a number is obtained which exceeds the original number by 29. Find the number.

10. A student takes a certain number of minutes to walk from his residence to the Normal School. Were the distance $\frac{1}{8}$th of a mile greater, he would need to increase his pace (number of miles per hour)

by $\frac{4}{5}$ of a mile in the hour, in order to reach the school in the same time. Find how much he would have to diminish his pace in order still to reach the school in exactly the same time, if the distance were $\frac{3}{55}$ of a mile less than it is.

Second Class Certificates, 1875.

1. Find the continued product of the expressions, $a+b+c$, $c+a-b$, $b+c-a$, $a+b-c$.

2. Simplify $\dfrac{a^3+a^2b}{a^2b-b^3} - \dfrac{a(a-b)}{b(a+b)} - \dfrac{2ab}{a^2-b^2}$.

3. Find the Lowest Common Multiple of $3x^2-2x-1$ and $4x^3-2x^2-3x+1$.

4. Find the value of x from the equation, $\dfrac{a^2-3bx}{a} - ab^2 = bx + \dfrac{6bx-5a^2}{2a} - \dfrac{bx+4a}{4}$.

5. Solve the simultaneous equations,
$$\frac{a}{x}+\frac{b}{y} = in,$$
$$\frac{c}{x}+\frac{d}{y} = n.$$

6. In the immediately preceding question, if a pupil should say that, when $nb = nd$, and $bc = ad$, the values of x and y obtained in the ordinary method, have the form $\frac{0}{0}$, and that he does not know how to interpret such a result, what would you reply?

7. Two travellers set out on a journey, one with $100, the other with $48; they meet with robbers, who take from the first twice as much as they take from the second; and what remains with the first is 3 times that which remains with the second. How much money did each traveller lose?

8. A and B labor together on a piece of work for two days; and then B finishes the work by himself in 8 days; but A, with half of the assistance that B could render, would have finished the work in 6 days. In what time could each of them do the whole work alone?

9. P and Q are travelling along the same road in the same direction. At noon P, who goes at the rate of m miles an hour, is at a point A; while Q, who goes at the rate of n miles in the hour, is at a point B, two miles in advance of A. When are they together?

> Has the answer a meaning when $m-n$ is negative? Has it a meaning when $m = n$? If so, state what interpretation it must receive in these cases.

10. P is a number of two digits, x being the left hand digit and y the right. By inverting the digits, the number Q is obtained. Prove that $11(x+y)(P-Q) = 9(x-y)(P+Q)$.

Second Class Certificates, 1876.

1. Divide $(1+m)x^3 - (m+n)xy(x-y) - (n-1)y^3$ by $x^2 - xy + y^2$.

> Shew that $(a + a^{\frac{1}{2}}b^{\frac{1}{2}} + b)^3 - (a - a^{\frac{1}{2}}b^{\frac{1}{2}} + b)^3$ is exactly divisible by $2a^{\frac{1}{2}}b^{\frac{1}{2}}$.

2. Resolve into factors $x^4 + 2xy(x^2 - y^2) - y^4$,
$a^2(b-c) + b^2(c-a) + c^2(a-b)$, and $25x^4 + 5x^3 - x - 1$.

3. If $x^3 + px^2 + qx + r$ is exactly divisible by $x^2 + mx - n$, then $nq - n^2 = rm$.

4. Prove that if m be a common measure of p and

q, it will also measure the difference of any multiples of p and q.

Find the G. C. M. of $x^4 - px^3 + (q-1)x^2 + px - q$ and $x^4 - qx^3 + (p-1)x^2 + qx - p$ and of $1 + x^{\frac{1}{2}} + x + x^{\frac{3}{2}}$ and $2x + 2x^{\frac{3}{2}} + 3x^2 + 3x^{\frac{5}{2}}$.

5. Prove the rule for multiplication of fractions.

Simplify $\dfrac{x^2-(y-z)^2}{(y+z)^2-x^2} \times \dfrac{y^2-(z-x)^2}{(z+x)^2-y^2} \times \dfrac{z^2-(x-y)^2}{(x+y)^2-z^2}$

and $\dfrac{a}{a^2+b^2} - \dfrac{a}{a^2-b^2} + \dfrac{a^3}{(a-b)(a^2+b^2)} - \dfrac{2a^3-b^3-ab^2}{a^4-b^4}$.

6. What is the distinction between an *identity* and an *equation*? If $x-a = y+b$, prove $x-b = y+a$.

Solve the equations $(2+x)(m-3) = -4-2mx$,

and $\dfrac{16x-13}{4x-3} + \dfrac{40x-43}{8x-9} = \dfrac{32x-30}{8x-7} + \dfrac{20x-24}{4x-5}$.

7. What are *simultaneous equations*? Explain why there must be given as many independent equations as there are unknown quantities involved. If there is a greater number of equations than unknown quantities, what is the inference?

Eliminate x and y from the equations $ax + by = c$, $a'x + b'y = c'$, $a''x + b''y = c''$.

8. Solve the equations—

(1) $\sqrt{n+x} + \sqrt[3]{n-x} = m$

(2) $3x + y + z = 13$
$3y + z + x = 15$
$3z + x + y = 17$

9. A person has two kinds of foreign money; it takes a pieces of the first kind to make one £, and b pieces of the second kind: he is offered one £ for c pieces, how many pieces of each kind must he take?

10. A person starts to walk to a railway station four and a-half miles off, intending to arrive at a certain time; but after walking a mile and a-half he is detained twenty minutes, in consequence of which he is obliged to walk a mile and a-half an hour faster in order to reach the station at the appointed time. Find at what pace he started.

11. (a) If $\dfrac{a}{b} = \dfrac{c}{d}$ then will $\dfrac{a^4+c^4}{b^4+d^4} = \dfrac{a^2 c^2}{b^2 d^2}$.

(b) Find by Horner's method of division the value of
$x^5 + 290x^4 + 279x^3 - 2892x^2 - 586x - 312$ when $x = -289$.

(c) Show without actual multiplication that
$(a+b+c)^3 - (a+b+c)(a^2-ab+b^2-bc+c^2-ac) - 3abc = 3(a+b)(b+c)(c+a)$

McGILL UNIVERSITY.

First Year Exhibitions, 1873.

1. The difference between the first and second of four numbers in geometrical progression is 12, and the difference between the 3rd and 4th is 300; find them.

2. Find two numbers whose difference is 8, and the harmonical mean between them $1\frac{4}{5}$.

3. Prove the general formula for finding the sum of an arithmetical series.

4. The differences between the hypotenuse and the two sides of a right-angled triangle are 3 and 6 respectively; find the sides.

5. Solve the equations

$$x^2 + y^2 = 25 \quad , \quad x + y = 1;$$

$$\frac{x}{x+1} + \frac{x+1}{x} = \frac{13}{6};$$

$$x + y + z = 5, \; x + y = z - 7; \; x - 3 = y + z$$

$$\frac{x+4}{3x+5} + 1\frac{1}{6} = \frac{3x+8}{2x+3}.$$

6. A cistern can be filled by two pipes in 24′ and 2′ respectively, and emptied by a third in 20′; in what time would it be filled, if all three were running together.

7. Shew that

$$1 + \frac{a^2 + b^2 - c^2}{2ab} = \frac{(a+b+c)(a+b-c)}{2ab}$$

8. Prove the rule for finding the greatest common measure of two quantities.

First Year Exhibitions, 1874.

1. The sum of 15 terms of an arithmetic series is 600, and the common difference is 5; find the first term.

2. Find the last term and the sum to 7 terms of the series
$$1 - 4 + 16 - \&c.$$

3. Find the arithmetical, geometric, and harmonic means between $3\frac{3}{8}$ and $1\frac{1}{2}$.

4. The difference between the hypotenuse and each of the two sides of a right-angled triangle is 3 and 6 respectively; find the sides.

5. The sum of the two digits of a certain number is six times their difference, and the number itself exceeds six times their sum by 3; find it.

6. Solve the equations:—
$$x - y = 1 \; ; \; x^3 - y^3 = 19$$
$$\frac{3x - 7}{x} + \frac{4x - 10}{x + 5} = 3\tfrac{1}{2},$$
$$x - \frac{1}{7}(y - 2) = 5 \; ; \; 4y - \tfrac{1}{3}(x + 10) = 3.$$
$$\frac{132x + 1}{3x + 1} + \frac{8x + 5}{x - 1} = 52.$$

7. A man could reap a field by himself in 20 hours, but with his son's help for 6 hours, he could do it in 16 hours; how long would the son be in reaping the field by himself?

8. Find the value in its simplest form of
$$\frac{x + y}{y} - \frac{2x}{x + y} + \frac{x^2 y - x^3}{x^2 y - y^3}$$

9. Find the greatest common measure of
$3x^3 + 3x^2 - 15x + 9$ and $3x^4 + 3x^3 - 21x^2 - 9x$.

First Year Exhibition, 1876.

1. Solve the equations

$$\sqrt{a+x} + \sqrt{a-x} = \frac{12a}{5\sqrt{a+x}},$$

$$\frac{x}{a} + \frac{y}{b} = 1 - \frac{x}{c}; \quad \frac{x}{a} + \frac{y}{b} = 1 + \frac{y}{c}.$$

2. Reduce to its simplest form the expression:—
$7\sqrt[3]{54} + 3\sqrt[3]{16} + \sqrt[3]{2} - 5\sqrt[3]{128}.$

3. Find the greatest common measure of
$2x^3 + x^2 - 8x + 5$ and $7x^2 - 12x + 5.$

4. Simplify $\dfrac{\dfrac{m^3+n^2}{n} - m}{\dfrac{1}{n} - \dfrac{1}{m}} + \dfrac{m^2 - n^2}{m^3 + n^3}.$

5. A number consists of two digits, of which the left is twice the right, and the sum of the digits is one-seventh of the number itself. Find the number.

6. Solve the following:—

$$\frac{x}{a} + \frac{y}{b} = +1, \quad \frac{x}{a} + \frac{z}{c} = 2, \quad \frac{y}{b} + \frac{z}{c} = 3;$$

$$\frac{1}{x} + \frac{1}{y} = 2, \quad x+y = 2.$$

7. Find the sum of n terms of the series 1, 3, 5, 7, &c.

(a.) Show that the reciprocals of the first four terms, and also of any consecutive four terms, are in harmonical proportion.

UNIVERSITY OF VICTORIA COLLEGE.

Matriculation, 1873.

1. What is the "dimension" of a term? When is an expression said to be "homogeneous"?

2. Remove the brackets from, and simplify the following expression:—
$$(2a - 3c + 4d) - \{5d - (m + 3a)\} + \{5a - (-4 - d)\} - \{3a - (4a - 5d - 4)\}.$$

3. Prove the "Rule of Signs" in Multiplication.

4. Multiply $a - \dfrac{a^2 + x^2}{a}$ by $x + \dfrac{a^2 - x^3}{x}$.

5. Divide $ax^3 + bx^2 + cx + d$ by $x - r$.

6. Divide 1 by $1 + x$.

7. Find the Greatest Common Measure of $6a^4 - a^2x^2 - 12x$ and $9a^3 + 12a^2x^2 - 6a^2$. $8x^5$.

8. From $3a - 2x - \dfrac{ax - x^2}{x^2 - 1}$ subtract $2a - x - \dfrac{a - x}{x + 1}$.

9. Given $\begin{cases} \dfrac{x}{8} + \dfrac{y}{9} = 42 \\ \dfrac{x}{9} + \dfrac{y}{8} = 43 \end{cases}$ to find x and y.

10. Divide the number a into four such parts that the second shall exceed the first by m, the third shall exceed the second by n, and the fourth shall exceed the third by p.

11. A sum of money put out at simple interest

APPENDIX.

amounts in m months to a dollars, and in n months to b dollars. Required the sum and rate per cent.

12. Given $x^2 + ab = 5x$, to find the values of x.

13. Divide the number 49 into two such parts that the quotient of the greater divided by the less may be to the quotient of the less divided by the greater, as $\frac{4}{3}$ to $\frac{3}{4}$.

14. Divide the number 100 into two such parts that their product may be equal to the difference of their squares.

15. Given $\begin{cases} x^2 + xy = 56, \\ xy + 2y^2 = 60, \end{cases}$ to find values of x and y.

16. A farmer bought a number of sheep for $80, and if he had bought four more for the same money, he would have paid $1 less for each. How many did he buy?

Matriculation, 1874.

1. Find the Greatest Common Measure of $2b^3 - 10ab^2 + 8a^2b$, and $9a^4 - 3ab^3 + 3a^2b^2 - 9a^3b$, and demonstrate the rule.

2. Add together $a - x + \dfrac{a^3 + x^3}{a + x}$, $3a - \dfrac{a^3 - ax}{a + x}$, $2x - \dfrac{3a^2 - 2x^2}{a - x}$, and $- 4a - \dfrac{a^3 + x}{a^2 - x^2}$.

3. Divide $\dfrac{1}{1 + x} + \dfrac{x}{1 - x}$ by $\dfrac{1}{1 - x} - \dfrac{x}{1 + x}$ and reduce.

4. Given $\frac{1}{3}(x - a) - 1.5(2x - 3b) - \frac{1}{2}(a - x) = 10a + 11b$ to find x.

5. A sum of money was divided among three persons, A, B, and C, as follows: the share of A exceeded $\frac{4}{5}$ of the shares of B and C by $120; the

share of B, ⅜ of the shares of A and C by $120; and the share of C, ⅔ of the shares of A and B by $120. What was each person's share?

6. Given $\begin{cases} x^2 + y^2 + xy(x+y) = 68 \\ x^3 + y^3 - 3x^2 - 3y^2 = 12 \end{cases}$ to find x and y.

7. Show that a quadratic equation of one unknown quantity cannot have more than two roots.

8. Given $\dfrac{2\sqrt{x}+2}{4+\sqrt{x}} = \dfrac{4-\sqrt{x}}{\sqrt{x}}$; to find the value of x.

9. There is a stack of hay whose length is to its breadth as 5 to 4, and whose height is to its breadth as 7 to 8. It is worth as many cents per cubic foot as it is feet in breadth; and the whole is worth at that rate 224 times as many cents as there are square feet on the bottom. Find the dimensions of the stack.

10. Given $\begin{cases} \dfrac{x+y}{2} = \sqrt{xy} + 5 \\ \dfrac{2xy}{x+y} = \sqrt{xy} - 4 \end{cases}$ to find x and y.

11. In attempting to arrange a number of counters in the form of a square it was found there were seven over, and when the side of the square was increased by one, there was a deficiency of 8 to complete the square. Find the number of counters.

12. Reduce to its simplest form
$$\dfrac{a^2 - (b-c)^2}{(a+c)^2 - b^2} + \dfrac{b^2 - (c-a)^2}{(a+b)^2 - c^2} + \dfrac{c^2 - (a-b)^2}{(b+c)^2 - a^2}$$

13. A and B can do a piece of work in 12 days; in how many days could each do it alone, if it would take A 10 days longer than B?

14. Given $\begin{cases} \dfrac{x}{y} = \dfrac{z}{w} \\ x - y = 4 \\ z - w = 3 \\ x^2 + y^2 + z^2 + w^2 = 62\frac{1}{2} \end{cases}$ to find $x, y, z,$ and w.

15. Find the last term, and the sum of 50 terms, of the series 2, 4, 6, 8, &c.

16. Write down the expansion of $\left\{ x - \dfrac{1}{x} \right\}^7$

17. How many different strains may be rung on ten different bells, supposing all the combinations to produce different notes?

ANSWERS.

Junior Matric. 1872. *Pass.*

1. $\frac{1}{6}x^4 - (\frac{1}{18}x^2y^2 - \frac{1}{2}xy^3 + y^4)$; $(a^2 + 9b^2)(a + 3b)$;
   $(x+a)^2 + (x+a)(y-b) + (y-b)^2$.   2. $a^2 + ap + q$
3. $(a), 1\frac{1}{3}$; $(b), \frac{34}{25}$; $(c), 4\frac{1}{2}$; $(d), \frac{1}{2}, \frac{1}{3}$.   4. 640, 660.

Junior Matric., 1872. *Pass and Honor.*

1. $\left\{z^{\frac{1}{2}} + (x^{\frac{1}{2}} - y^{\frac{1}{2}})\right\}^2 \left\{z^{\frac{1}{2}} - (x^{\frac{1}{2}} - y^{\frac{1}{2}})\right\}^2 =$
   $\left\{z^{\frac{1}{2}} - (x^{\frac{1}{2}} - y^{\frac{1}{2}})^2\right\}^2$; $a + 2b + 3c$.   2. We have
   $c^2 - pc + q = 0$ and $c^2 - p'c + q' = 0$, from which to eliminate c.

4. If β be one root, $\dfrac{b}{a} - \beta\left(1 + \dfrac{p}{q}\right)$, $\dfrac{c}{a} = \beta^2 \dfrac{p}{q}$,
   and, eliminating β, $\dfrac{b^2}{ac} = \dfrac{(p+q)^2}{pq}$.

5. $(a), 4, -7, \frac{1}{2}(-3 \pm \sqrt{277})$; $(b), 3, 2,; -3, -2$
   $\dfrac{7}{\sqrt{6}}, \dfrac{5}{\sqrt{6}}; -\dfrac{7}{\sqrt{6}}, -\dfrac{5}{\sqrt{6}}$. $(c), -3$
   $\pm \sqrt{2}$. (d), Divide through by x^2 and put y fo.
   $x + \dfrac{1}{x}$, and $\therefore y^2 - 2$ for $x^2 + \dfrac{1}{x^2}$, then $y = \dfrac{10}{3}$ or $-\dfrac{5}{2}$ and $x = 3, \frac{1}{3}, -\frac{1}{2}$ or -2

ANSWERS.

6. $\dfrac{1}{(1-r)^2} \cdot \dfrac{a^q - 1}{a - 1} \left\{ a^{\mu} + a^{\mu+\nu} + a^{\mu+2\nu} + \ldots \right\}$

7. $a(b_1c_2 - b_2c_1) + a_1(b_2c - bc_2) + a_2(bc_1 - b_1c) = 0$.

8. $\dfrac{2rm}{1+r}$.

Junior Matric., 1872. *Honor*

1. 8 and 6 miles.   2. Each of the first set of fractions may be shewn equal to

$2abc \dfrac{\dfrac{x}{a}}{b^2 + c^2 - a^2}$ or $2abc \dfrac{\dfrac{y}{b}}{c^2 + a^2 - b^2}$, or $2abc \dfrac{\dfrac{z}{c}}{a^2 + b^2 - c^2}$, which are therefore equal.

3. Multiplying the equations successively by y, z, x and z, x, y, we obtain $c^2x + a^2y + b^2z = 0$, $b^2x + c^2y + a^2z = 0$; thence $\dfrac{x}{a^4 - b^2c^2} = \dfrac{y}{b^4 - c^2a^2} = \dfrac{z}{c^4 - a^2b^2}$, and $x = \dfrac{\pm a(a^4 - b^2c^2)}{\sqrt{\{(a^4-b^2c^2)^2 - (b^4-c^2a^2)(c^4-a^2b^2)\}}}$

4. $a^2 + b^2 > 2ab, \therefore c(a^2 + b^2) > 2abc$, &c.

5. $3, 0; -2, -5; -3, 6; -8, 1$.   6. 90 and 240 mls.

Junior Matric., Honor.
Senior Matric., Pass. } 1874.

1. (a), From first $x = 2y$ or y, and then solutions are $3, \tfrac{3}{2}$; $-3, -\tfrac{3}{2}$; $\sqrt{21}, \sqrt{21}$; $-\sqrt{21}, -\sqrt{21}$.
(b), $\tfrac{3}{16}(41 \pm \sqrt{769})$, $\tfrac{1}{3}(-37 \pm \sqrt{769})$. (c), $\tfrac{1}{3}, \tfrac{1}{4}$; $-\tfrac{1}{3}, -\tfrac{1}{4}$; $\tfrac{1}{2}, \tfrac{1}{3}$; $-\tfrac{1}{2}, -\tfrac{1}{3}$. (d), 4, 18. 2. 3.

4. (a), $\dfrac{33x^2 + 61x + 10}{x + 2}$; (b), $\dfrac{x^2 + 3x + 2}{x^2 + 11x + 30}$

5. $\frac{1}{8}\left\{(-\frac{1}{2})^n-1\right\}$; $\frac{1}{x-1}\left\{x+1+(x-1)(1-x)\right\}$
$=\frac{x(3-x)}{x-1}$.

6. $x-2$ and $x+5$ are factors, and roots are, 2, -5, $\frac{1}{2}(-3 \pm \sqrt{35})$.   7. $7\frac{1}{2}$ gals.

8. 4.88 .....per cent.   9. 1 days.
He receives $3 every day the work continues; he returns nothing the first day he is idle, $1 the second, and so on, and the number of days he works is 10.

Junior Matric., 1876.   *Pass.*

1. a^2; $a^{m-n} b^{n-2} c^{-1} d$.   2. a^6-x^6; a^3+x^3.
3. $1+2x+3x^2+4x^3+5x^4+6x^5+\ldots\ldots$; rem. $7x^6-5x^7$. $9x^2-6x+1$.
4. $a^{2m}+(ab)^{m}{}^{n}+(ab)+b^{2m}$.
5. (1), 2. (2), 2, 5, 7 ; or $-2, -5, -7$.

Junior Matric., 1876.   *Honor.*

1. 35 mls.   2. (2), These quantities are in $H.P.$ if $\frac{b+c}{ab+ac+bc}$, &c., are in $A.P.$, i.e., if a, b, c are in $A.P.$

5. It may be shewn that the remainder at the nth decimal place is $2n$; hence if the nth digit be increased by unity, and the whole subtracted from 1, the remainder is the remaining part of the period.

6. $z=4, x=2$ or $-3, y=3$ or -2; $z=-1, x=2\pm\sqrt{10}$, $y=-2\pm\sqrt{10}$.

ANSWERS.

Junior Matric., 1876. Honor.

1. 121 and 400 yards.
2. $(a-b+c)(ab+bc+ca)(a^2+b^2+c^2+ab+bc-ca)$
3. Irrational roots go in pairs $\therefore 3-\sqrt{2}$ is a root; and other roots are $\frac{1}{2}(-1 \pm \sqrt{-3})$.
4. $x^{\frac{5}{3}} + x^2 y^{\frac{1}{3}} + x^{\frac{3}{2}} y^{\frac{2}{3}} + xy + x^{\frac{1}{2}} y^{\frac{4}{3}} + y^{\frac{5}{3}}$.
5. $\dfrac{ab}{b+(n-1)(a-b)}$. 6. 3 mls.
7. (1), Plainly $x+2$ divides both sides, and roots are $-2, 2\pm\sqrt{7}$. (2), $x=3, y=4$ or $\frac{1}{4}$; $x=-3, y=-4$ or $-\frac{1}{4}$.

Second Class Certificates, 1872.

1. $\left(\dfrac{a}{b}+\dfrac{b}{a}\right)^2 - 1 = \dfrac{a^2}{b^2}+1+\dfrac{b^2}{a^2}$.
2. $(a-b)-(a-4b) = 3b$.
3. $\dfrac{(x^2+\frac{1}{3})^2-(\frac{x}{2})^2}{(x^2-\frac{x}{2})^2-(\frac{1}{3})^2} = \dfrac{x^2+\frac{x}{2}+\frac{1}{3}}{x^2-\frac{x}{2}-\frac{1}{3}}$.   4. (67, −99).
5. $(a-b)(x-y)=0$; $\therefore$ if a be not $=b$, $x-y=0$, if $a=b$, $x-y$ may have any value.
6. $\dfrac{43-14m}{14m-13}$.   7. $\frac{2}{3}$, provided x be not $=-2\frac{1}{4}$; then fraction becomes $\frac{0}{0}$ and is indeterminate.
8. $\dfrac{1}{x-y}=1, \dfrac{1}{x+7}=\frac{1}{5}$; $x=3, y=2$
9. 13.   10. $\frac{3}{4}$ of a mile per hour.

ANSWERS.

Second Class Certificates, 1875

1. $2(a^2b^2 + b^2c^2 + c^2a^2) - (a^4 + b^4 + c^4)$.   2. $\dfrac{3a}{a+b}$.

3. $(3x+1)(4x^3 - 2x^2 - 3x + 1)$.   4. $\dfrac{2a(2b^2 - 5)}{4a - 3b}$.

5. $x = \dfrac{bc - ad}{nb - md}$, $y = \dfrac{bc - ad}{mc - na}$.

6. x and y are indeterminate: there is but one equation.   7. $88, $44.   8. 14 days, $11\tfrac{2}{3}$ days.

9. In $\dfrac{2}{m-n}$ hrs. $m-n$ negative means that they were together $\dfrac{2}{n-n}$ hrs. before noon. $m = n$, they are never together.

10. Each side equals $99(x^2 - y^2)$.

Second Class Certificates, 1876.

1. $(1+m)x - (1-n)y$.   2. $(x+y)^3 (x-y)$; $(a-b)(b-c)(c-a)$; $(5x^2 - 1)(5x^3 + x + 1)$.

3. Let the other factor be $x + a$; multiply and equate co-efficients; eliminating a, $nq - n^2 = rm$; other condition is $pn - mn = r$.   4. $x - 1$; $1 + x^{\frac{1}{2}}$.

5. $\dfrac{(x+y-z)(x-y+z)(y+z-x)}{(x+y+z)^3}$; $\dfrac{1}{a-b}$

6. $-\tfrac{2}{3}$; 1.

7. $a''(b'c - bc') + b''(ac' - a'c) + c''(a'b - ab') = 0$.

8. (1,) Cube, and $3(n+x)^{\frac{1}{3}}(n-x)^{\frac{1}{3}}(m) = m^3 - 2n$,
$\therefore x = \left\{ n^2 - \left(\dfrac{m^3 - 2n}{3m}\right)^3 \right\}^{\frac{1}{2}}$; 2, 3, 4

9. $\dfrac{a(c-b)}{a-b}$, $\dfrac{b(a-c)}{a-b}$.

10. 3 miles an hour

11. (a), See § 359. (b), 2,000. (c), Substitute successively $-b, -c, -a$ for a, b, c, in the left hand side, and it appears that $a+b, b+c, c+a$ are factors, and $\therefore$ expression is of form $N(a+b)(b+c)(c+a)$; putting $a=b=c=1$, we get $N=3$.

First Year Exhibitions, 1873.

1. 3, 15, 75, 375.   2. 9 and 1, or $\frac{8}{15}$ and $-\frac{7\frac{1}{2}}{15}$.   4. 9, 12.
5. (a), 4, -3; $-3, 4$. (b), 2, -3. (c), 4, -5, 6. (d), $-\frac{9}{8}$.
6. 40'.   7. $=\dfrac{(a+b)^2-c^2}{2ab}=$.

First Year Exhibitions, 1874.

1. 5.   2. $(-4)^6$; 3277.   3. $2\frac{7}{16}$; $2\frac{1}{4}$; $2\frac{1}{13}$.
4. 9, 12.   5. 75.
6. (a), 3, 2; $-2, -3$. (b), 7 or $-1\frac{2}{3}$. (c), 5, 3. (d), 14.
7. 30 hours.   8. $\dfrac{y}{x+y}$.   9. $3(x+3)$.

First Year Exhibitions, 1876.

1. $\dfrac{4}{5}a, \dfrac{3}{5}a$; $\dfrac{\frac{1}{a}-\frac{1}{b}-\frac{1}{c}}{\frac{1}{a^2}-\frac{1}{b^2}-\frac{1}{c^2}}$, $\dfrac{\frac{1}{a}-\frac{1}{b}+\frac{1}{c}}{\frac{1}{a^2}-\frac{1}{b^2}-\frac{1}{c^2}}$
2. $-12\sqrt[3]{2}$.   3. $x-1$.   4. m.
5. 21, 42, 63, or 84.   6. $a, b, 2c$; 1, 1.   7. ∞.

Matriculation, 1873

2. $11a - 3c - 5d + m.$ 4. $-ax.$
5. $ax^4 + (ar + b)x + (ar^2 + br + c) + \dfrac{ar^3 + br^2 + cr + d}{x - r}.$
6. $1 - x + x^2 - x^3 + \ldots\ldots$ 7. $3a^2 + 4x^2.$
8. $\dfrac{(a - x)(x^2 - 2)}{x^2 - 1}.$ 9. $144, 216.$
10. $\tfrac{1}{4}(a - 3m - 2n - p),$ &c.
11. $\dfrac{mb - na}{m - n}, \dfrac{1200(a - b)}{mb - na}.$
12. $\pm \tfrac{1}{2}\sqrt{ab}.$ 13. $28, 21.$
14. $50(\sqrt{5} - 1), 50(3 - \sqrt{5}).$
15. $x = \pm 10, y = \mp 10; x = \pm 4\sqrt{2}, y = \pm 3\sqrt{2}$
16. $16.$

Matriculation, 1874.

1. $a - b.$ 2. $\dfrac{4a^3 + a^2x - 2ax^2 + x^3}{x^2 - a^2}.$ 3. 1
4. $-5a - 3b.$ 5. $600, 480, 360.$
6. $2, 4; 4, 2.$ 8. 4 or $9\tfrac{1}{9}.$
9. $20, 16, 14$ ft. 10. $40, 10; 10, 40.$ 11. 56
12. $1.$ 13. 30 and 20 days.
14. $6, 2, 4\tfrac{1}{2}, 1\tfrac{1}{2},$ or $-2, -6, -1\tfrac{1}{2}, -4\tfrac{1}{2}.$
15. $100, 2550.$
16. $x^7 - 7x^5 + 21x^3 - 35x + 35x^{-1} - 21x^{-3} + 7x^{-5} - x^{-7}.$ 17. $1023.$

www.ingramcontent.com/pod-product-compliance
Lightning Source LLC
Chambersburg PA
CBHW051732300426
44115CB00007B/533